Construction Quantity Surveying

Construction Quantity Surveying

A Practical Guide for the Contractor's QS

Second Edition

Donald Towey MRICS

This edition first published 2018
© 2018 John Wiley & Sons Ltd

Edition History
Wiley-Blackwell; (1 edition July 23, 2012)

The right of Donald Towey to be identified as the author of this work has been asserted in accordance with law.

Registered Offices
John Wiley & Sons, Inc., 111 River Street, Hoboken, NJ 07030, USA
John Wiley & Sons Ltd, The Atrium, Southern Gate, Chichester, West Sussex, PO19 8SQ, UK

Editorial Office
9600 Garsington Road, Oxford, OX4 2DQ, UK

For details of our global editorial offices, customer services, and more information about Wiley products visit us at www.wiley.com.

Wiley also publishes its books in a variety of electronic formats and by print-on-demand. Some content that appears in standard print versions of this book may not be available in other formats.

Library of Congress Cataloging-in-Publication Data

Names: Towey, Donald, author.
Title: Construction quantity surveying : a practical guide for the
   contractor's QS / by Donald Towey, MRICS.
Description: 2nd edition. | Hoboken, NJ, USA : Wiley, [2018] | Includes
   bibliographical references and index. |
Identifiers: LCCN 2017015390 (print) | LCCN 2017023024 (ebook) | ISBN
   9781119312949 (pdf) | ISBN 9781119312956 (epub) | ISBN 9781119312901 (pbk.)
Subjects: LCSH: Quantity surveying.
Classification: LCC TH435 (ebook) | LCC TH435 .T594 2017 (print) | DDC 692/.5–dc23
LC record available at https://lccn.loc.gov/2017015390

Cover design by Wiley
Cover image: © teekid /Getty Images

Set in 10/12pt Warnock by SPi Global, Pondicherry, India

10 9 8 7 6 5 4 3 2 1

Contents

Preface

The contractor's quantity surveyor plays a pivotal role in the commercial management of construction projects due to the practical skills required to assist with the successful delivery of a scheme.

The aim of the edition of this book is to build on those skills by providing insight into the working practices of contracting organisations with regards how projects are secured and buildings delivered. This commences from the invitation to tender when the skills of quantifying and estimating the price of works is exercised whilst acknowledging the risks involved with the type of procurement and form of contract. This sets the framework for effectively running a construction project with this 2nd edition exploring recent industrial changes that includes:

- New rules of measurement (Nrm2)
- Types of contract including JCT 2016
- Impact of the Construction, Design and Management Regulations, CDM 2015
- The various procurement routes available for obtaining a building.

The book is set out in a logical manner with Tables, Figures, technical information and text that helps define the subjects in sufficient detail. This makes the book a suitable source of reference for students on appropriate courses as well as qualified and experienced professionals and those with an interest in the construction industry.

Enjoy.   Donald.

1

The Construction Industry and the Quantity Surveyor

1.1 Industry Overview

The construction industry is a generic term for a service industry forming part of the nation's economy that carries out the planning, designing, constructing, altering, refurbishing, maintaining, repairing and demolition of structures. It is a large dynamic and complex industry that plays an important role in the economy of which there are three sectors, that is, buildings, infrastructure and industrial. Building construction can be subdivided into two groups, residential and non-residential. The former requires no elaboration, while the latter encompasses commercial, institutional and government-owned/leased projects covering a range of building types such as hotels, banks, schools and hospitals. Infrastructure refers to highway and civil engineering structures, including large public works such as motorways, bridges and other transportation networks, utility distribution and water/wastewater treatment. Industrial includes chemical processing plants, warehouses, factories, power generation facilities, manufacturing plants and mills. The construction process commences with a planning stage stemming from early designs and includes financing and developing the designs for working purposes. This continues with a construction phase until the project is complete, which triggers the occupational phase when the building is operated as its intended use.

1.1.1 The British Construction Industry

The demand for new buildings and the refurbishment of existing is driven by available spending in the public and private sectors. Because of this, the construction industry is buoyant in terms of the demands it must meet, yet is susceptible to the mood of local economies and the national economy as a whole at any time. According to a House of Commons Briefing Paper entitled *Construction industry: statistics and policy* published during Q4 of 2015, the British construction industry amassed £103 billion in economic output during 2014. This represents 6.5% of the gross value added (GVA), which is the construction industry's economic contribution to the total value of the national accounts. The briefing paper advises that employment in the industry grew at a steady pace since 2010, with 2.11 million jobs filled during 2015. The paper predicts that a decade of future economic growth lies ahead based on the (then) coalition government's report *Construction 2025*, which was published during Q3 2013 and prepared from the guidance and support of the Construction Industrial Strategy Advisory Council (CISAC), an advisory body comprising members that seek, construct and issue

Construction Quantity Surveying: A Practical Guide for the Contractor's QS, Second Edition. Donald Towey.
© 2018 John Wiley & Sons Ltd. Published 2018 by John Wiley & Sons Ltd.

advice on buildings and infrastructure. This report predicts world economic output will grow at a rate of 4.3% per annum through to 2025, which will create changes in the international economy and provide new opportunities for the United Kingdom. To embrace these opportunities and be well placed domestically, the government has pledged to work with a range of industrial bodies with end goals for 2025 that aim to:

- reduce the initial cost of construction and whole life asset cost by one-third (2009/2010 levels);
- reduce by half the overall time it takes to acquire new/refurbished buildings from inception to completion (2013 industrial outputs);
- reduce by half greenhouse gas emissions in the built environment (based on 1990 levels); and
- reduce by half the trade gap between imports and exports for construction products and materials (based on a trade deficit of £6 billion in 2013).

With such challenges ahead, the industry must be ready for change and is indeed a giant in terms of the contribution it makes to the nation's economy which creates room for interesting careers and job security in the process.

1.1.2 Equal Opportunities and Diversity

The UK is a diverse society comprising people from multicultural and multilingual backgrounds, where everyone has something different to bring to society and the workplace. The construction industry is one that requires a variety of skills and abilities to function, which means it is important for people from different backgrounds, life experiences and abilities to be suitably employed to enable the industry to achieve the high levels of skills and deliverables needed. For this reason, employers, unions, service providers, service users and industrial bodies are encouraged to endorse integration regardless of age, disability, gender reassignment, marriage or civil partnership, pregnancy and maternity, race, religion and beliefs, gender, sexual orientation or socio-economic background. This requirement is also legislated under UK labour law with the Equality Act 2010 applicable in England and Wales and, in part, Scotland and Northern Ireland. The Act makes it illegal to discriminate against access to education, public services, private goods and services or premises and employment opportunities. Hailed by lawyers as the most significant development of equality legislation in decades, the Act harmonises and consolidates previous anti-discrimination legislation, and strengthens legal rights to equality. The Act's purpose was to replace a mass of disjointed legislation with more uniform, accessible and comprehensive rights. Following its introduction, it has succeeded in setting standards and raising awareness of rights to equality, as well as tackling discrimination and, in particular, the role of the public sector with regards achieving equality.

1.1.3 Global Construction

In a report entitled *Global Construction 2030* published by Global Perspectives Ltd during Q4 2015, the global construction market is expected to grow by an average of 3.9% per annum from 2015 through to 2030. This is comparable to the 4.3% prediction through to 2025 advised in the UK Government's report *Construction 2025*. According to *Global Construction 2030*, cumulative growth through to 2030 will surpass global domestic product (GDP) (the construction industry's economic contribution to the total value of a nation's accounts including taxes less subsidies) by one-quarter. This is primarily due to developed countries continuing to gather pace following a sustained period of economic stability, and the ongoing confidence of developing countries with industrialisation and reform. China is expected to be the largest construction market for most of the period, anticipated to level off by 2030, with the United States growing at a faster rate in second place with the financial gap narrowing during the period. India's economy is expected to surpass that of Japan to become the third-largest construction market by 2021, with the top three accounting for 57% of all growth. It is predicted that Japan's role will be notched down to fourth place by 2030 to be taken over by Indonesia.

For cultural reasons, countries tend to rely on home-grown companies to design, manage and construct projects with their residents/citizens incentivised under labour law to carry out services. This varies from country to country and region to region, and even with the local market tested, it is still possible for skill shortages to affect the servicing of projects. This is appeased with globalisation and the services of international recruitment and construction companies that seek candidates for project employers in host countries. The selection and suitability of such candidates can be endorsed with experience, qualifications and membership of trade and/or professional institutions that have reciprocal agreements with their counterparts in other countries, meaning the status can be obtained in more than one country at the same time.

Foreign recruitment and the investment in overseas schemes can lead to the expansion of a business and the opening of overseas branches. The integration of a new business with the construction industry of another country is indeed a challenge, and one that requires commitment to time and resources. A risk management strategy is therefore vital, which must be created by any business wishing to diversify its interests meaning the impact of the investment must be fully understood prior to making commitments. When appraising the possibilities for starting an overseas construction business, the investor must have an understanding of risks associated with any of the following:

- the need to invest, competition expected and the likelihood of securing contracts;
- referral from others that may have already ventured into the locality and their results;
- anticipated duration of the overseas investment (i.e. short- or long-term or permanent);
- financial stability of the overseas country;
- financial stability of the home country and foreseeable trends (e.g. currency exchange rates, existence of double taxation treaties, changes in legislation and tax breaks);
- performance of competitors on completed projects (i.e. what is normal and the quality expected);
- trade unions and their influence;
- health, safety and environmental attitudes;
- availability of suitable labour skills and material resources;
- political stability;
- cultural working practices;
- existence of corruption;
- legislation with regards planning at local and national level;
- existing industrial relations and building control;
- land and terrain;

- sources and status of infrastructure and utility service providers;
- terrorism and militants;
- communication methods, including any potential language barriers;
- climate and volatility of the scheme to natural disasters (i.e. earthquakes, hurricanes, etc.);
- decision to rent or purchase office space, including set-up costs and the need for financial loans;
- time involved to register the company and/or the need for sponsors;
- business development potential (time and money);
- relationship with local and central government regarding trade and employment restrictions (e.g. if trading overseas, the percentage of employees who must be nationals of the host country);
- procedure, processing time and availability of visas for employees who would relocate;
- cost to employ, including overheads, versus potential income;
- understanding usual credit terms which may be prolonged in the host country; and
- familiarity with law and the forms of contract used for service and services in the host country.

The above are drivers for realisation that, when combined with effective strategies, can mean involvement in an overseas investment is beneficial to a business. A fine example is the company Laing O'Rourke founded when R. O'Rourke & Son bought out John Laing Construction in 2001. Since formation, this company has grown internationally with offices in Europe (including the UK), Australasia, the Middle East and Canada that carry out an array of project deliverables including buildings, transport, power, water/utilities, mining/natural resources and oil and natural gas.

1.2   Parties Involved in a Construction Project

1.2.1   The Client's Team

A project client is an entity that seeks and pays for construction works, and is usually the party that enters into a contract with a contractor that carries out the works. A client may be an individual, partnership, group of persons, organisation or business from the public or private sector. Public sector means central and local government offices and/or facilities, and private sector means an individual(s), firm(s) as partnerships and companies that may be limited or unlimited in their legal business trading. Clients in the need of construction works stem from all walks of life, with some having none or limited dealings of the design and construction processes with others experienced enough to understand the importance of engaging teams. A client's role involves determining a suitable procurement method for obtaining a building while accepting or transferring risk in the process and for this reason, a client must make suitable team appointments for the successful delivery of a scheme.

1.2.1.1   The Design Team

The design team comprises design consultants from a range of disciplines that create concept and schematic designs from a client's brief and coordinate and develop the design so it is suitable for construction purposes. Details of the various design team members' involvement are discussed in Sections 1.2.2–1.2.7. The design team also writes and develops specifications that comply with legislation, prepares schedules and specifies the criterion required for the client's needs. With traditional procurement, each design team member forms an agreement with the client, usually with the parties agreeing to coordinate their designs for consistency. Alternatively, under a design-and-build arrangement, the project client enters into an agreement with a contractor to complete the design and deliver the project. Typically, this involves novating the design teams' early services to the design-and-build contractor, with the services reinstated under new agreements with the contractor.

1.2.1.2　The Construction Team
The construction team is the supply side of the industry, a term used to describe the appointment of parties necessary to carry out and facilitate the works on site. This includes:

- the main contractor (builder);
- subcontractors to carry out works for the main contractor;
- material suppliers (including manufacturers); and
- suppliers of plant equipment to assist with construction operations.

The combination of design team members and the main contractor is often referred to as the building team. Under a traditional procurement arrangement, building team members enter into individual contracts with the project client that generally excludes construction team members, except of course for the main contractor that is an integral part of both the construction and building teams.

1.2.1.3　The Development Team
The development team is an integral part of the client's team, separate from the design and construction teams, and comprises members that issue advice on marketing, technical, financial, legal and business planning matters to the project client. Members of this team can also include a parent or side company that expresses interest in a project for business reasons, such as being guarantor for prospective purchasers. Other members include building maintenance and facilities managers, politicians, heritage and conservation groups, local planning authorities and members of the client's own organisation that may be accountants, lawyers, real estate practitioners and coordinators of the development process. The development team usually initiate a project's viability by creating a client's brief, which is a document created as a frame of reference to determine reasons why the project should exist as well as the pathway for procuring the scheme and rewards the completed project is expected to deliver. Figure 1.1 demonstrates a hierarchy arrangement of design, construction and development team appointments.

1.2.1.4　Client's Agent
A client may be inexperienced with the design and construction processes, and may wish to outsource the management duties. This has its advantages as it permits a project client to focus on the day-to-day running of their business without distraction, avoiding the need to commit internal resources that may otherwise lack the expertise required to oversee a construction project. The manager may be an individual project manager or

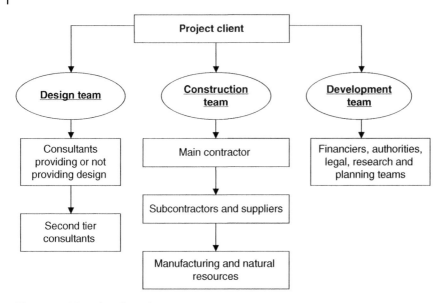

Figure 1.1 Hierarchy of appointments.

a project management company that becomes the client's agent in the process. Depending on the size of a project in terms of financial value, complexity and duration of the works, this can involve the mobilisation of a team skilled in project planning, commercial and contract management and the supervision of large-scale projects. Typically, the project manager/company becomes the single point of contact for the building team, with all communications to the client usually made via the client's agent. The client's agent may also be a member of the design team and possibly an architect or quantity surveyor, thus serving a dual role. In isolation, the development team might be sourced and engaged directly by the project client with the client's agent responsible to senior members of the development team. Where a project manager is appointed, various building team members' services are tendered. Upon receipt of tenders, a recommendation for each appointment is made by the client's agent to the client and, once appointed, each is contractually bound to the client and not the project manager/ client's agent. Each member is then expected to perform in a collaborative manner to service the project and project client, a matter usually expressed in the terms of each engagement.

1.2.2 Architect

A leading project architect's services involves the development and coordination of a building design to ensure it is compatible with other consultant's designs, and reports to a client on progression of the design and documentation until it is suitable for working purposes. Services can also extend through the construction phase where the architect responds to clarification requests, attends meetings and possibly amends the design to suit a client's revised needs. Architects usually operate from a professional practice consisting of design team members who create site and block plans and drawings that show elevations, plans and cross-sections of buildings as well as the finer details of construction.

A project usually takes its first breath with a feasibility study, the contents of which steer a decision to explore the viability of the project. Once a decision is made to build, development of the design is triggered from the contents of a client's brief that initiates a concept stage outlining the requirements in sketch form, traditionally produced with ink on paper or conventionally from computer-aided design (CAD) software. The design is further developed to a schematic stage that includes setting out information and technical details of building parts to smaller scale and developing the documentation to suit. Detailed drawings include cross-sections of walls, frames, roofs, etc. showing a vertical plane through the construction details including critical dimensions. At a later stage of design development, drawings demonstrate how information shown on large-scale drawings fit together as components in the final product, for example joinery items, junctions with ceilings and walls, etc. Prior to being released for construction purposes, the detailed design is audited and certified, confirming it as complying with health and safety standards and the building regulations. An architect's services also include reviewing and approving shop drawings that show how parts are to be installed in a building which are produced by fabricators and/or installers usually inherited from the master design. This involves checking drawings and/or schedules for accuracy and quality to enable their assembly and/or installation into the works, with the architect usually empowered to reject any information supplied and request resubmissions until such time they are suitable. At the end of a project, an architect's services can involve reviewing as-built information provided by the contractor stemming from approved shop drawings to ensure they reflect the installed works and approving maintenance and operating manuals also provided by the contractor. An architect/interior designer may also design loose furnishings etc. for this stage.

Separate to designing buildings, an architect may design external works to a building and issue drawings and specifications that show requirements for: ornamental screen walling; fencing; paths, paved areas and car parking; hard landscaping schemes, including public footways, cycle paths and lighting; soft landscaping schemes that encompass turf, shrub and tree planting; irrigation schemes; furniture such as benches, litter bins, planter boxes and bollards; signage; playground equipment; security (e.g. video surveillance cameras, pedestrian gates and barriers); and water features. Alternatively, some aspects of hard and soft landscaping schemes may be designed by a landscape architect appointed separately by the client.

Independent and commercial architectural practices are usually members of the Royal Institute of British Architects (RIBA), a professional body representing architectural designers of the built environment. This accreditation is also available to individuals who complete a recognised qualification and/or may be listed on the Architects Registration Board (ARB), a statutory body for the registration of architects in the UK. The RIBA provides accreditation to schools teaching architecture in the UK under a course validation procedure, and will also validate international courses not requiring ARB endorsement yet satisfying the RIBA's criteria. There are three parts of the RIBA education process. Part I is generally a three-year first degree course followed by one years' work experience in an architectural practice. Part II is generally a two-year postgraduate diploma or masters degree course, from which a further year out must be taken before the Part III professional examinations are sat. Overall, it takes a minimum of seven years before architectural students can seek chartered status that, when achieved, permits members to use the initials RIBA after their name.

1.2.3 Geotechnical Engineer

Geotechnical engineering is an arm of civil engineering and refers to the services of an engineer engaged to carry out a site investigation (SI) which culminates with a report into the site conditions above and below ground for advice. The SI recognises topography, existing buildings, structures and greenery; ascertains the type(s) of subsoil(s) through soil mechanics; advises on the water table depth; and tests for the presence of any contamination which, if found, includes a remedial action plan that when carried out successfully can endorse the land as suitable for development. The information provided in the SI permits structural and civil engineers to write specifications and design structures that the ground can withstand. Training leading to qualifications is through the civil engineering route (see Section 1.2.5 below). Suitably qualified engineers can become members of the British Geotechnical Association (BGA) through a scheme developed jointly and supported by the Institution of Civil Engineers, the Geology Society of London (Engineering Group) and the Institute of Materials, Minerals and Mining.

1.2.4 Structural Engineer

When a new building is designed by an architect, the design is issued to the structural engineer for a structural appraisal. As part of the appraisal, the engineer creates a structural design reciprocating the architectural proposals with profiles and detailed sections to provide the building with structural integrity and stability. The design is developed further to include product specifications that give criteria such as concrete strength in building components (e.g. foundations, walls, roof, columns, beams and slabs) as well as the quality of workmanship and testing requirements needed to ensure the building is suitable for the intended use. When considering a type of foundation/substructure design, the engineer refers to the ground conditions obtained in the SI report as this can influence the concrete strength and specification (e.g. a blended concrete mix to mitigate alkaline attack from subsoils). Above ground, architecturally designed elements are analysed to confirm their suitability for the building. This involves reviewing architectural information to enable the creation of suitable structural criteria. The review process considers the size, shape and use of a building, together with health and safety practices for construction purposes and the need to comply with the building regulations. After the structural appraisal is complete, the engineer issues the information to the architect that may include recommendations for architectural modifications to suit the permanent works. For example, steel members creating walls and floors may need to be wider and deeper than the architect's proposals to withstand structural stresses imposed by external factors such as wind loading/live loads on a building. In addition to permanent works, the engineer may design temporary structures such as shoring, which is a bracing system designed to stabilise surrounding structures and avoid collapse while new works are in progress. If a building is to be refurbished with the works affecting its structural integrity, the structural engineer will also assess stresses imposed on existing building elements and design permanent and/or temporary works to suit.

As with an architect, the structural engineer reviews shop drawings but of a structural nature, and checks and approves their suitability for a scheme. Furthermore, this engineer specifies testing requirements and reviews test results submitted, and has the authority to enforce compliance with the design and specification and instruct the removal and making good or replacement of any unacceptable works.

To qualify as a chartered structural engineer (MIStructE), the Institution of Structural Engineers in the UK requires the completion of key stages of education and training. The traditional pathway involves obtaining an accredited degree and then following a training programme to bridge any gap between the qualification and experience, known as a period of initial professional development (IPD). At the end of this period, the graduate attends a professional review interview (PRI) which must be passed together with an entry examination in order to obtain chartered status. Alternative routes apply for those who possess appropriate qualifications and have suitable experience.

1.2.5 Civil Engineer

Civil engineering involves the creation of new structures including roads, sewers and bridges and the maintenance of existing similar structures in the built environment. The title also covers non-structural works including excavations and remediation of contaminated land which can involve the services of a geotechnical engineer. When a new structure of a civil engineering nature is required, the civil engineer designs a scheme and writes a specification for a scope of works out of due concern for public safety and the environment. In addition, the civil engineer coordinates new design requirements with existing infrastructures, including roads and drainage as well as works by utility service providers (e.g. gas, water, electric and telecommunications) that may be part of master planning for a district, borough, town or city. Duties of the civil engineer include (but are not limited to):

- providing a topographical site survey and levels in relation to ground and construction items;
- issuing setting-out information for the works;
- assessing tenders from contractors for the works and making recommendations to the client;
- obtaining permits from local authorities;
- attending public meetings;
- liaising with utility service providers; and
- supervising works in progress.

During the construction phase, the civil engineer will enforce the contract requirements and can instruct defective works to be replaced so they comply with the approved drawings and specification. Once a new structure is complete, and if it is to be under the eventual ownership of someone other than the contracting parties such as a local authority, it can trigger the commencement of a maintenance period. This is a stated duration written into an agreement during which time the contractor remains responsible for defects until such time the structure is transferred to the adopting party. Under a separate arrangement to which the contractor may not be a party, the local authority may also adopt utility services forming an integral part of the works under contract on behalf of the utility services providers. For this reason, the civil engineer works closely with the contractor and adopting authority during the construction phase to ensure the design and constructed items comply with the approved design and specification, in order to aid a smooth transfer upon expiry of the maintenance period.

Where a project is engineered without the requirements for a building such as bridge or land remediation, the engineer takes the role of client/employer's agent under the title engineer or to that stated in the contract. In essence, the role of the engineer in an engineering contract is equivalent to the role of a client side project manager in a building contract.

To qualify and be employed as a civil engineer, the incumbent must possess an academic degree in civil engineering. Studies towards obtaining the degree take 3–5 years with the completed qualification being a Bachelor of Engineering (B/Eng) or Bachelor of Science (BSc) undergraduate degree. The curriculum generally includes courses in physics, mathematics, project management and design, plus specific topics of civil engineering. After taking basic courses in most sub-disciplines of civil engineering, candidates then specialise in one or more sub-disciplines at advanced levels and can obtain a masters degree (MEng/MSc) in a particular area of interest such as geotechnical engineering or façade engineering. A qualified civil engineer may be chartered, and a Member of the Institution of Civil Engineers (MICE) and must hold a degree in civil engineering which can also act as a stepping stone to other aspects of engineering.

1.2.6   Service Engineers

Service engineering includes methods of supplying, installing and commissioning systems that permit utility service providers to supply power, water and gas for distribution through a building. The term also encompasses drainage, fire protection, mechanical air systems, transportation, machinery and a range of specialist services used for fitting out a building. In multistorey buildings, building services are distributed via an infrastructure in rising mains or risers. These vertical risers run through the core of a building with horizontal branch connections at building floor levels and comprise specified pipework, cables, conduits and ducting, which are fitted off for a building's intended use. Smaller buildings such as residential properties are serviced in a similar manner without risers, but with the pipes, cables etc. secured to and through walls and/or structural timbers such as joists. The cost of building services generally runs at *c.* 30% of the total price of a project, and being this considerable, the contractor's quantity surveyor is encouraged to understand the scope driving this proportion.

1.2.6.1   Electrical Engineers

These engineers specify and design schemes that distribute electricity for power, lighting, security, heating, information technology and intelligent and communication systems in buildings. Moreover, the discipline includes the creation of design and specifications for artificially lighting external works and methods of obtaining power supplies to a building from the electric mains supply. In conjunction with an architect, electrical engineers may specify types of light fittings and design solar-controlled panels that produce energy from the sun for battery storage for use with a building's function.

1.2.6.2   Plumbing or Hydraulic Engineers

This engineering discipline designs and specifies water and gas supplies as well as heating and drainage systems within buildings. They may also seek the stamping of plans, which is endorsement of their designs from the water board and/or other authority that has an interest in the supply and distribution of water in buildings.

1.2.6.3 Fire Protection Services Engineers

These engineers author a fire-engineering report to identify potential fire, smoke and heat hazards in a completed building and/or design, and write specifications applicable to the works so the completed building complies with the fire-engineering report. Suitably designed schemes include active and passive measures with objectives of protecting the vicinity, structure, contents and building occupants from the effects of fire, smoke and heat. Active measures include sprinklers, fire blankets, hydrants, hose reels, portable extinguishers and air pressurisation systems along corridors, stairwells and lift shafts to suppress the effects of fire and smoke, while passive measures are architectural and include doors, partitions and escape routes that divide a building into parts, known as compartmentation. Hydrants for use by the fire brigade may fall under the category of water services with the design possibly part of the water supply design provided by the plumbing/hydraulics engineer. Pressurisation systems are dry systems driven by pressurised air which may fall under the scope of mechanical and air conditioning engineering (see following section). Other dry systems include clean agent fire suppression for use in equipped electrical and telecommunications rooms where gas in lieu of water is used to extinguish a fire or suppress smoke to mitigate damage to the equipment that would otherwise occur if using water.

1.2.6.4 Mechanical and Air Conditioning Engineers

These engineers provide a design and specification for naturally flowing and fan-assisted air systems to provide a building with a suitable atmospheric pressure as well as adequate heating, ventilation and air conditioning (HVAC). HVAC refers to technology that provides suitable air changes and thermal comfort of the internal environment of a building.

1.2.6.5 Transportation System Engineers

Engineers under this category create designs and specification for vertical, horizontal and inclined transportation to deal with a stream of people or products moving through or within buildings.

1.2.6.6 Other Engineers

A range of other engineers that provide designs and specifications for works of a specific nature that are often project specific includes:

- waste-disposal systems;
- solar heating;
- oil-fired heating systems;
- district heating for distributing mass-generated heat (gas, cogeneration or solar) from a source to a number of buildings simultaneously;
- district cooling where treated chilled water is provided at pressure to a building for use in HVAC systems from a central district cooling plant;
- types of mechanical plant for specific use (e.g. swimming pools); and
- building automation/management systems, which are intelligent-based systems that inform a building's facilities manager on the status of installed parts and their efficiency while a building is operational and if any malfunction has occurred; management involves monitoring from computerised central control rooms with alert notifications possible via smartphones and emails.

As with other engineering disciplines, authenticity of an engineer is gained by the successful completion of recognised courses that measures knowledge, competence and practical training.

1.2.7   The Client's Quantity Surveyor/Cost Manager

A client venturing into a construction project may engage a quantity surveying firm as a cost consultant who becomes an integral part of the design team in the process. One of the services these firms offer is *pre-contract* cost advice which involves estimating construction costs prior to the client entering into a construction contract. At the earliest stage, it is possible to offer cost advice without the need for a design, providing the client issues a statement of requirements. In order to estimate cost from such requirements, quantity surveying firms collect cost data from past and current projects where they are engaged as cost consultant to produce dynamic single-line construction costs, for example a 650 bed hospital (price per bed) or 300 pupil school (price per child), etc. Where a client provides a limited design or gives a statement of areas, cost advice can be more certain using either the functional building as a whole (e.g. by applying a rate per m^2 to the floor area of a hotel) or refining rates suitable to floor areas of parts of a building (e.g. hotel guest rooms and restaurants). Where an advanced design is provided and further cost data available, it is possible to estimate the costs of the functional elements of a building by applying rates to the areas of walls, floors etc. or suitable rates for the detailed components of each functional element (e.g. concrete per m^3, bar reinforcement per tonne and formwork per m^2 applicable to walls, floors etc.). Depending on the level of detailed design available, it may also be possible to create cost targets of the elements, the combination of which can produce a project budget. Once a budget is created, the quantity surveyor/cost manager can be engaged to monitor different stages of design development and advise the client's team of changes that impact the budget. Pre-contract services also include the preparation of trade bills of quantities for tendering purposes, once the design and documentation is fully developed; vetting main contractors' tenders; and cash flow forecast predictions by time to pay for the works.

Once a client enters into an agreement with a contractor for the works, it triggers the *post-contract* period. During this period, the quantity surveyor/cost manager may be engaged to cost-manage a construction project on behalf of the project client. Typical services include: recommending financial amounts as interim payments to the contractor while works are in progress; issuing monthly reports on the physical progress of the project; valuing changes to the works instructed by the client; assessing contractors' claims; and preparing and issuing a final account. In addition, some large-sized consultancies expand traditional quantity surveying services and offer project management and advisory services to a client that includes:

- recommending an appropriate procurement route for a type of project;
- risk management strategies, including identification and analysis of risks;
- due diligence reporting by vetting and confirming scopes of services in main contractors' tenders and the suitability of submitted offers with the client's expectations;
- selection of an appropriate form of construction contract (i.e. standard or purposely drafted);
- life-cycle costing on a building or parts of a building to demonstrate how an investment in the construction/supply and installation price and the price to maintain through

the occupational phase until replacement/removal can derive a rate of return, often called 'cradle to grave' assessments;
- advice on dispute resolution services if there is conflict on a construction project with the contractor;
- business feasibility studies to assess the viability of a scheme prior to investing in a design;
- acting as client/employer's agent under the title project manager;
- certification of buildings with Energy Performance Certificates (EPC); and
- Building Information Modelling (BIM) manager (see Section 1.8.2 below for details on BIM).

The quantity surveyor/cost manager may be a sole practitioner, in a partnership or operate as part of a large consultancy. To qualify, professionals need to hold an academic degree and/or are members of the Royal Institution of Chartered Surveyors (RICS). The RICS is the leading international body that regulates members and firms to ensure ethics and professional conduct are maintained. Professional members are termed 'chartered quantity surveyors' with the RICS having the largest network of quantity surveyors worldwide. The client's quantity surveyor/cost manager is a design team member and consultant, and may also be referred to as the professional quantity surveyor or PQS. This reference can mean a practising consultancy or individual(s) engaged on a project if the individual(s) is a chartered quantity surveyor or deemed suitably trained and holding a relevant degree. This is not to be confused with the contractor's quantity surveyor employed by the main contractor who is a commercial member for the supply side of the industry who, when accredited with a chartered building qualification or a degree holder with suitable experience, is also deemed professional.

1.2.8   Main Contractor

The main (sometimes called general) contractor carries out works in accordance with the agreement it has with the project client. The contractor will also adopt the title principal contractor when legally responsible for the health and safety duties on a construction project, as required by the health and safety regulations. The main contractor rarely carries out all of the works themselves and subcontracts, or sublets, a number of trade works in order to fulfil the obligations of the contract. Subject to the conditions of contract this is often without client input, which gives the contractor a main role to procure, manage and deliver a scheme. However, the contract may permit the client to have a say in the subletting/subcontracting process by naming or nominating certain trade contractors because of their reputation or skills, which a contractor must acknowledge if part of an agreement. Standard forms of contract generally omit the title 'main' and recognise the capacity as 'contractor' only with the main contractor's duties involving:

- establishing the site accommodation, including temporary offices and amenities for use by the contractor's staff, site visitors and operatives working on site, and any specific accommodation for the project client including members of the client's team if a requirement of the contract;
- managing health and safety procedures;
- coordinating, procuring, planning and supervising construction works;

- reporting periodically to the client and coordinating with the client's team where necessary;
- ensuring budgets are maintained; and
- implementing a method of quality control to ensure works are achieved in accordance with the drawings, specification and conditions of contract.

In order to deliver a project on time, at an agreed cost and of the expected quality, the contractor will need to commission a project team. The team must be suitably qualified and experienced to help deliver the scheme with personnel either based full time on site or are regular visitors that are assisted by the contractor's head office (e.g. accounts department). On a project valued at say £10 million which is straightforward in nature, with the design fully developed and the contractual provisions suitably captured in an agreement, the following personnel would normally be site based:

- project manager in charge, with some time spent at the contractor's head office and client's offices;
- site manager;
- structural and finishing trade supervisors;
- health and safety officer;
- quantity surveyor (possibly part time or in between other projects);
- general site operatives (e.g. labourers); and
- administration support staff and trainees who may visit site periodically.

A project of similar standing of half the financial amount may have the site-management requirements reduced by cancelling the need for finishing trade supervisors, leaving supervision to the site and project managers whose skills are appropriate. The role of the contractor's quantity surveyor involves dealing with post-contract duties on behalf of the contactor and, depending on the project in terms of value and/or complexity and the contractor's management structure, the role may require full-time commitment to a single scheme.

Normally, the contractor's quantity surveyor is answerable to a commercial and/or project manager for addressing commercial, administrative and contract matters including:

- regular cost reporting on works in progress (monthly, bi-monthly or quarterly);
- recommending awards to material suppliers and subcontractors and ensuring those in receipts of awards have binding agreements in place;
- ensuring project insurances are current and relevant;
- vetting health, safety and environmental submissions from subcontractors for compliance with the contractor's project health and safety plan;
- providing the flow of information to the contractor's supply chain (suppliers and subcontractors);
- assessing the price of contract variations and their submission for approval;
- assisting in administration of the construction contract;
- preparing applications for interim payments from the client;
- preparation of a project final account; and
- processing payments to the supply chain, administering and agreeing final accounts.

Team members may be qualified chartered building professionals and Members of the Chartered Institute of Building (MCIOB). The CIOB is the leading construction management voice in the construction industry, with its members representing a body that has knowledge managing the building process. Team members may also be chartered quantity surveyors (MRICS), although this is usually limited to commercial and project managers and contractors' quantity surveyors.

1.3   Legislation and Control of the Building Process

Whatever the type of building project undertaken, construction operations and the final building must comply with built environment legislation which is enforced by planning control and regulatory systems. Anyone wishing to pursue and pay for building works must be satisfied it is lawful, and for this reason it is necessary to obtain permission from the local authority before commencing operations to confirm that the design and works comply with the law. When a building is required to undergo a change of use from one classification to another (e.g. a residential property to commercial premises), planning approval is required which is legislated in England and Wales by the Town and Country Planning Act 1990, in Scotland by the Planning etc. (Scotland) Act 2006 and in Northern Ireland by the Planning (Northern Ireland) Order 1991.

Obtaining planning approval is usually the responsibility of the building owner and not the contractor carrying out the works. However, if works commence without a permit, claim of a lack of knowledge by either party to a construction contract on the need for a permit could set the pathway for conflict, as it may mean the works are stopped or cancelled by the local authority. For projects which are simple in nature, the approval procedure may be straightforward with the building owner possibly seeking permission themselves. However, with large projects in terms of floor area, value and complexity, the process can be time-consuming and the entity wishing to apply for a building permit may appoint a project manager to manage the process. Ideally, the party preparing the application should be conversant with local authority requirements which may be influenced by byelaws relevant to the location where the works are to be carried out. Byelaws are parochial powers granted from central government by an Act of Parliament that empowers local authorities to make decisions relevant to the community, and can be a deciding factor in the issue of planning approval.

The building regulations (sometimes called building codes) are separate from planning approval, and are a set of statutory requirements that seek to provide guidance and define standards for the purpose of designing and constructing buildings. They are contrived with skill and care to ensure a completed building is constructed with due consideration to the environment, health and safety of occupant(s) and the public at large. The regulations are modified from time to time to reflect changes in legislation, which may apply to any part at any time. The enabling act empowering them is the Building Act 1984 (England and Wales) that underwent change to become the Building Regulations 2000 (England and Wales). In Scotland, the driving legislation is the Building (Scotland) Act 2003 that steers the Building (Scotland) Regulations 2004, and in Northern Ireland the Building Regulations (Northern Ireland) Order 1972, amended 2012.

1.3.1   Planning Permission

A minor change to a building (usually for residential purposes) is termed 'permitted development' and is usually exempt from planning approval. However, a party seeking to change or modify a building should contact the local authority to confirm if works can proceed without formal approval before commencing the works. If approval is required, the process involves seeking clarification of the planning requirements and confirming that the design is compliant with the building regulations. In addition, whoever carries out the design and building processes must affirm a commitment to safe working practices as required by health and safety law.

The approval process commences with an applicant lodging a formal proposal to the local authority. The lodgement usually includes a set of building plans and elevations or other information as required to demonstrate the extent of the works, which activates an assessment procedure to arrive at a decision. In arriving at a decision, the local authority's assessment takes into account the building process and effect of the completed project on the built environment and existing buildings in the locality. Moreover, a decision will be influenced by any impact the proposed scheme will have on the Local Development Framework Plans as well as local amenities and infrastructure. Local Development Framework Plans outline planned changes to a district over a stated period, possibly up to 10 years (or more), which are in force at the time of receiving an application. Depending on the size of the proposed scheme in terms of building floor area, height and location, there may be a requirement to provide an environmental impact assessment (EIA); this assesses environmental consequences (both positive and negative) of the construction and occupational phases on the environment. A well-researched project will identify the need for an EIA at the earliest stage if approval of an EIA is a requirement for granting planning permission.

The length of time for issue of a response from the date of lodgement depends on the decision the applicant seeks, which can be influenced by the contents of the submission, the type of project, local authority policy and/or complexity of the scheme. If a submission is speculative and includes information which satisfies the category of the application, the decision may be to grant outline planning permission only. This means the local authority accepts the intent to develop in principle which is subject to further review, and advises what the review will entail (e.g. the submission and approval of a boundary walling design to a new residential estate). For vendors submitting a speculative application this decision provides a sigh of relief, as it means the process of developing the design can progress. An application seeking outline planning permission can be made by a landowner as a vendor that wishes to sell a parcel of land for development, meaning approval (and its conditions) would be relayed to the purchaser following the sale of the land, as without the permission the land may be worthless to a developer.

The second stage of approval is acceptance with reserved matters. This means the scheme is approved yet subject to a set of terms and conditions discharged over time, usually by the end of the construction phase. For example, an application may be submitted seeking permission to construct a high-rise office building in accordance with a set of building drawings, with the applicant failing to provide details of external works such as vehicle parking and landscaping which is not designed. Here, a local authority's response might be to grant permission for the office building to be practically complete within a stated time frame from the date of issuing the decision with a reserved matter for the whole of the works to be practically complete at the same time, which must be in accordance with an external works scheme that is to be lodged and approved.

The third stage is full planning or sometimes called detailed planning permission, which is approval to develop unconditionally or with matters the applicant can comfortably accept, which is of course the most favourable outcome for an applicant. If any type of planning permission is refused, the applicant may lodge an appeal that can only be heard if it relates to matters governed by legislation. These include:

- legalities involving restrictive covenants, for example the existing ownership of land, buildings or parts thereof not owned by the applicant that require discharging by the owner;
- a request to review granted outline planning permission not recognised by the local authority; or
- resolution of confliction between granted outline planning permission and any existing Local and Development Framework Plan.

Until an appeal is resolved, the applicant would be unwise to commence the works as the local authority would probably instruct the demolition of anything created that goes against the expressed interest of the deciding committee and may also impose fines.

1.3.2   Building Regulations

Building regulations are divided into 15 headings, each designated with a letter from 'A' to 'Q' (with the exception of 'I' and 'O'), and covers matters such as 'structure', 'fire safety', 'ventilation' and 'security' with each part accompanied by an Approved Document. The approved documents take the form of first stating the legislation and then providing a number of means which are deemed to satisfy provisions. These provisions detail methods showing the works required to satisfy the regulations through the use of text and illustrations. The building regulations are not created to stifle innovation, as there may be ways of complying with each part other than just using those set out in the deemed to satisfy provisions. The tendency by contractors however is to consider that innovative solutions may be too hard to validate, with most following the requirements literally and adhering to the approved working design so that the works can be completed on time. Updated versions of parts of the regulations are generally not applied retrospectively, and only apply to each new change or modification of a building which does not require the retrofitting of any existing elements.

1.3.3   Building Control

A main contractor must ensure the works it carries out complies with the construction contract (the contract) and building regulations/approved documents. In order to do this, a diligent and experienced contractor will ensure works in progress (instead of when complete) are carried out in accordance with the approved design, and cross-reference the detailed designs with observations made on site and act accordingly with any deviations. In addition, a level of independent control through periodic building inspections may be carried out by other parties that may be a condition of the contract and also of planning approval.

The discipline of building inspector takes one or more forms, with each qualified in their appropriate field to make professional judgments and issue notices of compliance or non-compliance of inspected works. These judgements are based on observations and test results that when satisfactory confirm the building or part(s) of a building as meeting the requirements of the building regulations/approved documents. Independent building inspectors may be local authority representatives that inspect public and private works, and have delegated authority to instruct the correction, destruction and rebuilding of any works to comply with the intentions of the building regulations. Depending on the discipline, an inspector may have self-regulated authority to enhance or modify an approved design in the interest of public safety (e.g. requirements of the fire brigade). Where applicable, self-regulation may be stated in a code of practice document relevant to the inspector's discipline that may also be endorsed in the planning approval, which can mean the contractor may be expected to construct something in excess of an approved design. Whether or not any change can be contested depends on the conditions of contract as too would be an increase in the price of the works that can only be addressed with a variation.

Under a separate arrangement, a clerk of works (who is usually from a trade background) can be appointed to check works in progress with the appointment independent of the contractor and other inspectors. The clerk of works is usually contracted to the client and has a duty to ensure the contractor carries out works in accordance with the contract. For this reason, the services of a clerk of works must be written into the contract and not be an afterthought. The role is not a legal requirement and one of inspector only, meaning the clerk cannot issue instructions to alter the works or certify any works as compliant with the contract. However, the clerk can enforce compliance with the building regulations/approved documents if a contractor fails in their duties to construct works that do not comply which may have gone unnoticed by a building inspector or local authority representative. Here, only the client's agent can issue an instruction to vary the works under contract, and may do so based upon advice given by the clerk of works. For example, the conditions of contract may permit the client's agent to instruct the contractor to open up covered works such as a backfilled drainage trench to see if the pipes are laid to correct falls and encased in concrete as specified, and can then direct the clerk to inspect. Any subsequent instruction by the client's agent based upon the clerk's findings is enforced by the contract because, as a matter of procedure, the parties acknowledge their legal obligations to comply with the building regulations/approved documents.

Enforcement of this requirement is not a variation, and is merely intended to make the contractor aware of their responsibilities. However, and depending on the conditions of contract, if opened up works demonstrate that they are compliant, the contractor would normally be entitled to a variation with costs and an extension of time to the project end date because they have been delayed by no fault of their own. In essence, the clerk of works inspects the works to ensure they comply with the contract, and independent inspectors or local authority representatives enforce the building regulations/approved documents. The contractor has a fiduciary duty to acknowledge these levels of building control, and has a legal obligation to comply with the conditions of contract it has entered into and the requirements of legislation.

1.3.4 CDM Regulations 2015

The adoption of a suitable and proactive health and safety system is important for the successful delivery of a construction project. The positive culture it creates has advantages to a contractor, including: improved productivity and quality of work; better working environment; lower staff absence and staff turnover; reduced insurance premiums; and promotion of a good corporate image.

A poignant piece of early legislation for the UK construction industry occurred with the Health and Safety at Work etc. Act 1974. Broadly, this Act encourages and regulates the duties of employers, employees, contractors, subcontractors and persons engaged in the workplace. The Act established the role of the Health and Safety Executive (HSE), an entity empowered to delegate authority to health and safety inspectors to ensure compliance with the law.

Health and safety within the construction industry received a further legislative boost in the 1990s following European Directive 92/57/EEC. In the UK, this saw the introduction of the Construction Design and Management Regulations 1994 (CDM 1994), revised and updated in 2007 (CDM 2007) and again in 2015 (CDM 2015). As per its predecessors, CDM 2015 has aims of improving the overall health, safety and welfare of those working in construction. The regulations affirm a responsibility on parties involved with 'construction work' and applies to those involved in the carrying out of building, civil engineering, engineering construction, home maintenance and improvement works. The regulations aim to ensure proactive health and safety issues are considered on construction projects with goals of reducing the risk of harm to those that construct, occupy and maintain buildings. CDM 2015 defines responsibilities according to particular roles for a project that affect the client, designer and contractor. Characteristics of CDM 2015 that affect the client include:

- 'construction work' which is planned to be carried out by more than one contractor and/or where the works will last longer than 30 working days with more than 20 workers working on site at one time during any part of the project, or if the project exceeds 500 person days in total (in this case the project client has the legal duty to notify the HSE of the project);
- the requirement for the client to take a lead role and appoint a principal designer to plan, manage and monitor the pre-construction phase and prepare a health and safety file;
- a requirement for the client to appoint a principal contractor for any project deemed to comply with the requirements of CDM 2015;
- the client is accountable for all aspects of health, safety and welfare of a project with the regulations imposing strict liability on the client for the performance of duty holders they appoint (e.g. the principle contractor); and
- the client is responsible for making health and safety arrangements for a construction project from inception to completion and through its whole life cycle where CDM 2015 is applicable.

Pragmatically, the principal designer will delegate responsibilities to the principal contractor for the construction phase, and in their duties the principal contractor must:

- ensure workers have the correct skills, knowledge, training and experience to fulfill their obligations;
- provide adequate site supervision and organisational capabilities;
- demonstrate how the flow of information is managed;
- create and maintain a construction phase health and safety plan; and
- develop and maintain (until the end of the construction phase) a post-construction/ occupational phase health and safety plan.

To fulfill these obligations, the contractor's quantity surveyor plays an essential role by ensuring the contractor's and subcontractor's health and safety submissions are to a suitable standard. It is not a normal role of the contractor's quantity surveyor to review, comment, reject or issue advice on the improvement of submissions. However, the role warrants assistance in seeking the information in the offset for the contractor's project manager and health and safety officer's use.

1.4   Industrial Bodies

There are a number of trade industrial bodies within the construction industry that aim to promote, support and inspire their sectors and provide best quality, for example: British Association of Landscape Industries (BALI) for those involved with landscaping and Federation of Piling Specialists (FPS) for piling works, etc. Two significant industrial bodies involved in professional construction matters are the Royal Institution of Chartered Surveyors (RICS) and Chartered Institute of Building (CIOB). The RICS and CIOB have their head offices in the UK with additional offices worldwide. Professional members of either enjoy the benefits of networking the industry at national and international levels, while agreeing to comply with the rules of membership.

1.4.1   Royal Institution of Chartered Surveyors (RICS)

The RICS was founded in 1868 and is a self-regulating body that recognises qualifications in land, property and construction. It has approximately 118,000 professional members worldwide (as of 2016), of which the largest number are members of the Quantity Surveying and Construction Professional Group. The institution has a further 16 Professional Groups including Building Control, Building Surveying, Project Management and Dispute Resolution associated with construction, with the remainder involved in land and property. Members belong to one of the following grade of classes:

- *Student*: which helps members to realise their potential and work to the highest standards at college or university;
- *Associate* (AssocRICS): suitable for those with work-based experience or vocational qualifications;
- *Membership* (MRICS): professional membership which demonstrates to colleagues, clients and peers that the Member is a Charted Surveyor; or
- *Fellowship* (FRICS): recognises an individual's achievement of being a professional member.

The traditional method for obtaining chartered status is along the graduate route, which requires candidates to complete a cognate degree and structured training programme combined with work experience. Traditionally, a postgraduate commences a structured pathway of training towards achieving the APC (assessment of professional competence), which is the measure of an acquired qualification linked with practical training and experience in a relative field of work (e.g. quantity surveying and construction). The structured training and work experience minimum timeframe is two years. At stated intervals during this time (usually every 6 months), the candidate records their training and experience in a logbook to demonstrate details of professional development. This is recorded online by the candidate for endorsement by a counsellor that must be a chartered surveyor who is usually a colleague at the candidate's place of work and qualified from the same pathway as that chosen by the candidate. Subject to the requirements of each RICS region throughout the world, the services of a supervisor may also be required who will have frequent encounters with the candidate. Professional development is measured against benchmarked levels of competence:

- *mandatory*: competencies that are a mixture of professional practice, interpersonal, business and management skills considered necessary for all professional members;
- *core technical*: competencies in stated subjects considered to have uppermost relevance to the chosen pathway; and
- *optional technical*: competencies in additional selected subjects considered of relevance to the chosen pathway.

Assessment of competence follows a linear arrangement of order: Level 1, subject knowledge and understanding; Level 2, application of subject knowledge and understanding; and Level 3, reasoned advice and depth of technical knowledge on subjects. This allows candidates to execute specified subjects with precision by doing and not just knowing as counsellors are skilled to observe when a candidate 'walks the walk' and is not just 'talking the talk'. Once the candidate, counsellor and supervisor (where applicable) are satisfied competency levels are met, the candidate completes and issues the RICS APC submission template. As part of the submission, an APC candidate must provide a written 3000-word case study of a recent project or projects (undertaken up to two years before the assessment) from which the candidate has been involved. Subject to the submission being accepted, the candidate attends a professional interview as a final assessment with an RICS panel from the relevant professional group to discuss the submission and test the candidate's understanding of professional practicing and ethics. The panel later completes their assessment with a recommendation for membership or deferral. If successful, the candidate is invited to enrol as a professional member who, if accepted and the fee paid, receives chartered status and is permitted to use the initials MRICS.

Individuals and companies may apply for chartered status and, once accepted, are bound by rules of conduct for maintaining ethical standards. The RICS responds to the needs of the profession and, as membership routes may change from time to time, those interested in seeking membership should be acquainted with current information found on the RICS website (http://www.rics.org).

1.4.2 Chartered Institute of Building (CIOB)

The CIOB has a national and international reputation for excellence in construction matters. The institute places particular emphasis on construction management and the sharing of knowledge with companies, members and clients, influencing the way the industry operates. It was founded in 1834 as the Builder Society in London, incorporated as the Institute of Builders in 1884, changed to the Institute of Building in 1965 and was granted Royal Charter in 1980. The total number of individual members exceeds 48,000 (as of 2016), of which 20% are registered outside the UK in over 100 countries. Members belong to one of the following classes:

- *non-chartered membership* (applicant or student status): this helps realise potential to the highest standards at college or university, with the CIOB endorsing the status on a case-by-case basis;

- *membership* (MCIOB): professional corporate membership which demonstrates a Member is a chartered builder or chartered construction manager; or
- *fellowship* (FCIOB): recognises an individual's achievement of being a professional member.

The traditional route to professional membership is along the graduate route where candidates follow an educational pathway that requires graduation from a cognate degree and training following a professional development programme (PDP). PDP is the measure of a candidate's educational qualifications combined with practical learning and experience to assess occupational competence. The CIOB has established a new branch for young professionals called Novus, which links student status with corporate membership by providing peer support, mentoring and forums. Novus also liaises with educational bodies in the UK and Ireland to promote activities of the CIOB and recruit new members. A candidate's education and occupational experience is measured within a framework of support involving a CIOB-approved assessor, who may be a peer support of Novus, and who reviews, advises and eventually endorses a worthy candidate's assessment. Satisfactory completion entitles the candidate to attend a professional review with a panel, where a candidate's industrial and management competence together with a commitment to professionalism is assessed. If successful, the candidate can apply for professional membership; when accepted and the fee paid, this permits the professional to use the credentials MCIOB.

Alternatives to the graduate route are available to anyone without appropriate qualifications (e.g. company directors, contracts managers and senior managers of appropriate companies, or those from a military background with the same level of education and experience) who agrees to follow a structured training programme. Other routes to membership are available for persons who are members of affiliated organisations with the CIOB, who hold a cognate/non-cognate degree or NVQ, and are industrial professionals who have been working at a senior level with a minimum of five years of management experience. Individuals and companies may apply for chartered status and, once accepted, are bound to the conditions of membership. Changes to membership criteria can take place regularly; for the most recent information see the website of the CIOB (http://www.ciob.org).

1.4.3 Benefits of Membership

Benefits of membership of the RICS and CIOB include:

- status and respect from clients, colleagues and employers;
- invitations to seminars to learn about current industrial and business trends;
- legal advice;
- eligibility for assistance from benevolent funds;
- career advice; and
- discounts on insurances, software and financial services.

1.4.4 Continuing Professional Development (CPD)

One of the requirements for professional membership of the RICS and CIOB is a member's commitment to updating knowledge and skills and remain competent with lifelong learning (LLL). Methods of carrying out CPD include private tuition and reading, attending courses and seminars through work, or participating in online web classes or e-learning offered by professional bodies and mentoring. To be effective, learning should aim to improve knowledge of subjects that a member considers are important to their employment and profession. Advantages of CPD include:

- the updating and refreshing of knowledge from educational courses that may be otherwise frozen in time;
- providing a catalyst for the learning of new subjects; and
- increased competence in business, which may provide enhanced employment prospects.

Variance of a work task within a normal working day is not normally considered part of CPD. However, skills gained through study or coursework to increase competence could be sufficient, for example training in the use of computer software for improved business use. The RICS makes it mandatory for members of all classes to achieve and record a minimum of 20 hours CPD per calendar year, 10 of which is informal and 10 formal. With formal CPD, the course provider may issue a certificate after an event stating the hours of CPD achieved which can be uploaded to the RICS website in the members' page as a record of completion. Apps are also available from the RICS to record and monitor CPD, meaning the process can be conducted on smartphones. In a similar fashion, the CIOB has a portal and range of subject matters for CPD events which members can complete for monitoring and recording. To create learning outcomes for CPD, it is wise to plan ahead with objectives and focus on methods of obtaining resources to achieve the objectives and logging the achievements once learnt. There are various methods of recording completed and planned activities, including the use of spreadsheets and word processing documents; recording online in the members' area section of the professional bodies' website; diaries and notebooks. To demonstrate, Table 1.1 illustrates the use of a spreadsheet for planning and logging CPD activities.

Table 1.1 Logging and goal setting for CPD events. Skill levels: 1, aware; 2, knowledge and understanding; and 3, able to apply knowledge and understanding. Learning methods: 1, day release; 2, evening course; 3, CPD event; 4, private study; 5, web class or e-learning online; 6, work base project; 7, employment training; and 8, other.

Item	Goal	Current skill level	Required skill level	Learning method	Start date	End date	Learning outcome	CPD hours
A	Understanding MS Excel for construction estimating (informal)	2	3	7				
B	Learn principles of teamwork (informal)	2	3	3				
C	Acquire knowledge of Microsoft Project workings relevant to project running costs (formal)	1	2	5				

1.5 Funding and Market Drivers

The construction industry relies on secured funding to cash flow projects which is derived from public and private sector investment. Public sector funding is generated from accrued local and central government reserves, which is obtained by income from various taxations, rates and sell-offs, whereas funds within the private sector are generated from loans, reserves, investments and windfalls. The various types of funding for spending in each are as follows.

- *Public*: government-backed schemes; issue of grants; defence and military projects; government building upgrades or new works; overseas grants (European Union/ International Monetary Fund, etc); and government stimulus packages.
- *Private*: cash reserves and equity; private and corporate loans and mortgages; company profit and investments; insurance works; profit withdrawal from the sale of stocks and shares; charities and lottery grants; religious organisations; investment strategies from businesses and individuals; land banks (e.g. reserved funds to acquire land to develop property); and government incentives.

Funding of construction projects in either sector is subject to change, which may be the result of an economic cycle or specific event impacting the national economy. When the economy expands the construction industry is usually the first to feel the impact, as the demand for building works usually increases. Conversely, it may be the first to witness decline if the economy retracts. When local authorities and central government play positive roles in the stimulus of the economy this influences output, creating interest from other industrial sectors in the process, which in turn increases socio-economic development. An increase in socio-economic development creates an abundance of funds which in turn increases spending and the output of the construction industry. However, neglect of this development leads to decline and has a reverse effect. In order for central government to monitor growth or decline, it relies on data provided by advisory bodies such as the Office of National Statistics (ONS). The supply of data from the ONS provides a snapshot of the industry at any one time, with the information generally regarded as being of a reputable source regarding the strength of the national economy as a whole. This information is often relayed in treasury reports that can act as a catalyst for the funding of public works that act as market drivers.

Market drivers are found at local and national levels that drive the supply and demand of services to the construction industry. When there is an abundance of contractors, tradesmen and professionals available without full consumer demand, this leads to lower prices and a surplus of employment requirements. By contrast, when consumer demand is high and the supply remains unaltered, prices are driven up with the need for additional resources (i.e. more employment). If the supply is increased to meet the demand, this creates equilibrium and control of prices.

Markets are driven by events or circumstances that influence the level of supply and demand available, which is linked to spending. One such event is an economic recession that triggers a reduction in spending, starving the industry of projects and giving rise to unemployment in the process. By comparison, an increase in demand and spending without sufficient supply creates over-employment which can be modified by a correction in the amount of spending (e.g. decreasing the number and value of works orders under the control of local and central governments). Recognition and suitable control of market drivers creates economic security, resilience to recession and employment stability, all of which are ingredients in the recipe for a strong economy.

1.6 Economic and Construction Cycles

Economies usually create repeat cycles, with most completing their course over a period of 7–9 years. When growth is sustained at the top of a cycle (boom), a slowdown eventually occurs which triggers a rise in interest rates leading to diminishing share values and commodity prices, tighter money and a depressed property market at the bottom of the cycle. At the bottom the economy is vulnerable, and if sustained because of a lack in consumer confidence, a recession (bust) may be activated when the economy deteriorates and retracts. A recession is normally for the short term, unless there are compelling circumstances to entrench the decline. No circumstance in recent times has been as severe as the global financial crisis (GFC) of 2008–2010 that entrenched a recession in various parts of the world, in particular developed nations. The GFC is an isolated case that had an impact on a massive scale. However, it demonstrates what can happen at the bottom of a cycle and how hard it can become for struggling businesses to survive. Low demand for building works during a time of recession can mean fierce competition between contractors, with many prepared to work to reduced profits or none at all. This creates high risk and uncertainty, with many businesses relying on short-term borrowing in an attempt to remain solvent. A recession is not permanent; in the aftermath, the climb out of recession starts with a fall in interest rates, followed by rising share values, better commodity prices, easier money available and eventual increase in property prices, back to the top of the cycle. A demonstration of this cycle is provided in Figure 1.2.

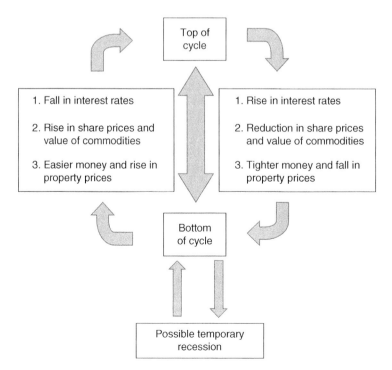

Figure 1.2 Economic cycle.

The above are characteristics of an economic cycle that impact the construction industry with the pattern of events driving or diminishing demand and the need for construction work. When interest rates fall, this encourages more lending and the possibility of an increase in tenders for construction works, with the opposite in force after a boom. Knowledge of these trends permits clients, contractors and design teams to be aware of the likelihood of changes over the long and short term to implement strategies for future planning (e.g. possibly refurbishing existing buildings instead of seeking new works during a downturn). Armed with this information, decisions can also be made regarding risks and opportunities in specific markets in order to recognise fashionable consumer demand at given times, meaning contractors must be open to diversifying interests in the type of works they undertake.

1.7 Development of Quantity Surveying

1.7.1 Background

Henry Cooper, the son of a Master Builder, set up Henry Cooper and Sons in Reading, England in 1785, and in 1799 opened a London office to deal with the costing of building works. The mid-nineteenth century saw the use of 'measurers' or 'master tradesmen' who were called upon to assess the amount of materials and labour required for building operations. At this time, clients would employ an architect to design a building and invite tenders from builders for the works, a traditional approach still in use today. The 'measurer' prepared schedules from various designs to which competing builders applied rates to create a price for the works. However, this was only suitable when the design was complete, with clients often wanting to know what price to expect before tenders were received that paved the way for the introduction of the independent quantity surveyor. This independent role involved measuring, quantifying and assessing the cost of works at different stages of the design process, and providing cost advice on changes instructed by the client during the construction phase. The demand for the service escalated during the late-nineteenth century, with the role being set up within the RICS and recognised as a client-facing profession. In the aftermath of this recognition, and mainly due to the migration of professionals from England to Commonwealth nations, the profession has manifested on the world stage, establishing quantity surveying institutions in many countries in the process. With the passage of time, and due to growth in the number of procurement routes now available to project clients, the requirement for the skills of a quantity surveyor is not restricted to client side and is also adopted by main and trade contracting businesses as part of their service to a project.

The core skills of quantity surveying have remained unchanged since the days of bolstering the profession in the nineteenth century, and are based upon a thorough understanding of construction technology, the competence to measure and quantify scope, and the ability to assess chargeable rates to determine a marketable price for the works. However, it is incorrect to perceive modern quantity surveying as a pioneering profession restricted to traditional core skills, as the role warrants additional services sought by contractors and project clients that see quantity surveyors as competent in administering contracts and managing the commercial aspect of construction projects. Moreover, and due to the growth of innovative financing, public sector clients have the opportunity to lease or mortgage new buildings or structures that a contractor builds, owns and operates for a fixed term with the quantity surveyor involved in the whole life-cycle cost (i.e. from design through to handover and the occupational phase), with quantity surveying skills sought by end-user clients and contractors.

1.7.2 Traits and Skills of a Quantity Surveyor

For those with knowledge of the design-and-build processes or employed in a management role in the construction industry, the role of the quantity surveyor is generally understood and recognised. However, outside the industry, and indeed beyond the shores of the UK, Ireland and Commonwealth countries where the role is firmly established, it is not so familiar. So, what influences someone to become a quantity surveyor? The answer could rest with family influence from a caretaker or parent who is a quantity surveyor or other construction industrial professional. It could also stem from advice given by those aware of the role who are entrusted to provide career opinions. It may also stem from career options provided in school or college, or from information given by a friend or network of friends employed in the industry who considers it a suitable role for a certain individual. Whatever the influence, personal traits can assist individuals to cultivate the necessary skills required for quantity surveying, illustrated in Figure 1.3.

In summary:

- *Traits*: (a) A natural prerequisite to enjoy the building process as well as wanting to learn about matters that eventuate on construction projects. This can mean building techniques, management systems, as well as contractual and commercial matters that arise in the day-to-day running of operations. (b) Thriving on being an integrated team member and relied upon to perform as expected. (c) Being systematic and orderly with the ability to understand and operate systems applicable to the management of projects.
- *Skills*: (a) Good literacy and concise communication skills, including an effective command of oral and written techniques and the ability to use information technology systems. (b) An ability to concentrate for long periods and accepting the office as part of the working environment, which applies on or off the building site. (c) A flair for mathematics, figures and geometry which provides understanding of the measurement techniques and calculations used for determining quantities and prices. In addition, problem solving and logical reasoning will be of benefit.

In addition to the above, the quantity surveyor requires an understanding of contract law including reasons a construction contract exists, which is to create a formal instrument of agreement that sets out the rights, risks and obligations of the parties in order for them to understand the contract they have undertaken.

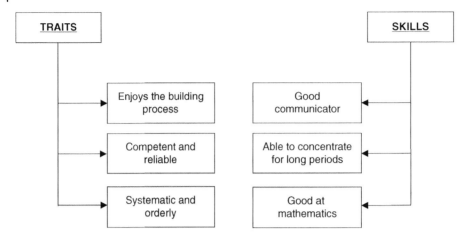

Figure 1.3 Quantity surveying traits and skills.

1.7.3 Education and Training

To acquire suitable education and training, students typically enrol and attend an undergraduate degree course in a university following a curriculum lasting four years. Early semesters involve the study of core subjects and, depending on the curriculum, typical subjects include: construction technology; industrial relations; management; building materials including composition and application; contract law; quantity surveying/estimating; and services in buildings. The combination of these provides an understanding of construction and the industry at technical level. Thereafter, training becomes in-depth dealing with the specifics of the core subjects including subtopics that leads to a series of examinations. Students may be content with completing the degree or may seek chartered status in addition to the qualification. There are abundant cognate degree courses available in the UK accredited by both the RICS and CIOB as stepping stones towards achieving chartered status. If the aim of a student is to obtain RICS or CIOB chartered status through a traditional route of education and training, students need to be aware of the accredited degree courses because a decision to transfer or leave during a semester to commence an appropriate course could be costly in both time and money. Methods for achieving chartered status with the RICS or CIOB through the traditional route and other routes are discussed in detail in Section 1.4 in this chapter.

1.8 Construction Innovation and the Contractor's Quantity Surveyor

Since the dawn of the twenty-first century, the most salient features influencing the construction industry are probably the growth of information technology and changes in environmental attitudes. As these subjects change regularly, it is in the interest of the quantity surveyor and others engaged in the construction industry to remain innovative. This is required in order to keep abreast of the continuous growth in these subjects as they influence the way we work and the buildings that are produced.

1.8.1 Information Technology (IT)

Traditional communication methods for relaying information such as post, fax, meetings minutes, courier and telephones are still in use today and will remain so in the future. However, with the growth of IT, modes of communication have expanded rapidly and in the process modified the way we live and work. The use of electronic systems means modes of communication and the methods of transmitting data can be carried out in a safe and efficient manner, expediting the flow of information on construction projects.

1.8.1.1 Document Communication Systems

Collaborative document transmission involves the distribution of project management information such as drawings, minutes of meetings, general correspondence, etc. through a host network. The process involves a sender uploading information to a selected network, with the network advising authorised recipients via e-mail when the details are available for retrieval. Recipients then log on to the host network using a password to access the information for downloading. Features of the system include: tracking and audit trails; search engines by category; calendars; diaries; contact details; and the management of formal correspondence, all with 24-hour access. The adoption of the system by a project client is discretional, and can depend on the duration and financial value of a scheme to warrant the investment. If a project client considers a collaborative system is viable, the name of the host provider is usually mentioned in the invitation to tender documents. This advises potential duty holders of the client's intention to include a communicative system for the project under tender, and for tendering companies to provide staff training in order to be competent with the arrangement, the cost of which is to be included in the tendered sum. For this reason, a collaborative system is suitable for projects where the cost of training is minor in proportion to the value of the project. A range of host providers can be found on the following websites: http://www.aconex.com; http://www.viewpoint.com; http://www.dochosting.co.uk; and, with Oracle's Primavera applications, http://www.oracle.com.

1.8.1.2 Project Management Planning

Construction project managers make use of software programmes designed to assist the planning of time, resources, budgets and tasks on a project. Probably the most commonly used format is a Gantt chart (named after Henry Gantt, American engineer and inventor of the chart), a horizontal bar chart used to identify project activities and the length of time it takes to start and complete each task. The chart can also include a critical path following the duration of each activity which must be started and completed before the next activity can start. For example, a programme of works may state a requirement for electrical wiring run within partitions and walls of a building. On the critical path, the programme may demonstrate that the wiring must start and be complete before wall boarding can commence and stipulate the start and completion dates (or number of days) of the wiring works, that is, creating the duration of the critical path.

Programmed information can be distributed to parties via hard copy or email by attaching a scanned copy of the document. A better solution for producing higher quality is to issue the programme as an electronic file created from the software programme and, depending on the file size, can be issued via email, compact disc or USB. To be effective, recipients must have access to the same software as it cannot usually be opened through any other source.

1.8.1.3   Estimating and Cost Management Systems

At the heart of producing tenders is competent estimating which involves the development of an estimate of cost to carry out the works which, in the process, relies on estimating software to produce results. There are a number of off-the-shelf estimating software systems available for purchase with varying degrees of sophistication to suit the organisation's size and value of works undertaken. A powerful top-end system has provisions for manually inserting dimensions to calculate quantities with add-on features allowing *.pdf or AutoCAD designs to be imported for scaling using polylines that determines quantities, and viewports that show the measured design results in coloured format. Other features include a viewport to calculate chargeable unit rates, with the option of importing price lists and a facility to produce a variety of reports.

Cost management systems are different from estimating systems and are used post-contract to produce reports on the status of project expenditure and cost forecasts which are recorded in the system under a series of cost codes. Main contractors use these systems as part of project management reporting which are produced with the assistance of the quantity surveyor, usually on a monthly basis.

1.8.1.4   Webcams

Webcams can be used to video-record construction activity with cameras placed in strategic locations on construction sites for optimum viewing. The use of webcams for office use has become outdated due to the high quality of modern video and teleconferencing equipment becoming available.

1.8.1.5   E-tendering

E-tendering is an electronic method of procurement that commences with an invitation to tender and concludes in an award of contract. In general, the process involves an administrator uploading tender design and documentation to a host website for retrieval and downloading by competing companies. Tender responses may be either uploaded to the host website or delivered in hard copy to a physical address, with the preference usually stated in the conditions of tender. Various levels of security and sophistication exist with e-tendering, with access to the information similar to a document communication system used in project management. A main difference between the host web provider of a project management system and e-tendering is that e-tendering is usually conducted via the client's own website instead of a collaborative online system.

E-tendering has advantages over a paper system as it is sustainable and can help reduce tender periods. It is also spontaneous and depletes the need for manually inputting receipt of tenders, saving administrative time in the process. To maintain procedural integrity, tendering companies may be requested to pay a fee or non-refundable deposit to an administrator to cover the cost of the service. The public sector is an advocate of the process, possibly because the use is seen as a long-term investment which improves services and reduces costs, benefitting the public purse. Private sector clients that frequently call for tenders may also benefit from the arrangement. However, this sector tends not to use the system as much as the public sector, probably because of the expense involved in setting up the arrangement and ongoing management responsibilities.

1.8.1.6 Cloud Computing

As a business develops, it creates data which must be stored in a compact and secure manner, which may not always be possible if saved on individual computers that are prone to system failure. Central servers mitigate this problem, yet require allocation of physical space and cooling systems to deal with generated energy, which can be costly to acquire and operate. Cloud computing is seen as the solution to this which frees up space and operates by uploading data to a cloud provided by a host service provider where information is stored in a sustainable and efficient manner. The term 'cloud' is internet-based and a metaphor that represents a cloud from which a list of providers, such as Google or Microsoft, can be accessed. To be implemented, the provider installs software and hardware in the hirer's office and upgrades existing systems as necessary. The hirer then makes use of storage space in the cloud by uploading data, which is charged by the provider under the terms of an agreement. A luxury of the system is the ease of use, as amendments to the hirer's requirements can be altered by the host with the storage upgraded or downscaled to suit. The system is also available for mobile computing including laptops, portable computer devices and smartphones.

1.8.2 Building Information Modelling (BIM)

BIM is the AutoCAD digital representation of a building in model form using three-dimensional images (3D). These images create visualisation of a building that can also be created with greater AutoCAD dimensions applicable to the design, construction and occupational phases of a building, for example:

- 4D refers to the linking of 3D components or assemblies with time or schedules of time-related information in real-time mode. This enables visualisation of the entire duration through a series of events displaying progressive activities through the construction phase.
- 5D links the components of 4D while also adding cost-related information.
- 6D refers to the linking of 3D components or assemblies to the aspects of a project's life cycle as advice for facilities managers and occupants during the occupational phase.

Tools used for BIM during the design phase set the pathway to create a computerised system storing everything from standard 2D drawings to 3D designs, specifications and the finer details of product components. This involves BIM simulating construction processes enabling various issues to be addressed, for example sustainability, including how to reduce waste and choose the most cost-effective schemes. Images and presentations include animations and a walk through of the building at different phases of construction, showing virtual construction methodology and business activities once the building is occupied and operating. For effectiveness, BIM involves logging information and data to illustrate where components are located in a building for identification and maintenance purposes. This is to aid facilities management and assist with the allocation of budgets for maintenance and life-cycle replacement costs.

Where BIM is specified as a requirement, a BIM manager provides advice to the project client and design consultants on the requirements of the system and the features it can provide. For the system to operate, suitable software is installed to each duty holders' IT system that may involve overhauling existing hardware, which is carried out by the BIM manger or an IT consultant. These changes or upgrades can be expensive, with staff training necessary in order to understand and gain the benefits of the system. BIM has distinct advantages over traditional design approaches, as virtual reality demonstrations means there is no doubt regarding the images and function of the building once it is complete. This helps to avoid any misunderstandings if viewing traditional flat 2D drawings, which may provide misperceptions of what a building or part of a building may look like once it is complete.

The benefits of BIM to the industry and project client are: it enhances health and safety; improves communications; is collaborative and reliable; reduces the chance of errors; and mitigates design and construction risks.

Disadvantages include: the size of electronic files and storage required to retain the information; a general lack of understanding to how BIM operates, which requires training for use of the system; and incompatibility of required software with existing systems, unless users consider the cost to upgrade as a financial investment.

Advocates of BIM consider that the benefits for designers outweigh the disadvantages because, with the passage of time, financial outgoings of an early investment can provide dividends for the future if the intent is to be involved in future schemes.

1.8.3 The Environment

The impact of legislation and due concern for the environment has created a new thought process to the way buildings are designed, constructed and managed through their operational life. This attention to detail has created language that would otherwise be obsolete from industrial vocabulary such as conservation and sustainability, which refers to the capacity to protect, endure, maintain and support resources through the building process and occupational life of buildings. Poignantly, the impetus for change has been at a political level following the discovery of a hole in the Earth's ozone layer, brought about from the use of chlorofluorocarbons (CFCs) used by many industries. Fortunately, this has been controlled by suitable legislation and regulations with a degree of success. However, focus on the situation has since digressed due to the increase in global warming as a result of greenhouse gas emissions (gas, mainly carbon dioxide, in the atmosphere that absorbs and emits radiation) created by human activities, in particular industrial use. The source and quantity of these emissions and the effect on global warming has prompted radical thinking from a range of industries with the consensus to act responsibly and show consideration to the environment. The construction industry has recognised this problem, and initiated improvements by proactively promoting changes in organisational practice and customs. A number of customary influences akin to the environment and the construction industry are discussed in the following sections.

1.8.3.1 Green Business
This term expresses a company's policy of committing to cancel any negative environmental impact during its business operations at a local and global level with a view to safeguarding the community, business and society.

1.8.3.2 Green Certification
In the UK, the use of Green certification in buildings is steered by the Housing Act 2004 and Housing Act (Scotland) 2006 that implements the Energy Performance of Buildings (Certificate and Inspections) (England and Wales) Regulations 2007 and the Home

Information Pack (Nr 2) Regulations 2007. This legislation makes it mandatory to display an Energy Performance Certificate (EPC) in a new residential building (or any type for sale or rented unless used for less than 4 months per year) and a Display Energy Certificate (DEC) for any public building with a useful floor area in excess of $250 \, m^2$. Certificates are graded as A (best energy efficiency and lower running costs) to G (least efficient and higher running costs), with the average being D. Compliance with this legislation underlines the prior accomplishments of BREEAM (Building Research Establishment Environmental Assessment Method). BREEAM was established by the Building Research Establishment (BRE), and is a voluntary measurement assessing the environmental impact on a range of building types for use by clients, developers, designers and those interested in the environmental aspect of buildings. BREEAM assessments include suggested methods for reducing running costs through the whole life cycle of buildings and providing innovative assessment tools for guidance. Certification is possible with the use of templates that have benchmarks with a scoring system for ideas and standards that can be included in the design of buildings.

Green certification schemes also exist in Australia under the control of the Green Building Council, in Qatar as the Qatar Sustainability Assessment System (QSAS) and in the USA (plus a number of others countries) with LEED (Leadership in Energy and Environmental Design). LEED was developed by the US Green Building Council, and is a private non-profit trade organisation promoting sustainability in buildings regarding design, the construction process and occupational use.

1.8.3.3 Sustainability

This broad term refers to the activity of any business participating in and promoting green business. The concept of sustainability is one that ensures work processing and product manufacturing address environmental factors while the organisation maintains a business profit. The objective here is that sustainability meets the 'triple bottom line' that refers to people, planet and profit as discussed in the Brundtland Report, published by the United Nations World Commission on Environment and Development.

1.8.3.4 Sustainable Materials and Buildings

A part of green business involves promoting product specifications that seek to control the use of natural materials used in the construction of buildings and consider recycled products instead. Recycling is possible with a number of building materials such as aggregates, metals, glass and paper that, when recycled, produce new products seen as alternatives to products manufactured from natural resources. For example, crushed and graded demolition material can be used as aggregate in the production of concrete instead of quarried products which is a drain of natural resources.

Buildings designed to be constructed with sustainability in mind are constructed to specifications that have the aims of:

- improving insulation in order to reduce heat and sound emissions;
- making use of grey water from washing machines, etc. for use as irrigation; and
- providing efficient heat exchange with heating systems by installing solar panels to create and store energy and reduce demand on the electric grid.

The additional cost for sustainable measures in new buildings compared to those that exclude the requirement is within the range of 2–5%. However, this additional cost is variable and influenced by specification criteria and the project, as logically, the higher the price to construct a building, the more attractive sustainable buildings become due to the add-on prices becoming less significant.

1.8.3.5 Life-Cycle Costs

A client venturing into a construction project may express a need to satisfy environmental matters, yet have concerns regarding the additional cost of constructing a building and the running costs to be incurred once it is occupied. It is possible to address this concern with a life-cycle cost assessment, which is usually carried out by a consultant who may be a quantity surveyor, and is the critical analysis of a full building design or part thereof. A completed assessment will demonstrate how cost premiums included in the construction can provide benefit through the life cycle, giving a return of investment. For example, suppose lighting options are being considered for back-of-house areas to a new building undergoing design development (e.g. basement-level car parking, corridors, etc.) with the option to use either fluorescent or light-emitting diode (LED) fixtures, the latter being a more sustainable option. This could be appraised by assessing the cost of installation of each option plus the cost to maintain and replace over a given duration using the data provided by the manufacturers. In this example, a period of 25 years is assumed as the duration the areas will be used as back of house. See Table 1.2 for an analysis.

Life-cycle cost assessments provide an indication of the value of a building (or part thereof) beyond a price to build, and provide a financial awareness that can be linked to environmental issues as well as the rewards that can be delivered. If life-cycle factors are a consideration in a design, options should be addressed as early as possible because afterthoughts involving redesigns may cost time and money.

1.8.3.6 Waste Management

It is estimated that the UK construction industry produces one-third of all physical waste generated in the country as a result of demolition and the disposal of excavated material and surplus products created from the construction processes. To mitigate this and make the creators of waste aware, a landfill tax is levied by the government on waste disposal fees with the funding generated used to pay for long-term plans dealing with the impact on the environment. The levy is charged by weight or volume with different rates applying to the type of waste disposed, that is, inactive (natural soils and building fabric materials such as concrete and glass) or active (natural resource products such as wood and chemically produced goods such as plastics). For this reason, there are benefits to a contractor for managing and reducing waste during a construction project which are twofold. Firstly, a reduction in the waste allowance included in a contractor's tender will save the contractor expense on disposal fees. When a contractor submits a tender, an amount of material waste in comparison with installed quantities is allowed as the risk due to cutting to size and damage that varies with the type of product (e.g. additional 5% for wood, 15% for bricks, etc.). If a contractor purchases materials to values that exceed the allowances, it will result in a financial loss which is not usually recoverable from the client. However, it may be possible to avoid the burden by implementing waste management strategies with aims of reducing waste and saving disposal fees. This management style involves a policy of ordering materials as 'just in time deliveries' where goods are installed as soon as practically possible after delivery to site.

Table 1.2 Life-cycle cost estimates for different internal lighting options (CAPEX: capital expenditure; OPEX: operating costs).

Investment period: 25 years			Maintenance cost			Life-cycle replacement cost			CAPEX
Asset nr	Level	Installation cost (£)	Preventative	Corrective	Total (£)	Replacement frequency (years)	Nr replacements	Replacement cost (£)	Total (£)
Option 1: fluorescent fittings including lamps and ballasts (35 W)									
1001	0–3	40,000	0	2,500	62,500	4	6	240,000	342,500
1002	4–7	10,000	0	500	12,500	4	6	60,000	82,500
1003	8–11	10,000	0	500	12,500	4	6	60,000	82,500
Total CAPEX		60,000			87,500			360,000	507,500
Total OPEX/operating cost, say 1000 fittings for 180,000 hrs over the investment period = 6,300,000 kWh @ 0.10									630,000
Total CAPEX and OPEX over investment period									1,137,500
Option 2: LED fittings including lamps and drives (30 W)									
1001	0–3	50,000	0	1,000	25,000	7	3	150,000	225,000
1002	4–7	15,000	0	300	7,500	7	3	45,000	67,500
1003	8–11	15,000	0	300	7,500	7	3	45,000	67,500
Total CAPEX		80,000			40,000			240,000	360,000
Total OPEX/operating cost, say 1000 fittings for 180,000 hrs over the investment period = 5,400,000 kWh @ 0.10									540,000
Total CAPEX and OPEX over investment period									900,000

Notes:
(1) The above tables demonstrates that a higher investment for the installation can provide benefits over the long term.
(2) Figures exclude inflation or deflation in prices over the investment period.
(3) Risk from defects and damage to lights is greater at lower building levels due to the frequent use by the public and vehicular traffic. The risk for LEDs is however considered less.
(4) No provision for preventative maintenance is needed, as routine inspections deemed unnecessary.

For maximum effect, the provision of adequate storage and protection of unfixed materials is required as well as the careful planning of site activities. Secondly, the inclusion of a waste management plan on a project demonstrates commitment to the environment, helping to raise corporate image.

The contractor's quantity surveyor must show an interest in waste management as it can affect budgets and profit margins. An effective measure for limiting the disposal of waste includes compacting and breaking up bulky items and placing them into skips, thus reducing voids in bulk waste and reducing the number of skips required. Another effective measure is dealing with excess spoil generated from earthworks operations where it is possible to accommodate the material on site by spreading and levelling around low levels of land as fill material. This process is used by developers during new works on green (virgin) or brown (reclaimed) land, where the topography permits innovation with the levels. If existing levels on land are not suitable for filling, an option is to raise the level of the building(s) by a nominal height and filling the areas created. This option requires the quantity surveyor to carry out a cost exercise by assessing the additional cost for varying the construction works which can be offset or mitigated by the reserved budget allowance for loading, hauling and disposing spoil off site. If proving viable, the filling method will require approval from a person in authority to ensure any increase in building height is practical and complies with planning approval. Furthermore, any filling works are subject to approval by the engineer and contractor who must consider handling of the material and capping the fill with inert material if required.

1.8.3.7 Lean Construction

Adopted by the International Group for Lean Construction, the lean approach strives to continuously improve standards, minimise cost and maximise value while maintaining the project client's needs. The strategies commence at design stage and continue through the construction phase, with intent to limit maintenance works after a building is occupied. This can be carried out with precision by following the lean construction theme of limiting waste with the use of proactive management that actively seeks flawless behaviour by improving communication and using procurement systems with strong supply chains. It has the same aims as a 'master builder' concept, is not restricted to environmental issues and applies a lean theme for minimising time and effort applicable to the design, construction and occupational phases of construction projects.

1.9 Prospects for the Contractor's Quantity Surveyor

People from a quantity surveying background may seek to work in affiliated roles under the guise of other titles where the training and qualifications are comparable. To be effective, suitably qualified individuals wishing to specialise in certain fields of work or seek roles in countries other than the country where their education and qualification was acquired must commit to obtaining the required skill sets. In order to obtain these skill sets, there may be a need to undergo additional training; before doing this, it is necessary to understand the roles as each discipline carries different levels of responsibility.

1.9.1 Contracts Administrator

Contract administration refers to post-contract activity dealing with the commercial, contractual and cost management of projects. In the UK construction industry, qualified quantity surveyors can be engaged as contract administrators by cost consultancies, main contractors and large-sized subcontractors, responsible to a project manager or team leader for their duties. Incumbents are usually employed full time to service one or more concurrent schemes under the title of quantity surveyor, contract administrator or project administrator. For a range of duties applicable to this role, see Sections 1.2.7 and 1.2.8 earlier in this chapter.

In commonwealth countries outside the UK and Ireland, the title quantity surveyor is often restricted to client-side quantity surveying. However, procurement strategies in these countries usually exclude a project bill of quantities as a contractual document and, where produced, are used for reference only. As a result, the consensus is that measurement skills and the production of a bill of quantities is specific to client-side activities and an integral part of quantity surveying degree courses with the training included to a lesser extent in construction management and other related degree courses. Moreover, the overseas title contracts administrator is adopted by main contractors and large-sized subcontractors, which in essence is the UK's equivalent of the contractor's quantity surveyor.

Contract administration can also mean administering the contract between the project client and main contractor, which carries a greater degree of responsibility than the aforementioned and involves complying with and enforcing the terms and conditions of the agreement, as well as documenting and agreeing any changes that may arise during the period of the agreement. For the client side, the position may also be known as superintendent, client's agent, employer's agent/representative, project manager or other lead discipline suitable for the type of project (e.g. architect for buildings and engineer for works of an engineering nature). The contractor carrying out the works will also appoint a contract administrator who may be the contractor's project manager or quantity surveyor and has an interface with the client-side contract administrator. A fundamental characteristic of contract administration is that whoever administers the contract is bound by the rules, meaning the conduct involved must be free of personality. They must therefore abide by the terms and conditions of a legally executed contract, whether considered fair or not, as departure will constitute breach of contract.

1.9.2 Contracts Manager

The title contracts manager can mean contracts administrator with the added duty of being involved with the pre-contract activities of a contractor's business where the terms and conditions of any construction contract provided with an invitation to tender is reviewed to identify risks and responsibilities which may influence the price to tender. The duties of a contracts manager may extend to the post-tender period, where the terms and conditions of a pending contract are negotiated to ensure the contractor understands the risks, rights and obligations of an agreement it may elect to enter into. Contract manager's skill sets revolve around an understanding of the structure of various standard forms of contracts and the reasons for the creation of edited standard forms and purpose-made agreements as permitted by law. These managers also need to understand the process of receiving and issuing Letters of Intent, Letters of Acceptance and Preliminary Agreements, as well as their wording. Such letters and agreements are issued in the absence of a formal contract, where a start on site is required before the formality can be concluded, and the contract manager must understand their affect until the contract is executed and the contents of the letter or agreement discharged.

1.9.3 Commercial Manager

Commercial managers have varied roles that focus on the commercial activities of a business, including:

- marketing and business development for company expansion;
- contract negotiations, including reviewing the terms and conditions of pending awards and the price involved;
- property management;
- supply chain management, including vetting and administration of their contracts;
- cost managing projects; and
- managing business overheads.

Large companies undertaking a number of concurrent projects may engage one or more commercial managers to oversee a group of people. When employed in this capacity they are usually responsible to a director for commercial activities and may have legal training in commercial law and/or contract law. With smaller companies, the commercial manager would normally be responsible for tasks themselves and be aware of the strategic functions of the business.

1.9.4 Project Manager

A project manager engaged by a client acts as an agent of the client and addresses matters required for the successful delivery of a project. On large-scale projects, a client may engage a project management company under the control of a project director. The project director does not carry the legal status of a company director for a project because a project is not usually a business. However, a project director may be a director of a project management company that supervises a team to oversee a scheme. The scope of services provided by a project management company includes:

- applying for planning permission;
- preparing a project brief of key requirements;
- tendering and recommending consultant appointments;
- setting budges for defined scopes of work;
- recommending suitable procurement routes;
- monitoring design development;
- vetting contractors and inviting tenders for works;
- negotiating the terms and conditions of a construction contract and advising the client;
- overseeing construction of the works to completion; and
- acting in capacity of contract administrator and administering the contract on behalf of the project client/employer.

A client-side project manager involved with pre-contract activity can also be referred to as contracts manager or procurement officer tasked with the responsibility of managing the scheme for the client, and at the same time be the client's agent.

Project managers employed by contractors provide working programmes, supervise staff and accept overall responsibility for delivering a scheme on behalf of the contractor. They also provide feedback from committed projects to senior management in the contractor's business to identify risks that influence commercial decisions for works under tender.

1.9.5 Cost Engineer

The title quantity surveyor is adopted by the UK and is widely recognised overseas, in particular in Ireland and Commonwealth countries. However, outside these countries, the title cost engineer is often used instead, the USA being an example. This is modified further to 'engineer' or 'eng', as used in the Middle East, placing it on a par with engineering professions, a presumption being the qualification of a quantity surveyor is comparable with a technical engineer.

1.9.6 Estimator

It is worth understanding the difference between estimating services provided to a client for advice and estimating carried out by a contractor to secure work. The client's quantity surveyor/cost manager provides a cost planning service for a client in the capacity of consultant. In this capacity, an estimate of probable cost for the proposed works is issued with little or no design information being available, with the estimate modified at intervals of design development. This is a cost management role advising the client's team of cost forecasts until the design and documentation is suitable to invite tenders for the works from competing contractors. A main contractor's estimator prepares a cost estimate based upon the tender design and documentation, and gains a test of market pricing for the works under tender. Once the estimate is complete, a sum is added to cover business overheads and profit to convert the estimate into a tender and legal offer to carry out the works. Ideally, the client's quantity surveyor/cost manager's final estimate of probable cost should be similar to the tendered amounts. If there is a large discrepancy, it normally has nothing to do with the main contractor's estimator and is a matter for the client and client's team.

1.9.7 Independent Roles

People may seek flexibility and diversity in their career if not wishing or are unable to work full time, which can be accommodated by employers offering short-term contracts. These contracts benefit employers when tasks require completing in order to fulfil short-term needs which may suit independent contractors. For example, a main contractor may require a quantity surveyor to prepare variations or assist with the award of trade packages when permanent staff members are on leave. Individuals may also seek supplementary fields of work and make use of short-term employment opportunities to provide income while undergoing training for roles, such as in:

- dispute resolution, including being on a panel of arbitrators, expert witness or dispute boards;
- contract and commercial aspects of facilities management;
- BIM manager;
- BREEAM assessors of quality assurance certification; or
- book writing.

The key to the success of independent working is an individuals' reliability, effective communication skills and self-organisation, together with qualifications, proven experience and marketing skills. The downside is a lack of permanent job security, no entitlement for holiday leave and no career progression within the company.

The role of quantity surveying has expanded from one of traditional core skills involving measurement and the pricing of building works to one of broader involvement in the construction industry. There is a demand for quantity surveyors and the requirement to diversify into various roles as discussed in this chapter, and individuals are encouraged to seize the opportunity and work in their chosen field. The construction industry changes at a fast pace and, with the growth in information technology, environmental issues and management arrangements, there is a need to share ideas that benefit employees, employers, peers, clients and the industry. Although the profession has strong traditional values that have stood the test of time, it is encouraging the future appears to warrant the continuous need for the quantity surveyor, albeit in a variety of forms.

2

Measurement and Quantities

2.1   Measurement Guides and Coverage Rules

Measurement and the quantification of trade works is at the core of quantity survey-ing and has long been the basis of the construction industry's tendering systems. The traditional document used to show this information is a trade bill of quantities (BQ, or sometimes referred to as BOQ) prepared by a project clients' quantity surveyor. If a client wishes to include a BQ as a tender document in order to obtain a lump-sum price for the works, the BQ and stated quantities will become binding on the con-tractor and client if included in a contract. Once becoming a contractual document, the stated quantities take priority and can only be adjusted with a variation order authorised by the client's agent, or title named in the contract with the authority to issue a variation.

A BQ should be formatted in accordance with a set of coverage rules obtained from a measurement guide and comprise quantities alongside descriptions of the works. This permits the contractor's estimator to understand the requirements and charge rates to given quantities inclusive of labour, plant and materials to determine a price. When preparing a BQ, the design and specifications must be sufficient and should represent the final information because the objective of a BQ is for compet-ing contractors to provide a fixed lump-sum price based upon firm quantities. If a client wishes an early start on a project with the design partially complete, a bill of approximate quantities can be prepared instead. Here, quantities are ascertained from the design and documentation available to which contractors apply rates to permit the works to commence at a provisionally agreed price. Subsequently, the actual installed works are re-measured, quantified and valued in accordance with the rates creating a revised price in the process.

Clients may decline from including a BQ as a tender document because of the alternative procurement routes available that make it possible to transfer risk for ascertaining quantities to main contractors. Where applying, main contractors, and in turn subcontractors and suppliers, tend to adopt the term BQ generically or refer to it as builder's quantities or measured schedules of works. However, the use of a traditional BQ as a tender and contractual document refuses to become obsolete with project clients, contractors and cost consultancies recognising the advantages it provides, which is summarised as follows:

Construction Quantity Surveying: A Practical Guide for the Contractor's QS, Second Edition. Donald Towey.
© 2018 John Wiley & Sons Ltd. Published 2018 by John Wiley & Sons Ltd.

- It is a valuable source of reference for obtaining tenders that reduces contractor's tendering costs, meaning greater competition between contractors which in turn benefits the client.
- A competitive market is created as contractors and subcontractors are more willing to price work from a BQ rather than prepare quantities themselves.
- Negotiation periods following receipt of tenders are rapid, with a quick start on site possible due to the scope identified in the BQ.
- It assists contract administration to identify and value works in progress.
- Rates in the BQ can be used as a basis for pricing variations.
- It assists with the preparation of a final account as it sets the basis of the contract sum.

To standardise the format of a BQ, the industry adopts standard guides that define the rules of measurement and the methods used to describe works for a range of project types. The various guides available are discussed in Sections 2.2 and 2.3 below.

2.2 RICS New Rules of Measurement (NRM)

The RICS 'new rules of measurement' (abbreviated as nrm) is a suite of documents developed by the Quantity Surveying and Construction Professional Group of the RICS. The purpose of the documents is to underpin a set of measurement rules in a series of comprehensive guides. The guides can be used by cost consultancies, contractors, subcontractors and others involved in the measurement, estimation and cost management of projects that require a measurement guide for the presentation of information within their business.

2.2.1 NRM1

The first document of the nrm suite, nrm1, is entitled 'Order of cost estimating and elemental cost planning'. It was published in 2009 and republished in 2012 as the 'Order of cost estimating and cost planning for capital building works', and is a publication that sets the guidelines for the preparation of cost estimates and cost plans. Divided into four parts, the publication provides advice on the logical preparation and arrangement of works for the presentation of cost plans, and provides a set of measurement rules using terminology understandable to those involved in a construction project. Although written for the UK construction industry, the rules have an underlying philosophy for use anywhere in the world. Part 1 addresses the appropriateness of the rules with the Royal Institute of British Architects (RIBA) Outline Plan of Works and the (then) Office of Government and Commerce (OGC) Gateway review. (When it existed, the OGC produced guidance about best practice in project management. The OGC guidance has now been archived; however, its use is still seen as a comprehensive set of guidance for public sector projects.) Parts 2 and 3 address principles and measurement rules for cost estimating and cost planning, and Part 4 tabulates the measurement rules into group elements, including appendices of templates to satisfy the needs of any project.

2.2.2 NRM2

The second volume of the suite, nrm2, is entitled 'Detailed measurement for building works'. The first edition was published in 2012 and, according to the RICS, 'provides a uniform basis for measuring and describing building works'. The RICS publishes a number of documents which are categorised as practice statements, codes of practice, information papers or guidance notes to which nrm2 is classified as a guidance note with its status defined as 'Recommended good practice'. Moreover, the RICS considers the publication a fundamental document for use as a measurement guide and a suitable source of reference for quantifying and describing building works when preparing a bill of quantities (BQ) or quantified schedule of works. The guide is also considered a suitable source for assisting the development of a bespoke schedule of works where quantities cannot be affirmed. Nrm2 is of particular interest to the contractor's quantity surveyor, as it is likely at some time the quantity surveyor will deal with a BQ/schedule prepared from the publication. The scope of nrm2, including examples, is dealt with in more detail throughout this chapter.

2.2.2.1 Composition

The document comprises three parts plus a series of appendices. Part 1 is a 'General' section addressing the context of the guide and its relevance to the RIBA Plan of Works and OGC Gateway process. Other contents address the purpose of nrm2, its use, document structure, symbols for use with quantities, abbreviations and a glossary of definitions. Part 2, 'Rules for detailed measurement of building works', outlines the purpose and benefits of detailed measurement with a focus on the advantages, preparation and contents to include in a BQ/schedule including guidance on how to deal with:

- contractor's preliminaries, including overheads/profit and design responsibilities;
- project credits;
- how to address project specific information;
- defining risk allocation;
- critical information required to produce a BQ/schedule; and
- suggestions on the coding of trade bills.

Part 3, 'Tabulated rules of measurement for building works', comprises a formatted list of project overheads and trades required for the procurement of building to include in a BQ/schedule, with the works tabulated for easy identification.

The appendices comprise templates for use with the parts, including:

- guidance for the preparation of a BQ/schedule;
- expanded and condensed pricing schedules for a projects' preliminaries;
- expanded and condensed pricing summaries for building elements that can be used as a tender return form;
- templates for the separation of provisional sums, risks and credits from works; and
- work breakdown structure (WBS) for code identification.

2.2.3 NRM3

The third volume in the series, nrm3, is entitled 'Order of cost estimating and cost planning for building maintenance works'. The first edition was published in 2014, and follows the framework and premise of nrm1 but with the theme of cost estimating and cost planning for the maintenance of buildings. Nrm3 has the aims of recognising items associated with the cost of maintenance not usually reflected in the scope of maintenance work descriptions but which incur indirect expense, for example contractor's management and administrative charges, risk allowances, design consultant fees and contractor's overheads and profit. Nrm3 is intended to be a reliable and trustworthy document that assists with the creation of schedules necessary to clarify and quantify works for the purpose of evaluating costs once a building is occupied. It is therefore suitable for use by facilities management consultants, contractors and those with an interest in the cost of maintaining buildings.

2.3 Other Measurement Guides

In addition to the NRM suite of documents, there are a number of other standard measurement guides that provide coverage rules for the preparation of a BQ/measured schedule applicable for a range of project types.

2.3.1 Standard Method of Measurement of Building Works (SMM)

The first SMM was produced by the RICS and appeared in 1922 with the aims of providing advice to quantity surveyors on how to measure building works for the production of a BQ. The guide helped define the methods needed for creating a BQ that clearly details the work requirements necessary for obtaining tenders for building works. The last edition (SMM7), re-published in 1998, was replaced by RICS nrm2 (see Section **2.2.2** for details).

2.3.2 Civil Engineering Standard Method of Measurement (CESMM)

The Institution of Civil Engineers published their first Standard Method of Measurement in 1976 to provide those responsible for pricing civil engineering works with advice regarding the rules of measurement. It has since been updated and is currently available in a fourth edition (CESMM4), published in 2012.

CESMM4 differs from nrm1 and 2 due to the nature of civil engineering works. Fixed scopes of works within a civils BQ are described as composite items that list inclusions, whereas building measurement guides such as nrm1 and 2 consider the finer details of building construction. The nature of civil engineering works means some works carry a greater degree of risk with price certainty when compared with building works, with the project client often retaining the lion's share of the risk. For this reason, a civil works BQ based on CESMM4 may include provisional sums and/or provisional quantities in addition to a quantified scope of work as well as a schedule of plant items for pricing by the contractor, for example mobilisation to site, running costs and demobilisation. The assessment of works in progress for payment purposes is based upon specified operational plant mobilised to site at the time of the valuation, plus the value of completed works ascertained from the priced BQ. Any authorised and completed works involving plant operations which are extra to the contract are reimbursed in accordance with the running cost schedule of rates.

2.3.3 Manual of Contract Documents for Highway Works (MCHW)

This manual comprises a number of volumes of documents dealing with the specifications, design and management of highways. It includes model forms of contract, with one volume dedicated to the preparation of a BQ including a Method of Measurement for Highway Works and guidance notes.

2.3.4 Standard Method of Measurement for Industrial Engineering Construction

This guide was produced as a result of an authorised agreement between the RICS and the Association of Cost Engineers (ACostE). ACostE is a body representing the professional interest of those tasked with the responsibility of predicting, planning and controlling resources and the cost of activities associated with engineering, manufacturing and construction works. ACostE has also produced the Standard Method of Measurement for Industrial Engineering Projects, dealing with measurement principles for the management of industrially engineered projects.

2.3.5 Overseas

There are a number of guides similar to nrm2 used in counties outside the UK which are culturally developed for their own use. In Australia, the Australian Standard Method of Measurement of Building Works (sixth edition) as published by the Australian Institute of Quantity Surveyors (AIQS) is widely used. Similar guides are also used in Canada and Hong Kong, and in Ireland the Agreed Rules of Measurement version 4 (ARM4) authored by the Society of Chartered Surveyors Ireland is seen as an alternative to nrm2. In the Middle East, the Principles of Measurement (International): For Works of Construction (POMI), as published by the RICS, is widely used and available in English, Arabic, French and German.

2.3.6 International Construction Measurement Standards (ICMS)

The ICMS was formed in 2015 as a coalition of professional bodies representing cost management professionals in building and infrastructure in key global markets. It comprises a Standards Setting Committee (SSC) that has the aims and objectives of the ICMS:

- to agree what is included and excluded from construction costs for reporting at both project and national levels;
- to create a framework for a standard system of project costing to allow for cost comparisons on a like-for-like basis between countries; and
- to let governments and international bodies compare project costs.

The ICMS framework comprises four levels: Level 1, project category (e.g. buildings, bridges etc.); Level 2, cost categories including capital and associated costs; Level 3, cost groups; and Level 4, subgroups. Levels 3 and 4 are for populating with data. The arrangement can be an alternative to RICS nrm1 for use when preparing a pre-contract high-level cost forecast for a project, except it is not restricted to building works. In parallel to the nrm suite, the framework of ICMS has aims of transferring information to the construction phase so the works/items can be migrated for schedule production and cost management.

2.4 Arrangement of Project Information

A contractor is responsible for producing a schedule of quantities for pricing if wishing to submit an offer for works where the terms of an invitation to tender exclude the supply of a bill of quantities. Before commencing BQ/schedule production, it is necessary for the contractor to be in receipt of relevant designs, documentation and project information. This criterion is listed in a document register that usually categorises each discipline as a series and subseries for easy identification, for example AR-1000 to 1010 architectural plans, AR-2000 to 2020 enlarged architectural plans, SE-1000 to 1010 structural engineering plans, SE-2000 to 2020 enlarged structural engineering plans, etc., including titles, numbers, revisions (if applicable) and dates of each. When the information is distributed it is usually accompanied with a document transmittal, which is a copy of the document register listing the information transmitted and the date of issue. The mode of distribution is as stated in the terms and conditions of the invitation to tender, either: hard copy by hand, courier or post; soft copy by compact disc or USB; or electronically by email (subject to the consignment size), collaborative system or via e-tendering. For more information on e-tendering refer to Chapter 1, Section 1.8.1.5.

2.4.1 Drawings

Drawings relay working information in dimensional and illustrative forms to create the image of a building. A project at an advanced stage of design development and suitable for preparing a BQ/schedule will include survey drawings based on a measured or ordinance survey, including site and block plans as well as the outline of any existing buildings and/or location of existing utility services. Such drawings will include:

- demolition plans (if applicable);
- architectural floor plans, elevations and building sections;
- specific construction details showing details to a smaller scale than shown on other drawings;
- detailed joinery requirements;
- ceilings and partitions layouts if not already a specific construction detail;
- door, windows and ironmongery schedules;
- wall and floor finishes;
- structural engineering drawings and schedules (e.g. columns and beams);
- details of any temporary works (e.g. façade retention);
- building services engineering drawings;
- civil engineering drawings and schedules (e.g. inspection chambers);
- external works layouts with any specific details for fencing and hard surfaces;
- landscaping layouts;
- specialist trade works that may be read in conjunction with any of the above; and/or
- schedules of plant and equipment.

The number of drawings issued depends on the size and complexity of a project. For BQ/schedule preparation, the drawings need to be clear (including to scale), accurate and suitable for the purpose.

2.4.2 Specifications

A specification is a technical document prepared by a consultant that details the standards required to fulfil a project's needs. It is an important document as it defines the quality and standards expected so that a completed building is suited for the intended purpose. There are two types of specification, namely functional and descriptive. A functional, or sometimes called performance specification, is the expressive detailing of the requirements of a building necessary for it to serve the intended use. This may apply to the whole or part(s) of a building and is drafted in the early stages, possibly in the absence of a design to identify the outline needs of a project. This type of specification is dynamic and lateral and provides a snapshot of the requirements, permitting a designer to build a platform from which to create a concept design. An example of a functional specification is as follows.

> New mainstream primary school constructed over two storeys to provide accommodation for 300 children including learning resources, assembly hall and dining area, staffroom, storage, toilets, circulation, outdoor physical education area and external car parking for staff and visitors. Target building area 1,500 m^2 on a site area of 2,500 m^2.

Descriptive specifications are a later version of the functional/performance specification and segment the requirement into specific components, producing numerous specifications in the process. Wording in these specifications is definitive, comprising descriptions that break down the functional requirements into finer details in order to clarify and regiment the purpose. The defined details comprise sections formatted to specify scope necessary for a project that includes material manufacturer's names, brands and schedules; trade workmanship standards; general items; codes of practice; quality standards; and legal requirements (including reference to the building regulations, health and safety and the environment). This generally follows the requirements of the National Building Specification (NBS), the UK's standard system of specification writing which is used by architects and other building professionals for the description of materials, standards and workmanship for a construction project. For the purpose of measurement and BQ/schedule preparation, the descriptive specification is of interest to the contractor as it defines the requirements for inclusion in a BQ/schedule. To demonstrate, Table 2.1 provides a descriptive specification for moisture-resistant barriers and insulation applicable to new roofing works.

2.4.3 Reports

Dependent on the type of project, the documentation provided may include reports that have relevance to the project under tender. For example, a hazardous material report would be prepared for a refurbishment project to identify the presence of any asbestos. If identified, this material poses a risk to the health and safety of operatives when disturbed; if the report is supplied, it must be mentioned in the BQ/schedule as there is a cost for its removal. In addition, any presence of this material in a building together with the method of handling and disposal is required as part of a health and safety plan that competing contractors must supply with their tender.

2.4.4 Planning Production Schedules

With small-sized projects uncomplicated in scope, sufficiently designed and valued up to say £3 million, the contractor's estimator and assistant will usually prepare the BQ/schedules. Beyond this, the contractor's business is normally of a sufficient size to manage the process and employs a suitable number of technical staff to carry out the tasks. Alternatively, the contractor may elect to outsource a BQ/schedule from a cost consultancy. If outsourcing, it is not unusual for competing contractors to pool together

Table 2.1 Moisture-resistant barriers and roof insulation

1 Quality

1.1 Scope

The scope of work for this trade includes:

1) Supply and installation of water-vapour-permeable membrane.

2) Supply and installation of thermal insulation boards to new roofing.

1.2 Cross-references

General

Conform to the general requirements of the specification.

Related work sections as follows:

- ROOFING
- SEALANTS
- PLYWOOD AND TIMBER COVERINGS

1.3 Environmental Management System (EMS)

Refer to the general requirements section that outlines the EMS. Products used are to comply with the lowest environmental impact in accordance with ISO 14001.

1.4 Standards

Installation of mineral wool insulation

Comply with Health and Safety practices in accordance with the Construction Design and Management Regulations (CDM) 2015 and Control of Substances Hazardous to Health (COSHH) Regulations 2002.

2 Inspection

2.1 Inspection

Undertake and record inspections of the following to ensure compliance with the contract requirements:

- Sarking-type material before being covered up or concealed.
- Thermal insulation boards before they are covered up or concealed.

3 Materials and components

3.1 Sarking material

Sarking: Spun-bound polypropylene composition with polyolefin coating. National Building Specification P10: sundry insulation/proofing work, subcode P310: vapour-control layer at rafter level to BS 5250

Water vapour permeability: Not exceeding 20 g/m^2 per day to BS 4177

Nail tear resistance: 260 N

Flammability: DIN 4102: B2

Head of water sustained without penetration: to 2 m as BS 20811

3.2 Thermal insulation boards

Boarding: Polyisocyranate foam core with aluminium foil composite to both sides for warm and cold roof pitch construction. National Building Specification P10: sundry insulation/proofing work; subcode P140: insulation fitted at rafter level to BS 4841 Part 5.

Boarding strength: Compressive strength to BS EN 826

Fire protection: SAA rating to BS 476 Part 3 2004

Thermal resistance: 'R' value 2.61 (m^2 K/W)

4 Execution

4.1 Roof insulation and moisture-resistant barriers

- *Sarking material*: Secured between battens in accordance with the manufacturer's recommendations.
- *Thermal insulation boards*: Secured to timber battens and plywood in accordance with the manufacturer's recommendations.

and share the cost for preparing a common BQ/schedule procured from a cost consultant. This has advantages to tendering companies as it could become a shared tender cost while also placing contractors on an equal footing, as in effect they should be pricing the same quantified scope of works. Of course, the cost consultancy must agree to this arrangement and will usually require one contractor to be a point of contact that agrees to receive and pay for the BQ/schedule, which is proportionally reimbursed by the recipient's competitors.

Following nrm2 as a guide, there are three approaches to managing the process of creating a work breakdown structure (WBS) for the arrangement of items to include in a BQ/schedule, which are discussed in Sections 2.4.4.1–2.4.4.3 below.

2.4.4.1 Elemental Breakdown

This breakdown divides the project into 12 parts (e.g. preliminaries, facilitating works, substructure, etc.). Once created, each elemental breakdown is subdivided into a series of group elements with the works so described and quantified.

2.4.4.2 Work Section BQ/Schedule

Here, nrm2 provides a list of 41 works sections suitable for any project described as building components/items. This format is preferred by contractors as it aids the pricing of trades and similar products which are grouped together, for example in-situ concrete for the in-situ concrete works section, whereas an elemental breakdown will allocate in-situ concrete to wherever it is required (e.g. substructure foundations, superstructure walls, external planter box walls, etc.).

2.4.4.3 Work Package Breakdown

The third choice involves the use of 23 recommended work package bills. This is similar to the work section BQ/schedule but with a reduced number of sections to suit uncomplicated work packages. The work package breakdown can be divided into client/employer, quantity surveyor/cost manager or contractor-defined packages, the selection of which may be made by the contractor. The format is considered beneficial for the contractor, as it is compatible with the procurement of individual work packages that a contractor may award in sequence with the construction programme. For example, groundworks package Serial number 4 includes in-situ concrete works for ground beams and pile caps, and concrete works package Serial number 6 includes in-situ concrete works for the superstructure frame, core, shear walls and stairs.

During the measurement process, there may be situations where the contractor might need to raise queries with the client's agent regarding the design and documentation provided. However, before a query is raised it would be wise to examine the information supplied as the answer may not be obvious but nonetheless available in the information provided. For this reason, if there is doubt on any content supplied it would be wise to ask a colleague for a second opinion; this may produce the answer without the need to raise any query that would otherwise be a waste of time.

When it is necessary to initiate a query, the usual approach is to issue a request for information (RFI) which should be in writing and recorded in a register. The recording is necessary because delays in a response may jeopardise timeframes for the BQ/schedule preparation. The request should be addressed to the client's agent and/or appropriate consultant or other entity stated in the tender documents. An RFI should be concise and refer to drawings, specifications, reports and location(s) in a building or elsewhere if applicable. This is to enable the client's agent or consultant to quickly identify the nature of the query, as there may be a large amount of information to sieve through in order to locate the problem. The RFI should also include a timeframe for a response which should be to a reasonable period, say seven working days, unless of an urgent nature.

The register should record the date the RFI was issued, the mode of issue, who it was sent to, description of the query, date of receiving a response and a note stating if the response is conclusive or inconclusive. The exact nature of what is considered inconclusive must be noted, for example if a response states there will be no answer until the architect is back from holiday. Once conclusive, a completed tick box on the register will confirm the matter as closed for there to be no doubt. The creation of an RFI register permits a supervisor to understand the status of RFIs at any time to enable the production of the BQ/schedule to meet the deadline. Should the timing of a conclusive RFI response create a genuine delay for the preparation of a tender, the estimating manager should request an extension to the tender submission date. Normally, this is granted if the consensus of competing contractors is to make similar requests, with any granted extension being the most commonly requested to ensure competing contractors are treated on an equal basis.

2.5 Measurement Terminology

The use of jargon with measurement terminology uttered by quantity surveyors and estimators requires to be understood by project managers and those involved with the commercial management of construction projects, as it forms an integral part of estimating the cost of works under tender. Furthermore, such jargon is also used when assessing the price of work changes initiated from client change requests during the construction phase. The sections below discuss the fundamentals of such jargon and its use in the process.

2.5.1 Taking-Off

Taking-off is a term used to describe the method of determining quantities for inclusion in trade work sections of a BQ/schedule. It involves scaling or transferring critical dimensions of construction components from drawings to take-off paper or computer software for converting to quantities. When taking-off, applied dimensions are measured and quantified net as fixed in position. This means that resulting quantities derived from transferred dimensions exclude allowances or methods a contractor may use to carry out the works (e.g. waste and shrinkage from mixing materials such as concrete with water; cutting materials such as timber from standard lengths that produce off-cuts), which is a risk the contractor must consider within the chargeable rate. Traditional take-off paper comprises a series of columns as provided in Figure 2.1, which also explains the purpose of the columns.

The method of inserting dimensions and quantities into columns A, B and C is a straightforward process. However, what is more involved are the methods of determining waste calculations entered into columns A and B. For this reason, column D requires notes on the calculations as they may need referring to at a later date if the works change and the take-off needs revisiting. Measurements and quantities require recording in a format compatible with the rules of nrm2 that use metric dimensions and conversions.

A	B	C	D	A	B	C	D

Notes: (A) Multiplication column for multiplying dimensions entered in Column B. (B) For recording measurements 'taken-off' drawings. The sequence of measuring involves entering figures in order of length (first) width (second) followed by depth. (C) Squaring column or working-up column, the calculated quantity from information entered in columns A and B. Squaring of calculations is rounded to two decimal places. (D) Description of works, also part of the final BQ/schedule. This column is also used as a source of reference to show how methods for assessing dimensions in Column B (known as 'waste calculations') are derived and how multiplications in Column A are calculated.

Figure 2.1 Traditional take-off lined paper.

Figure 2.2 lists examples of the most frequently used units of measurement and the methods of presenting dimensions and calculations.

With taking-off there is a need for accuracy and, as a general rule, quantities resulting from measures in a take-off should be within 1% to allow for the rounding of calculations in the waste and squaring columns.

2.5.2 Centre Line Calculation

When measuring trade works to the perimeter of a building, the mean girth dimension provided by a designer is a suitable benchmark from which to base other measurements and determine quantities. This includes strip foundations, substructure works and perimeter wall construction including façade coverings and finishes. However, and for practical reasons, construction drawings seldom show this information as it is generally unnecessary for construction purposes; it is however advantageous for taking-off purposes. To assess a mean girth, it is necessary to calculate the centre line which is ascertained from given or scaled lengths of the external or internal dimensions of a building. This is demonstrated in Figure 2.3.

Using the example in Figure 2.3, and in order to calculate the centre line, it is necessary to recognise the relationship of the wall faces BC and CD with AB and AD; this is the wall thickness, 255 mm. The centre line is assessed by calculating the girth of the building and adjusting the length by the number of external corners twice (BC and CD), multiplied by half the wall thickness (AB and AD). For example, if a building is rectangular or square on plan, a deduction of 1020 mm ($4 \times 2 \times 0.5 \times 255$ mm) is taken from the external girth dimension. If internal dimensions are provided, 1020 mm would be added to the internal girth dimension instead. Once ascertained, the centre line is advantageous as it forms the basis from which additions and deductions can be made to calculate quantities for the various perimeter construction components.

2.5.3 Descriptions of Works

The take-off must comprise clear descriptions of the works for inclusion in the BQ/ schedule for there to be no doubt of the scope and intention. The choice and style of descriptive wording will vary from surveyor to surveyor; however, the salient contents to include should be in accordance with a suitable measurement guide. In the case of

	Item	*Item*		
	Item	*Item*	Locating underground services, electric, maximum depth 600 mm below existing ground level (school playground)	A scope of works can be described as an item which is singular by default. By underlining the item or figure entered in the dimension and squaring columns, it will confirm it is restricted to the adjacent description.
	4	*4*	Removing trees, girth 500– 1500 mm, filling voids with inert material obtained on site	Quantified single measures are numerated (nr).
	50.00	*50.00*	Concrete kerbs, 125 × 254, bullnosed, on and including haunched 20 N/mm² in-situ concrete foundation, 175 × 325	Where the unit of measurement is in metres, it is stated as linear (m).
5/	15.00	*75.00*		When a given quantity requires multiplying, a multiplication factor with a forward slash is entered.
5/2	10.00	*100.00*		
3/2/2	5.00	*60.00*		
1./2	5.00	*15.00*		If extra quantities are required, they are 'dotted' to indicate an additional amount. In the examples opposite, an additional 1 and 4 are required.
4./3/2	8.00	*80.00*		
	70.00 10.00	*700.00*	Clear site of vegetation overgrowth and dispose off site	Where the unit of measurement is superficial, the area is stated (m²)
3/2	12.00 1.00 2.00	*144.00*	Extra over excavation for excavating in unstable ground	Where the unit of measurement is cubed, the volume is stated (m³)
6/	5.00	30.00	Plain member, universal beam UB310, overall 165 × 304, 40.4 kg/m	Steel is measured by length and converted to weight in tonnes as determined by the conversion factor where 1000 kg = 1 metric tonne.
		1.20	$\dfrac{30 \times 40.4 \text{ kg/m}}{1000}$ = **1.203 t**	
4/	20.00 2.45	196.00	Painting to general surfaces, exceeding 300 mm girth, internal	Where a list of dimensions is for the same description, wording is linked to each set of dimensions to save duplicate writing. In the example opposite, painting to the same specification is required for two different areas of a building.
3/2	30.00 2.70	486.00		
		682.00		

Figure 2.2 Take-off presentation.

	45.00 0.60 1.00	27.00	Foundation excavation, not exceeding 2 m deep, working around piles & Filling obtained from excavated material adjacent foundation excavations, 650 mm deep in maximum 250 mm thick layers	There may be occasions where identical dimensions are suitable for other descriptions. To save duplicate writing of the dimensions, the symbol **&** is used to link descriptions to calculations, known as 'anding on'. In the example opposite, the same calculated volume for the foundation excavation is required for site filling works.
	30.00 30.00	900.00	Lifting turf by mechanical means for preservation, 30 mm thick & Deposit in heaps on site, average 50 m from excavation $\times\ 0.030\,mm = 27.00\,m^3$	'Anding on' may also be used for different units of measurement as shown.
	30.00 45.00 60.00	NIL 60.00	Painting to general surfaces, not exceeding 300 mm girth, internal	If errors occur and a set of dimensions for quantifying is not to be included in the BQ/schedule, by inserting **NIL** in the multiplication column it will ensure the quantities are excluded. In the example opposite, the lengths of the first two dimensions are ignored.
	10.00 10.00 2.50	250.00	*Bulk excavation, basement, over 2 and not exceeding 4 m deep* Bulk excavn, basement, over 2 n.e 4 m dp	The length of a description may be considerable, and to save time in writing or word processing, the wording is abbreviated. A good understanding of abbreviations is therefore required to ensure they are interpreted correctly, hence the reason to use a standard set of abbreviations.
	Item	*Item*	Prepare timber substrate of stairs to receive new finish, including removal of discolourations and staining; fill cracks and holes with suitable filler; knot and stop defective timbers; apply caulking to visible gaps between staircase and wall; fine sand all surfaces; clean down areas with sharp solvent before base staining; apply one coat of base stain, base coat and top coat at recommended intervals to: newel post (1), handrail (1), treads and risers (13), stringers (1), balusters (12), all as per Drawing AR-5000 (Stair 1)	A spot item is used in refurbishment or alteration works, and is a description defining the practical aspects of a task. It is used when the design and/or documentation outlines the scope yet lacks detailed requirements that may only become apparent when carrying out the works. The information provided must be sufficient for it to be understood in a BQ/schedule and can be edited from the information obtained on a drawing which can also be cross referenced.
2/	1	2	Base cabinet 4350 × 915 × 600 mm; comprising polished finish marble slab type ST-14 countertop; wood veneer in stained full fill hi gloss buff finish type WD-03 caseworks; incl fixings and hardware as Specification Section 06 4023; and Drawing Nr AR-5555	A composite item is a description of an item of new work in a building. As with spot items, descriptions must address the item sufficiently enough for it to be understood and may be cross-referenced to a drawing. Where a number of drawings are available showing arrangements and amplified construction or fixing details, it is usual to make reference to the arrangement drawing only.

Figure 2.2 (Continued)

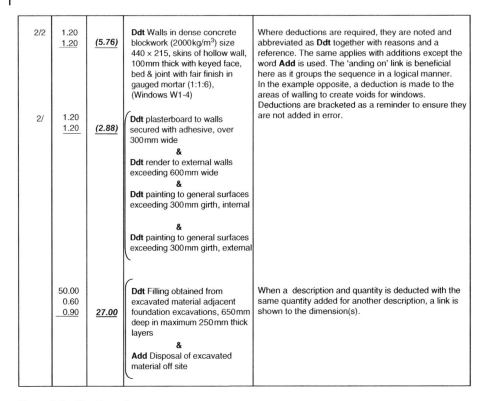

2/2	1.20 1.20	*(5.76)*	**Ddt** Walls in dense concrete blockwork (2000kg/m^3) size 440 × 215, skins of hollow wall, 100mm thick with keyed face, bed & joint with fair finish in gauged mortar (1:1:6), (Windows W1-4)	Where deductions are required, they are noted and abbreviated as **Ddt** together with reasons and a reference. The same applies with additions except the word **Add** is used. The 'anding on' link is beneficial here as it groups the sequence in a logical manner. In the example opposite, a deduction is made to the areas of walling to create voids for windows. Deductions are bracketed as a reminder to ensure they are not added in error.
2/	1.20 1.20	*(2.88)*	**Ddt** plasterboard to walls secured with adhesive, over 300mm wide **&** **Ddt** render to external walls exceeding 600mm wide **&** **Ddt** painting to general surfaces exceeding 300mm girth, internal **&** **Ddt** painting to general surfaces exceeding 300mm girth, external	
	50.00 0.60 0.90	*27.00*	**Ddt** Filling obtained from excavated material adjacent foundation excavations, 650mm deep in maximum 250mm thick layers **&** **Add** Disposal of excavated material off site	When a description and quantity is deducted with the same quantity added for another description, a link is shown to the dimension(s).

Figure 2.2 (Continued)

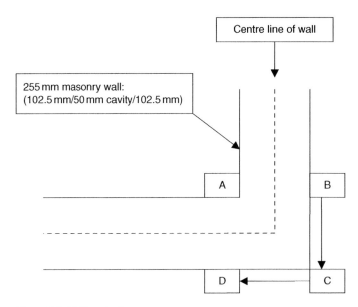

Centre line of wall

255mm masonry wall:
(102.5mm/50mm cavity/102.5mm)

A B

D C

Figure 2.3 Wall centreline.

building works and nrm2 this is found in Part 3, 'Tabulated rules of measurement for building works', which comprises 41 sections of building components (e.g. preliminaries, waterproofing, masonry, carpentry, etc.). These components are main headings that need to form part of the take-off and subsequent BQ/schedule which precede subheadings that divide each building component into parts (e.g. structural metalwork divided into structural steelwork and structural aluminium work), which list items or works to be measured in sequence. Naturally, the list of these items and works is exhaustive as they are intended to cover any scenario; to be effective, only the relevant parts are included in the take-off. To assist, nrm2 categorises items or works to be included in a take-off by levels, defined as:

- Level 1: critical dimensions or sizes to be noted in a description;
- Level 2: prioritised features to include in a description; and
- Level 3: method of fixing or installing when not at the discretion of the contractor, including the nature of the substrate or background.

To demonstrate, Table 2.2 lists extracts from the waterproofing and drainage aboveground sections of nrm2 and a take-off for each for the defined scopes of work. The relevant criteria to include in the take-off and subsequent BQ/schedule is shaded.

Where works are of a special nature, such as excavating in tidal conditions or works to existing buildings, and are not covered by the tabulated rules, nrm2 advocates that such works be described and collated with other similar types of work. This can mean allocating works to appropriate building components/items and creating bespoke descriptions cross-referenced to specifications, drawings, reports and employer requirements where available. This is to ensure the abnormal working conditions of a complicated project are suitably recognised in the BQ/schedule.

2.6   Measurement Example

Let us say a contractor is preparing a tender and a BQ/schedule is required to price the substructure of a new building with the preferred arrangement being an elemental breakdown. The design is shown in Figure 2.4 for this example, which shows an in-situ concrete ground-floor constructed over pile caps and piled foundations. Prior to commencing the take-off, it is necessary to study the design and specifications to become familiar with the requirements. A list of trade items required for the works as identified in the tabulated work sections of nrm2 should thereafter be tabled. An example follows.

Element: Substructure
Trade: Piling (nrm2: Tabulated work section item 7)
Subheading *Bored piling*: Bored piles (Item 2); Reinforcement to in-situ concrete piles (Item 8); Dispose of excavated materials (Item 10); Tests (Item 12).

Trade: Excavating and filling (nrm2: Tabulated work section 5)
Subheading *Excavations*: Foundation excavation (Item 6).
Subheading *Disposal*: Disposal of excavated material off site and groundwater (Item 9); Cut off tops of piles (Item 20).
Subheading *Fillings*: Imported filling (Item 12).
Subheading *Membranes*: Damp-proof membrane (Item 16); Insulation boards (Item 18).

Table 2.2 Nrm2 tabulated work sections

Waterproofing (building component)

Subheading: Flexible sheet tanking or damp proofing

Item or work to be measured	Unit	Level 1	Level 2	Level 3	Notes, comments and glossary
1 Coverings exceeding 500 mm wide	m²	1 Horizontal 2 Sloping pitch stated 3 Vertical 4 Curved: radii stated	1 Underlays 2 Insulation 3 Finish to exposed surface 4 Protection	1 Nature of base 2 Number of coats or layers	1. The area measured is that in contact with the base 2. No deduction is made for voids not exceeding 1.00 m²

Waterproofing

Flexible sheet tanking or damp proofing

30.00			Coverings exceeding 500 mm wide, horizontal to concrete base in single
25.00	750.00		layer, high-density polyethylene film, 1.6 mm thick, cold applied laid in accordance with manufacturer's guidelines

Drainage above ground (building component)

Subheading: Foul drainage installations

Item or work to be measured	Unit	Level 1	Level 2	Level 3	Notes, comments and glossary
1 Pipework	m	1 Nominal diameter	1 Straight, curved, flexible 2 Extendable 3 Method of jointing	1 Method of fixing to background	

Drainage above ground

Foul drainage installations

20/	15.00	300.00	Pipework, straight, 110 mm diameter PVC-U pipe including socket jointing with rubber seals secured to concrete walls with screwed brackets, 2 m vertically and 1 m horizontally

Trade: In-situ concrete works (nrm2: Tabulated work section 11)

Subheading *Reinforced in-situ concrete*: Horizontal work (Item 2).

Subheading *Surface finishes to in-situ concrete*: Trowelling (Item 8).

Subheading *Formwork*: Edges of horizontal work (Item 14); Sides of upstand beams (Item 19).

Subheading *Reinforcement*: Mild steel bars (Item 33); Mesh (Item 37).

Trade: Masonry (nrm2: Tabulated work section 14)

The brick/block wall trade measure including damp-proof course (DPC) is excluded from the exercise as it overlaps with the superstructure and will be measured separately under the superstructure masonry trade.

While carrying out the take-off, it would be wise to keep highlighter pens or coloured pencils to hand in order to mark hard-copy drawings and specifications to demonstrate

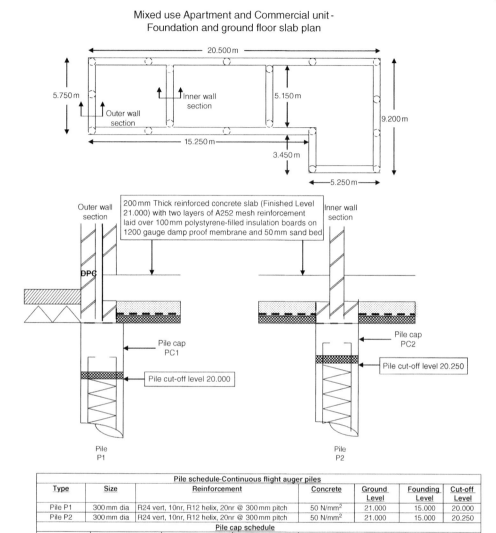

Mixed use Apartment and Commercial unit -
Foundation and ground floor slab plan

	Pile schedule-Continuous flight auger piles						
Type	**Size**	**Reinforcement**		**Concrete**	**Ground Level**	**Founding Level**	**Cut-off Level**
Pile P1	300 mm dia	R24 vert, 10nr, R12 helix, 20nr @ 300 mm pitch		50 N/mm²	21.000	15.000	20.000
Pile P2	300 mm dia	R24 vert, 10nr, R12 helix, 20nr @ 300 mm pitch		50 N/mm²	21.000	15.000	20.250
	Pile cap schedule						
Type	**Size**	**Reinforcement**	**Concrete**		**Notes**		
Pile cap PC1	300 mm × 600 mm	150 kg/m³	50 N/mm²		50 mm sand blinding below pile cap		
Pile cap PC2	300 mm × 350 mm	150 kg/m³	50 N/mm²		50 mm sand blinding below pile cap		

Figure 2.4 Plan of foundation and ground-floor slab.

the relevant criteria as transferred to the take-off paper, word processor/spreadsheet or commercial software if used. The take-off is shown in Figure 2.5.

2.7 Builder's Quantities

The original format for the presentation of a BQ is to BS3327:1970, endorsed as a seven-line column bill, but has since been withdrawn to permit the use of alternative bill formats including builder's bills of quantities (or simply referred to as builder's quantities). However, the original format remains unsurpassed with subtle differences made by quantity surveyors and computer software writers. Builder's quantities are flexible and informal in use, and comprise quantities and descriptions that define a scope of works from a builder's perspective. Descriptions include working or operational methods that can be in full accordance with nrm2 or abbreviated versions in order to create quantified schedules of works for pricing and bespoke schedules for rating only.

			New building - mixed use Apartment and commercial unit	
			General note: Works measured in accordance with RICS nrm2. Tabulated work section building component number shown in brackets.	
			Element: SUBSTRUCTURE	
			Trade: PILING (7)	
			Bored piling	
			(nrm 2 Items 2, 8, 10 & 12)	
		Item	Mobilisation of piling rig to site including setting up and demobilisation upon completion of the works	Piling is a type of building foundation created at deep level, used where soil conditions close to the surface are unsuitable to form a traditional concrete strip footing. For this example, the specification calls for a continuous flight auger (CFA) piling system, a soil replacement type of pile formed to a predetermined depth or on bedrock. The formation of the piles involves boring the ground to the required depth which is encased with lining for stability. Caged bar reinforcement is then inserted along the length with concrete poured down the shaft to form the pile.
	18	**18**	Setting out of piles	
18/	6.00	**108.00**	Bored augured piles, 300 mm dia, max concreted length 6.0 m, comprising boring, casing and 50 N/mm² concrete (18 nr) (P1, P2)	
				Due to the weight and size of the piling rig, a level platform to avoid ground collapse is required. The platform is usually a bed of compacted hardcore laid at ground level to permit the rig to manoeuvre within the footprint of the building and between piles. Any haul road from the entry point of the site to the footprint of the building for the rig to travel will be used for general site access. For this exercise, it is assumed the site is graded to the finished ground level with the haul road and piling platform considered temporary works to be constructed by the main contractor and the construction, maintenance and removal considered part of the Preliminaries (main contract), nrm2 Tabulated work section 1.
18/10	6.00	1,080.00	Mild steel reinforcement to in-situ concrete piles, R24, vertical in prefabricated cage including welding and connecting to helix binding bars	
		3.83	1080 × 3.55 kg/m / 1000 = **3.83 t**	
20/180	0.30	1,080.00	Mild steel reinforcement to in-situ concrete piles, R12, helix, 300 mm pitch in prefabricated cage including welding and connecting to vertical bars	The piling rig mobilisation, demobilisation and pile setting out is so noted or can be covered within the Preliminaries (works package contract) nrm2, Tabulated work section 1. Bored piles are described as per Item 2 of the Tabulated work section and reinforcement as Item 8 with the weight of bar lengths (kg/m) obtained from suppliers tables. Disposal is measured as per Item 10 and testing as Item 12. Level 2 of Item 12 requires the timing of tests to be stated.
		0.96	1080 × 0.89 kg/m / 1000 = **0.96 t**	
18/22/ 7	0.15 0.15 6.00	**7.64**	Dispose of materials arising from piling works to a tip offsite	P1 - Outer wall -14 P2 - Inner wall - 4
	18	**18**	Allow for integrity testing of each pile carried out within 7 calendar days of completion of the works with results submitted to the Engineer within 7 calendar days of carrying out the tests	
				P1 P2 Grd level 21.000 21.000 Form level 15.000 15.000 Length 6.000 6.000
			End of piling	

Figure 2.5 Take-off measurement example.

Trade: EXCAVATING AND FILLING (5)

Excavations

nrm2 Item 6

	58.20		Foundation excavation, commencing level 21.000, ne 2 m deep, working around piles (PC1) (PC2)
	0.30		
	0.65	11.35	
2/	5.15		
	0.30		
	0.40	1.24	
		12.59	**&**

The design shows a series of levels co-related to a fixed level in a position selected by the land surveyor or engineer. When a site is first surveyed, a fixed level point is required as a benchmark from which design levels relate. This point is a station and the level it creates a datum from which design levels are noted. This means design levels are either above or below the datum that commences from the station.

PC1 - Building perimeter

	2/ 20.500	41.000
	2/ 9.200	18.400
		59.400
Less corners, 5 −1 offset = 4		
	4/2/₁/₂ 0.300	−1.200
	Centre line (CL)	58.200

Disposal

nrm2 Items 9 & 20

Item	**Item**	Dispose of materials arising from foundation excavation to a tip offsite
18	**18**	Disposal of groundwater encountered up to 1 m below ground level
		Cutting off tops of piles, 300 mm dia and dispose of debris

The depth of excavation is calculated as the pile cap depth plus blinding. Foundations are described in accordance with Item 6 and disposal of arising materials, groundwater and cutting piles as per Items 9 & 20. Sand filling falls within Item 12, membrane Item 16 and insulation boards Item 18. Support to excavations is only measured where there is a design. For this exercise, there is no design with the works at the discretion of the contractor that must comply with CDM/H&S requirements, which is a normal condition of contract.

Fillings

nrm2 Item 12

	20.00		
	8.70	174.00	Imported filling, sand blinding bed, 50 mm th, laid level below slab and pile caps
	15.25		
	3.45	(52,61)	Ddt inset
2/	5.15		
	0.30	(3.09)	Ddt inner walls
	58.20		
	0.30	17.46	Add PC 1
2/	5.15		
	0.30	3.09	Add PC2
		138.85	

Building length	20.500
Less Walls 2/ 250 mm	−0.500
	20.000
Building width	9.200
Less walls 2/ 250 mm	−0.500
	8.700

Figure 2.5 (Continued)

			Membranes	
			nrm 2 Items 16 & 18	
	20.50 9.20	188.60	1200 gauge damp-proof membrane laid horizontally, over 500 mm wide	The membrane extends up the face of the inner blocks and an adjustment is required.
	15.25 3.45	(52.61)	Ddt inset	CL 58.200 Less half cavity & block $4/2/_{1/2}/$ 0.115 0.460 57.740
	57.74 0.15	8.66	Add inner face	
2/2/	5.15 0.15	3.09		
		147.74		
	20.00 8.70	174.00	Rigid polystyrene filled insulation boards, 100 mm th, over 500 mm wide	
	15.25 3.45	(52,61)	Ddt inset	
2/	5.15 0.30	(3.09)	Ddt inner walls	
		118.39		
			End of Excavating and Filling	
			Trade: IN-SITU CONCRETE WORKS (11) **Reinforced in-situ concrete** **50 N/mm^2**	
			(nrm 2 Item 2)	
	20.20 8.90 0.20	35.96	Horizontal work, in structures, floor slab, ne 300 th	Concrete pours are into formwork without mention noting the structure created and thickness range as required by Levels 1 and 2. Where concrete pours are not into formwork, the type of substrate would be noted.
	15.25 3.45 0.20	(10.52)	Ddt inset	

2/	5.15
	0.30
	0.20
		(0.62)	Ddt inner walls	Building length 20.500
				Less outer wall and cavity 2/150 mm −0.300
				                  20.200
				Building width 9.200
				Less outer wall and cavity 2/150 mm −0.300
				                   8.900

		24.82		
	58.20 0.30 0.60	10.48	Horizontal work, in structures, pile caps, over 300 mm th (PC1)	
2/	5.15 0.30 0.35	1.08	(PC2)	
		11.56		

Figure 2.5   (Continued)

			Surface finishes to in-situ concrete	
			(nrm2 Item 8)	
	20.20 8.90	179.78	Trowelling to top surfaces (Slab)	Horizontal surfaces are to have a trowel finish for follow-on trades
	15.25 3.45	(52.61)	Ddt inset	
2/	5.15 0.20	(2.06)	Ddt inner walls	
	58.20 0.60	34.92	Add PC1	
2/	5.15 0.60	6.18	Add PC2	
		166.21	**Formwork**	
			(nrm2 Items 14 & 19)	
2/	20.20	40.40	Plain formwork, edges of horizontal work, ne 500 mm high, 250 mm wide (slab edge)	Formwork is required for retaining concrete while it cures and a plain finish is suitable here as the faces of the slab edge and pile caps are not exposed. No edge formwork is required at the slab edge to the cross walls on PC2, as the blocks will act as a form.
2/	8.90	17.80		
		58.20		
2/2/	5.15 0.35	7.21	Plain formwork, sides of upstand beams, over 500 mm high, regular shape, rectangle (pile caps)	
	58.20 0.60	69.84		
		77.05		
			Reinforcement	
			(nrm2 Items 33 & 37)	
2/	20.20 8.90	359.56	Mesh reinforcement, A252 3.95 kg/m² to BS4483 with 150 mm min side and end laps	Mesh to the slab is described as per Item 37. Reference is not necessary for tying wire and fixing accessories for reinforcement as methods of fixing is deemed at the contractor's discretion.
2/	15.25 3.45	(105.23)	Ddt inset	
2/2/	5.15 0.20	(4.12)	Ddt inner walls	
		250.21		
			Mild steel bar reinforcement to pile caps including tying into pile reinforcement (PC1 & PC2): by ratio	Reinforcement to pile caps is by ratio to the volume of in-situ concrete.
		11.56	$\dfrac{\times\ 150\,\text{kg/m}^3}{1000}$ = **1.73 t**	
		1.73	**End of in-situ concrete**	

Figure 2.5   (Continued)

2.7.1 Preambles

Preambles are a set of coverage notes that precede described and quantified works in a BQ/schedule. The intention of including preambles is to identify scope not specifically mentioned under trade headings or descriptions of works for which it needs to be made obvious that they are included in the price. The notes comprise a list of items worded in sequence that define good practice and, where appropriate, cross-reference to project specifications, drawings, reports, employer/project requirements and the measurement guide used for preparing the BQ/schedule.

A BQ prepared by a cost consultancy for a project client will include trade preambles as a matter of course, as the document is usually prepared as part of an invitation to tender that will pass through the hands of competing contractors. However, experienced main contractors and trade contractors responsible for preparing BQ/schedules often omit the references in their schedules. This may be because they consider themselves experienced enough to be aware of inclusions or that the BQ/schedule is for internal use with the reminder unnecessary. A more responsible approach, and to avoid oversight, is for a contractor to create trade preambles and store the information in a database which is updated from time to time to reflect industrial changes. However, the use of preambles obtained from a database requires caution as they may be generic or project-specific, with the danger of transferring notes that could have irrelevant coverage for a project undergoing a tender. For this reason, any preambles included in a BQ/schedule must be reviewed and edited as necessary to ensure they serve the intended purpose. To demonstrate Table 2.3 lists preambles for excavating and filling works obtained from a contractor's past project, edited for the purpose of builder's quantities production.

2.7.2 Measured Works

Measured works is the largest section of a BQ/schedule that arranges trades in sequence as laid down in the measurement guide. Once quantities are ascertained, they are reconciled to account for adjustments (e.g. deductions to wall quantities to create voids for doors and windows). Thereafter, descriptions and quantified measures are entered into the BQ/schedule and rounded to the nearest whole unit (e.g. 25.4 m becomes 25 m; 211.50 m^2 becomes 212 m^2; and 33.785 m^3 becomes 34 m^3). Steel is entered by weight in tonnes with calculations rounded to two decimal places (e.g. 22.6788 t becomes 22.68 t; 28.4429 t is 28.44 t). The process may be carried out by hand for producing in a final word-processed BQ/schedule, or automatically created from suitable software stemming from the take-off. To demonstrate, Table 2.4 lists the format of builder's quantities that represent the take-off from Figure 2.5.

When creating builder's quantities, contractors tend to bypass the golden opportunity to include the full provisions of nrm2. Reasons for this may be a lack of time available for preparing the document; the informal use of builder's quantities warranting them for the contractor's use only; skill shortages of those competent to measure and quantify works; and the possibility a contractor is concerned it could be liable to subcontractors if electing to issue the information for advice. Nevertheless, nrm2 is an invaluable tool for clarifying and quantifying the scope of designed building works, and also acts as a checklist to ensure such scope is captured in any tender.

Table 2.3 Trade works preambles

Item	Description	Quantity	Unit	Rate	£	p
	<u>Substructure</u>					
	<u>Preambles</u>					
	<u>Excavating and filling</u>					
A	<u>SPECIFICATION AND DRAWINGS</u>					
	Refer to the specification and drawings for a full description of the works, materials and workmanship and allow full compliance					
B	<u>METHOD OF MEASUREMENT</u>					
	These builder's quantities have been prepared using RICS nrm2 as a guide for preparing each trade bill					
	These builder's quantities are issued for the purpose of assisting with the tender for the construction of [*insert name of project*] and are not intended to form part of any contract. Prices in this section and chargeable rates in the measured schedules are deemed to include allowances for waste, cutting and fixing					
C	<u>MEASUREMENT COVERAGE AND PRICES</u>					
	<u>Excavations</u>					
	It is assumed reduced-level excavations and disposal of material will be carried out prior to commencing trench excavations and stockpiled material will not hinder operations					
	No allowance has been made for bulking of the material after excavation and the chargeable rates in the measured schedules are deemed to include such allowances					
	Commencement levels of excavations are noted on the drawings with works expected to be carried out in a logical sequence. Allow here for any out of work sequences or alternative methods					
	The contractor acknowledges receipt of the Site Investigation (SI) report and includes for costs associated with handling, transporting and disposing of the subsoil as per the classification table with associated costs deemed part of the chargeable rates included in the schedules. This is to include all statutory and legislative fees for licensed disposal.					
	The contractor acknowledges receipt of the Environmental Impact Assessment (EIA) for the project under tender. Allow here for complying with provisions applicable to the construction phase, submitting proposals and observing all rules and regulations					
	Allow here for erosion and sediment control to prevent surface run-off from leaving the site, as specified					
	<u>Hardcore and filling</u>					
	Filling material shall be excavated material to the trenches and imported hardcore elsewhere					
	Prices for installing hardcore in layers shall include adequate tamping and compaction to ensure the material does not exceed the maximum moisture content permitted for hardcore beds beneath the ground-floor slab					

(Continued)

Table 2.3 (Continued)

Item	Description	Quantity	Unit	Rate	£	p
	All filling material is measured net in position without allowances for bulked material that compacts to a lesser volume. Chargeable rates in the measured schedules are deemed to include such allowances					
	Allow for the provision of imported hardcore samples as specified and subsequent approval by the Engineer prior to procurement and placement					
	Working space and maintaining faces					
	Allow here for a working space including additional excavation and disposal of material as well as maintenance and repairs to comply with the Health and Safety Plan and CDM Regulations 2015					
	Preambles: Excavating and filling total				£	

Table 2.4 Builder's quantities

Item	Description	Quantity	Unit	Rate	£	p
	Mixed use apartment and commercial unit					
	Substructure					
	Piling					
	Bored piling					
1	Mobilisation of piling rig to site including setting up and demobilisation upon completion of the works		Item			
2	Setting out of piles	18	Nr			
3	Bored augured piles, 300 mm dia, max concreted length 6.0 m, comprising boring, casing and 50 N/mm^2 concrete (18 nr) (P1, P2)	108	m			
4	Mild steel reinforcement to in-situ concrete piles, R24, vertical in prefabricated cage including welding and connecting to helix binding bars	3.83	t			
5	Mild steel reinforcement to in-situ concrete piles, R12, helix, 300 mm pitch in prefabricated cage including welding and connecting to vertical bars	0.96	t			
6	Dispose of materials arising from piling works to a tip off site	8	m^3			
7	Allow for integrity test of each pile carried out within 7 calendar days of completion of the works with results submitted to the engineer within 7 calendar days of carrying out the tests	18	Nr			
	Piling – Total	To summary			£	

Table 2.4 (Continued)

	Excavating and Filling		
	Excavations		
8	Foundation excavation, commencing level 21.000, ne 2 m deep, working around piles	13	m^3
	Disposal		
9	Dispose of materials arising from foundation excavation to a tip off site	13	m^3
10	Disposal of groundwater encountered up to 1 m below ground level		Item
11	Cutting off tops of piles, 300 mm dia and dispose of debris	18	Nr
	Fillings		
12	Imported filling, sand blinding bed, 50 mm th, laid below slab/pile caps	139	m^2
	Membranes		
13	1200 gauge damp proof membrane laid horizontally, over 500 mm wide	148	m^2
14	Rigid polystyrene filled insulation boards, 100 mm th, over 500 mm wide	118	m^2
	Excavating and Filling - Total	To summary	£
	In-situ concrete works		
	Reinforced in-situ concrete 50 N/mm²		
15	Horizontal work, in structures, floor slab, ne 300 mm th	25	m^3
16	Horizontal work, in structures, pile caps, over 300 mm th	12	m^3
	Surface finishes to in-situ concrete		
17	Trowelling to top surfaces	166	m^2
	Formwork		
18	Plain formwork, edges of horizontal work, ne 500 mm high, 250 mm wide (slab edge)	58	m
19	Plain formwork, sides of upstand beams, over 500 mm high, regular shape, rectangle (pile caps)	77	m^2
	Reinforcement		
20	Mesh reinforcement, A252, 3.95 Kg/m² to BS4483 with 150 mm min side and end laps	250	m^2
21	Mild steel bar reinforcement to pile caps including tying into pile reinforcement (PC1 & PC2) - by ratio	1.73	t
	In-situ concrete works - Total	To summary	£
	Substructure: Summary		
	Piling: Total		
	Excavating and filling: Total		
	In-situ concrete works: Total		
	Preambles: Total		
		Substructure: Total	£
		To grand summary	

2.7.3 Non-Measurable Works

Nrm2 recognises that with some projects it is not always possible to measure and quantify works due to a lack of design, the nature of the project or situations where the contractor will have design responsibilities involving risks and credits as discussed below.

2.7.3.1 Risks

In order to identify risks in a construction project and the finance involved, nrm2 recognises the matter in three parts: transfer of risk to the contractor; retention of risk by the client; and shared risk between the client and contractor. To understand where risk should rest, nrm2 recommends the use of a 'schedule of construction risks' for inclusion in a BQ/schedule for completing by the parties to a contract. The schedule is aimed to help identify the status of a project without stating the party considered most suitable to accommodate the identified risks that may only be determined upon receipt of contractors' tenders. For example, a site investigation (SI) report for a new build project may evidence rock below ground within the footprint of the proposed building. Here, a scenario could exist where the project client agrees a fixed price with the contractor for the building works with a condition of contract to reimburse the contractor on a schedule of rates for rock removal. On the flip side, the client may ask the contractor to fix the building price to include rock removal, thus transferring the risk in the process. Here, the schedule of construction risks could form part of an invitation to tender to competing contractors or, if not supplied, it could mean the client seeks full commitment from the contractor. For clarification, and where a risk schedule does not form part of an invitation to tender, the contractor could include a completed risk schedule with a tender, or remain silent by pricing and including the risk to comply with the terms and conditions of the invitation to tender.

2.7.3.2 Credits

Credit provisions can be provided in a schedule as part of an invitation to tender or created by a contractor and included in a tender submission. They are often suitable for demolition works when building fabric items considered of value can be removed and salvaged for resale by the contractor (e.g. stone, slate and some metals). The credit provision can also apply to the permanent removal of dispossessed assets given up by a project client prior to the refurbishment or demolition of a building. Nrm2 provides a template for such provisions in appendix F which, if completed by the contractor and included in a contract, provides the client with the opportunity to pass on full ownership of the goods to the contractor. If forming part of an invitation to tender, the schedule may or may not be quantified, and where not quantified, it means the project client expects competing contractors to state which materials they consider could be salvaged and offer a suitable credit. Where priced, the schedule reduces the financial value of the works; when accepted by the project client, the schedule can form part of a contract. These credits must not be confused with items included in the measured works section of a BQ/schedule under demolition for decanting, such as remove and store goods for later installation or remove goods and hand back to the client for reuse (e.g. a hospital that is to be demolished and newly constructed, where salvaged equipment is removed and temporarily stored for later reuse in the new building).

2.7.3.3 Annexes

Annexes are options to include in a BQ/schedule that comprise of schedules of project-specific items for including in an invitation to tender, which can mean price fluctuations where a fixed price cannot be guaranteed because of the economic climate at the time of inviting tenders for the works. Fluctuations may also apply when the duration of a project is considered too long to fix the price or the form of contract for the project allows fluctuations. Where a fixed price is required and no fluctuations clause exists in the proposed contract, the BQ/schedule should have a separate provision for 'main contractor's fixed price adjustment' and 'work package contractor's fixed price adjustment' to which suitable allowances are made for price certainty. Alternatively, the priced schedules may account for price fluctuations, cancelling the need for the annex.

Nrm2 also makes note of the inclusion of a director's adjustment that may be included in the preliminaries within each trade building component/item of the BQ/schedule or as an annex. This adjustment follows normal commercial practice of a contractor's business and is the correction of a price following a commercial review by directors of the company prior to being included in an offer to carry out the works.

2.7.4 Preliminaries

Preliminaries are the onsite overheads and sums of money a main contractor incurs for running a project that are both time- and fixed-cost-related. Time-related charges may include the following:

- supervision of the works, including contract administration;
- mechanical plant and equipment including fuel consumption, excluding operational plant such as excavators as their cost is based on productivity and included in measured work trade sections;
- other management and support staff if not part of supervision;
- security arrangements, including site watchmen and protective fencing;
- access and lifting equipment such as scaffolding, tower cranes and hoists;
- temporary services (e.g. water, power, etc.) for the site accommodation and building operations;
- potable water for consumption;
- wild air requirements, applicable in hot climates where temporary air conditioning is installed in an enclosed building to create a thermally controlled environment for the benefit of operatives working inside;
- cost to run the site accommodation, including maintenance and cleaning;
- temporary works;
- dewatering;
- site photographs (non-commercial use);
- staff training for use of collaborative management systems;
- attendance on employer engaged contractors;
- waste disposal; and/or
- health and safety purchases (e.g. personal protective equipment or PPE).

Fixed charges for lump-sum/stage payments may include the following:

- insurances;
- specific conditions of contract (e.g. tree protection);

- performance guarantees and bond charges;
- site signage;
- samples and mock–ups;
- minor plant hire purchases;
- building cleans prior to handover;
- surveying and setting out of the building;
- mobilisation and demobilisation of site accommodation and operational plant;
- connection and disconnection of temporary services; and/or
- supply of as-built information, operating manuals and the client's health and safety file.

Section 1 (Part A) of the 'Tabulated work sections of nrm2', entitled 'Information and requirements', comprises 12 main heading items that form the main contract preliminaries. These main headings have a theme of project reference that requires populating with information including project particulars, drawing numbers, employer's requirements and so on. Under each main heading is a series of subheadings in addition to a list of 'information requirements' and 'supplementary information/notes' that act as a prompt for including in a BQ/schedule. For example, under main heading 1.5 'The contract conditions', the subheadings prompt the form of contract to be stated, which could mean mentioning the main contract works as being procured under a bespoke/purpose-made contract.

Section 1 (Part B) is the 'Pricing schedule' and comprises a comprehensive list of fixed and time-related chargeable items for the main contract applicable to setting up the site, construction management during the building phase and demobilisation upon completion. Naturally, the contents of Part B are exhaustive and suitable for any construction project. A contractor would therefore be wise to create a database of all preliminaries items recognised under nrm2 as a source of reference. This could be applicable to any schedule which can be imported and edited for a particular project for the purpose of producing builder's quantities to ensure the relevant items are included in the price of the works.

A characteristic of Section 1 of nrm2 that differs from its predecessor SMM7 is the inclusion of works package contractors' preliminaries. This is distinct from the main contract preliminaries and comprises Part A and B, entitled 'Information and requirements' and 'Pricing schedule', respectively, that is, mirroring the main contract. For practical reasons, a number of components cross-reference to the main contract, as in essence many of the 'Information requirements' have relevance to the main contract requirements (e.g. allowance for main contractor's management and staff and allowance for works package contractors' management and staff). The exceptions to this applies to the likes of a projects' duration, for example a contract that could run for 104 weeks that includes an electrical works package to run for, say, 52 weeks.

2.7.5 Client-directed Sums

Where directed, an invitation to tender may include a list of financial sums or priced schedules for inclusion in a tender which must be recognised by the contractor in its price to carry out the works.

2.7.5.1 Contingency

A contingency is a nominal monetary sum to cover the cost of additional expenditure released at the discretion of a project client any time during the construction phase or defects liability/rectification period. Where stated, the terms and conditions of the invitation to tender will advise competing contractors of the amount and if it is to form part of the tendered sum.

2.7.5.2 Provisional Sums

Where it is not possible to quantify or schedule works because there is no design/specification, or the scope is not possible to identify and quantify in full until the works commence, a predetermined amount as a provisional sum can be included in the invitation to tender documents. The amount is usually client driven to safeguard the bottom-line price of a tender so that it incorporates all of the scope of works. Where a provisional sum amount is client driven, it is naturally a risk to the client; it is not the contractor's responsibility to reassess the amount in a BQ/schedule, even if an arbitrary value.

There are two types of provisional sums. The first is an undefined provisional sum and an approximation to the worth of the works which is broad in context without fine logic, with the intent the allowance will be sufficient. When logic is applied with substance and basis to a calculation, it becomes a defined provisional sum. For example, let us say a hard and soft landscaping scheme is not designed and the site area is known, and the project client wishes an amount to be included in contractor's tenders. This could be calculated using commercial rates, for example:

$$5,000 \, m^2 \, @ \, £150.00 \, per \, m^2 = £750,000.$$

The calculations used to create a defined provisional sum are not usually advised to competing contractors, with only the amount stated. Once a client-directed provisional sum is included in an executed contract, it becomes binding on the parties and is adjusted and revalued as the works proceed.

A variance to the provisional sum concept can occur when a client instructs a provisional quantity to be included in a tender for certain works for rating and pricing by the contractor to create a provisional price. This is subsequently re-measured and valued as the works proceed, creating a firm price in the process. Provisional sums/prices usually exclude value-added/goods and services taxes and can either form part of the appropriate trade section or an annex to the BQ/schedule that collates all provisional sums/prices.

A provisional sum can also form part of a contractor's tender if the contractor is of the opinion that certain scope is lacking in design detail or the commercial risk in tendering is too high. Such action may place the contractor at a disadvantage as it may go against the terms and conditions of tender, especially if competitors are willing to take the risk and submit an unconditional tender. Where electing to include a provisional sum of its own making, the tender must state it is not a fixed price and advise the provisional sum amount, what it is for and cite the reasons why the contractor has taken the decision.

2.7.5.3 Prime-Cost Sums

Where the final selection of a product is unknown, the terms and conditions of an invitation to tender may advise on a price per purchased unit as a prime-cost price (PC) or PC sum for advice. Traditionally, PC sums are created for budgetary reasons and are usually based upon negotiations an architect/engineer may have had with suppliers during the design development stage. The continuation of these negotiations is to include a PC sum in the tender documents as it provides certainty to the cost, with the intention the main contractor will procure the supplies and/or services from the nominated source. Where the documents refer to a PC price/sum the appropriate BQ/schedule must include the reference, for example 'Include the PC price of £300.00 per thousand for the supply and delivery of facing bricks to site' which in this example instructs bidding contractors to allow this amount for the purchase of facing bricks. When a PC price/sum is for the supply and delivery of goods only, and unless stated otherwise, it excludes the price of labour for installation/constructing or fixing and the price of fixing materials needed to carry out the works (e.g. mortar for laying facing bricks). It also excludes any specific attendances required such as design fees and the involvement of auxiliary works or works in attendance (e.g. chasing walls for fixing and securing electric cables). In this scenario, the attendance is called 'builder's work in connection' (abbreviated BWIC), with the main contractor making an allowance for the attendance in its price. Specific attendance attracts sums of money to cover the main contractor's company overheads and profit which is added as a separate amount. Nrm2 recommends that a building component/item be included in a BQ/schedule specifically for BWIC applicable to mechanical, electrical and transportation installations. Any general attendance incidental to the works, such as site storage which is not considered part of the preliminaries, is so noted and described in the BQ/schedule, usually alongside the specific attendances under BWIC.

2.7.6 Dayworks

A building project can require the need for dayworks, and is a method of reimbursing a contractor on a do-and-charge basis for works not included in measured trade sections of the BQ/schedule. In practice, this involves recording the hours of trade labour needed to complete an item of work as well as proof of the cost of materials purchased and price of any items of hired plant. Daywork is a suitable method of reimbursement to a contractor when the scope is unknown or is not possible to quantify, and may be used to evaluate the expenditure of a provisional sum. The concept is widespread in building refurbishment and some new works, where the consensus is that daywork is the most practical method of carrying out certain works where a system of logging real time and the constituents of expense is the most practical. When daywork is a known requirement and to be included in a contract, a daywork bill is prepared. This is to permit competing contractors to insert percentage additions to the prime cost of employing labour, and the cost of materials and plant hire which includes company overheads and profit. Where dayworks are not stated in an invitation to tender, the contractor can elect to include dayworks as part of its offer; Table 2.5 shows a schedule that could be annexed to a BQ/schedule.

Table 2.5 Dayworks schedule

A **Dayworks**	£	p
Allow for the following DAYWORK percentage additions for unpriced works instructed by the architect prior to practical completion. The additions include overheads and profit to the prime cost of labour and the purchasing and hiring of goods and services to complete the works		
Labour - Prime Cost PLUS _____ %		
Materials - Cost PLUS _____ %		
Plant - Cost PLUS _____ %		

2.7.7   Client-Engaged Contractors

A project can be procured with scopes of work that are intentionally excluded from the agreement between the main contractor and client, where the client engages and reimburses contractors directly. This occurs when a project client has an existing working relationship with a business for its services, or for works carried out by businesses that are not usually involved with the construction industry (e.g. ornamental work, supply of sculptures and artwork, date plaques, etc.) or works that are the normal business of the client. The intention of including reference to client-engaged contractors in an invitation to tender is for the client to have a greater degree of control with specialists and enter into direct agreements for their services. When applying, the invitation to tender documents should include a list of the works and preferably names of the businesses that will carry them out. In usual circumstances, their involvement on site commences and finishes during the construction phase with the contractor coordinating the works, acting as supervisor and providing any specific and/or general attendances. The BQ/ schedule for builder's quantities will therefore need to list each client-engaged contractor's scope and define the attendances required so the main contractor can create a price. The price of each client-engaged contractor's works requires no mention as it is of no concern to anyone except the client, which makes it distinct from a PC amount or provisional sum. The list can be included as a single trade bill encompassing all client-engaged contractor's works, as an annex to the BQ/schedule or may be included within the preliminaries.

2.7.8   Cost Centres

Cost centres are a series of codes written by a software provider for use with a cost management system for inclusion in a BQ/schedule in order to identify a project's facets. The codes are authored by a software provider, the format of which may be influenced by a purchaser that may specify certain codes or by default coding from the software seller. Coding for use as cost centres are abbreviations of items or works and usually comprise a mixture of numbers and letters, making it possible to include in builder's quantities to divide prices into parts. If used, these parts can be considered budgets created during the tender period. To demonstrate, if a project referenced A1234 is to have a BQ/schedule and builder's quantities prepared using work package formats obtained from nrm2, details for tracking the allowance for say the internal suspended ceilings could be coded as follows:

Project →	Package nr →	Trade →
A1234	Serial nr 14	Ceiling works (01); works package contractor's preliminaries (02)

In the above example, the ceiling works code could be A1234/14/01 and works package contractor's preliminaries A1234/14/02. The cost of the ceiling works could also be subdivided if a project involves different ceiling finishes, for example A1234/14/01.1 for suspended plasterboard ceilings, A1234/14/01.2 for suspended prefabricated timber ceilings, etc. This arrangement has advantages to a contractor as it sets the benchmark to the price of the works that can be used as a source of reference when awarding works packages.

2.8   Software Systems

This chapter has focused on measuring techniques using a traditional approach, which in the author's opinion is an excellent method for understanding the process of describing and quantifying construction works for pricing purposes. Furthermore, it provides knowledge of construction technology and industrial mannerisms as well as an understanding of the scope of works. However, a pitfall here is the absence of a system for abstracting quantities and descriptions derived from the take-off for inclusion in a final BQ/schedule. With a small project, the abstract can be carried out by hand where items are billed direct to paper, a word-processed document or a spreadsheet with descriptive schedules linked to tabs that include the take-off and quantities.

The downside of this type of abstracting is however twofold. The first is with formatting, as lined paper and tables in word-processed documents is time-consuming to create and spreadsheets are not designed to produce take-offs and BQ/schedules. The second is with errors that can occur with omissions, additions and calculations which, at worst, can go unchecked and included in the final document. A solution to this is with the use of industrial commercial software, which comprises intelligent, swift and accurate systems used for the preparation of trade BQ/schedules. The use of these systems means users can have confidence in the quality of formatting and the resulting calculations as they are self-produced during the taking-off process. It is for this reason a competent quantity surveyor experienced with pre-contract activities should be skilled in the use of at least one industrial software system in addition to proficiency with taking-off, quantifying and BQ/trade schedule production. There is respite here for the contractor's quantity surveyor not involved with pre-contract work, as these systems can be used during the post-contract phase as part of project administration for pricing variations. However, their implementation depends on each contracting organisation's business and in-house IT arrangements. Construction estimating and contract administration software systems can be purpose made for the larger contracting organisations when the purchase investment is for the long term. The downside here, however, is they are time-consuming to design and implement and expensive to purchase. A parochial solution is to purchase an 'off-the-shelf' system available from companies that offer a range of software systems with varying degrees of sophistication. Companies offering these systems operate at local, national and international levels, often branding their names in the process, and include 'Cubit' and other types of software by Buildsoft (Australia; http://www.buildsoft.com.au) and 'Cost-X' by Exactal (http://www.exactal.com).

When using these systems, electronic design drawings such as pdf or AutoCAD can be uploaded and attached to a project file created in the software. Designs can then be viewed to predetermined scales and zoomed for expanded view or reduced for taking-off dimensions, and usually include facilities to create location points which are colour-coded to trace the take-off. Designs may be viewed in 2D or 3D and displayed with a high level of accuracy for measuring large areas, elaborate planned shapes and irregular depths. To demonstrate, Figure 2.6 shows a 2D digital presentation of a take-off for the construction of a new school.

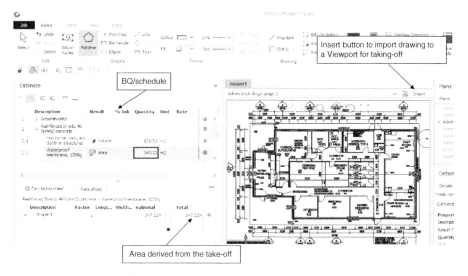

Figure 2.6 Digital take-off screen (BuildSoft).

2.9 Alternative Bills of Quantities

There are a number of alternative formats used by construction companies for the presentation of works in a schedule for pricing purposes other than trade sequence, as found with builder's quantities/BQ schedules. Factors to consider when selecting a format include who the schedule is for, the time available to prepare the scope and the quality of the design and documentation available.

2.9.1 Operational Bills

These bills were developed by the Building Research Establishment (BRE; a former British government establishment, now a private organisation that carries out research, consultancy and testing for the construction and built environment sectors in the UK) and describes works in terms of the labour, plant and materials required for each physical operation. When preparing these bills, operations are scheduled as labour requirements per gang (or gangs of tradesmen) plus the materials and items of plant required to carry out specified works (or operations) to which rates are applied to arrive at a price. Operational bills have advantages to a builder as they provide effective cost control during the construction works. In addition, they assist with sequential estimating for pricing a project because of the division of components that make each working operation easier to identify. The bill format is suitable for subcontractors and builders that carry out small projects, as the schedules provide detailed information which is advantageous for the ordering of materials. It has disadvantages on large projects however because it is a bulky document, making it time-consuming and expensive to produce.

2.9.2 Activity Bills

These bills are a modification of operational bills and involve measuring, quantifying and arranging trade works in order of site activities instead of operations. The bills take into account activities as a network and programme of works using a series of codes by location for reference. They have similar values as operational bills because they provide the impetus for effective construction management as well as the suitable distribution of materials, labour and plant. These bills are suitable for projects that are small in financial value, simple in nature or where the works are spread over a large site area. However, they are not normally used by contractors or subcontractors involved with large-scale works or projects of high value or durations longer than a few months.

2.9.3 Annotated Bills

Some main contracting and subcontracting organisations produce bills with annotations unique to a project. The wording in these bills refers to specifications which may be technical in nature and includes notation on the physical location of the works. As the wording is specific, localised and informal, they are not suitable for large projects that require a BQ/schedule adopting the use of normal industrial jargon and mannerisms.

2.9.4 Elemental Bills

When a price for construction works is required during the early stages of design development, the building and project requirements can be segmented into elemental parts (e.g. upper floors, external walls, etc.). These parts can then be broken down as composite items in elemental bills that have an all-inclusive list (e.g. 300 mm thick suspended reinforced in-situ concrete floor slab, total floor area 500 m^2), to which a suitable composite rate is charged to arrive at a price. This is part of pre-contract estimating, carried out before works are traditionally tendered, and used by quantity surveyors in cost consultancies as part of a cost planning service to a project client. Their use can also be adopted by competing contractors when preparing cost-planned tenders for a design-and-construct project when the design information is limited, which contactors accept as a risk to accommodate in their prices. The RICS new rules of measurement nrm1 includes guidelines that define the constituents of building elements for use with these types of bills. Elemental bills can also mean a formatted BQ/schedule with the building facets included in a summary (e.g. substructure, upper floors, etc.), with the exception the descriptions of works in the schedules are broken down in more detail using nrm2 as a guide.

2.9.5 Approximate Quantities

These are similar to builder's quantities and used by cost planners in cost consultancies as a pre-contract activity for cost checking a building's portions at key stages of design development. To create approximate quantities, the design must be advanced and involves preparing a schedule that, when priced, becomes the penultimate order of cost estimate prior to preparing the final BQ/schedule. Here, components of a construction element (e.g. excavation, concrete, etc.) for the substructure are measured, quantified and rated to produce a price. A schedule of approximate quantities lacks the final details of a BQ/schedule prepared in accordance with nrm2, yet has sufficient information to determine an approximate price (e.g. excavate foundation tranches, 600 mm wide, maximum depth 1000 mm, length 200 linear metres). Approximate quantities described in a schedule should not be confused with a trade bill of approximate quantities, which is a tender and contractual document prepared to obtain a provisional price for an early start on site with the works re-measured and re-valued once the design is complete.

3

Working with the Main Contractor

3.1 Contracting Organisations

The smallest size contracting organisation is a jobbing builder, which is a business usually run by a sole proprietor who is a self-employed tradesman based from an office in either the proprietor's place of residence or commercial premises. The jobbing builder carries out minor building works such as domestic alterations and property extensions and is assisted by subcontractors, with the proprietor also possibly carrying out some trade works. The proprietor usually deals with job estimating and the running of the business and is supported by a small number of staff. Other contracting businesses are tiered (an unregulated concept considered a means of reference only) within a range stemming from low to mid and high (or 3, 2 and 1, with 1 being the highest) and the grading referring to the size of projects undertaken in terms of financial value, complexity and duration and not usually the types of building works.

3.1.1 Tier 3 Small-Sized Contractors

Tier-3-sized contractors tend to be hands-on in their approach to securing work with the businesses usually managed by at least one director, often relying on regular clients for repeat work; this can pose a risk to their business if they do not have equity or receive regular payments. In these types of companies, one or more directors may act in the capacity as business manager and/or estimator and appoint other mangers to organise and run construction projects. They may also appoint office support staff to oversee accounts and administration and may engage an estimating assistant. The administrative procedures of a small contractor may not be as robust as their larger counterparts, with the accounts and estimating systems usually managed with computer spreadsheets instead of industrial software which may not be considered a worthwhile investment. For a trainee quantity surveyor or estimating assistant engaged by a small-sized contractor, the experience can be rewarding as the day-to-day tasks may not be as formal as found with the medium and larger-sized businesses. For this reason, a quantity surveyor can be invaluable to a small business by creating procedures to improve efficiency. Moreover, the culture of small-sized contracting organisations can provide advantages to employees as it exposes individuals to industrial practices at a grassroots level, providing variety in the process instead of fixed autonomous roles that can be found with the larger contracting organisations. For example, a role may involve estimating, measurement and contract administration for a number of projects, each valued up to say

Construction Quantity Surveying: A Practical Guide for the Contractor's QS, Second Edition. Donald Towey.
© 2018 John Wiley & Sons Ltd. Published 2018 by John Wiley & Sons Ltd.

£250,000, running concurrently. Anybody wishing to learn the ropes of the industry can gain invaluable experience from a small-sized contractor, acting as a stepping stone in career progression.

3.1.2 Tier 2 (Medium-Sized) and Tier 1 (Large-Sized) Contractors

Tier-2 and -1-sized contractors operate independently or under the guise of one or more satellite or regional offices that report to a central office on their profitability. These contracting organisations may be floated on the stock exchange where investors buy shares in the business and can reap handsome rewards when they operate with a profit. In this capacity, the central office is answerable to shareholders and does not usually involve itself in the day-to-day running of each regional office. The Tier-1-sized contractors usually have multiple satellite or regional offices, each governed by a board of directors with offices usually established in specific locations for a reason, for example marketing and managing project(s) when there is need for the contractor's presence in the area. Some contracting organisations with regional offices can operate in the same type of business, a good example being national house builders. However, some may be linked to a side or parent company providing different services, for example civil engineering, infrastructure and demolition. Each regional office and the projects they undertake usually have identical corporate identities (i.e. colours and logos) so their commercial identity remains uniform. The largest Tier-1-sized contractor may also have offices located in more than one country.

Advantages for employees engaged by medium- and large-sized contractors include:

- employment security;
- opportunities to learn working practices of other regional offices with the possibility of national and international travel;
- exposure to projects of high value;
- project variety;
- interaction with other professionals;
- possibility of being part of a joint venture (JV) scheme where contractors merge to collectively deliver a project while remaining separate businesses, or possibly creating a new business identity in the process;
- recognised as being part of an established and reputable business name;
- receiving benefits, including bonus schemes and supply of a company vehicle;
- chances for career progression including promotion; and
- exposure to a variety of procurement routes for acquiring buildings.

Disadvantages include:

- parochial role;
- can be competitive when seeking employment and/or to obtain promotion;
- being part of a business that can literally change overnight due to company takeover or a change in business policy;
- bureaucratic; and
- vulnerability of the business at times of economic recession, with compulsive redundancies.

These companies usually comprise a minimum of three departments: technical, including quantity surveying, procurement and estimating and (possibly) legal; construction, including quality assurance, health and safety, site management and procurement (if not part of the technical department); and accounts, encompassing payable and receivable transactions, IT systems, reception and marketing. Each department is usually under the control of a manager, with each manager reporting to a general or operations manager who oversees the business. Hierarchy above this role is at director board level.

The titles of employment positions that a company creates for its staff are industrial yet can be cultural, with employees possibly permitted to select their own job title subject to approval by the appropriate director. The directors of the business also steer the business management decisions and selection of the type of work undertaken in terms of a project's duration, complexity and financial value. To demonstrate, Figure 3.1 shows an organisation chart of a Tier-1-sized building contractor.

3.1.3   Premises and Assets

A typical contractor's premises comprises a corporate office and possibly an attached yard to store assets, including mechanical and non-mechanical plant that may be owned outright or purchased through a loan. As with any business contractors incur company overheads, which are sums of money to pay for the running and operating of the business. Contractors refer to these as offsite overheads; these are different from onsite overheads, which are the contractor's preliminaries and direct costs for running a project. It is important for the quantity surveyor to distinguish between these two items; they are equally important for estimating purposes, yet have different meanings with regards to project cost management.

Like any business, contractors have a financial year set to dates they consider appropriate. At the end of each financial year, the company accountant and directors review company turnover (income received during the financial year) and forecast a cost for the offsite overheads for the forthcoming year. A list of items to include in the forecast includes the following.

- *Salaries and benefits*: director's salaries; office-based staff salaries; employer's national insurance contributions for office-based staff, including directors; private health cover for staff; contributions to staff retirement schemes; the cost of insurance for staff, including third parties if not an onsite charge; staff training workshops; and entertainment expenses.
- *Office building*: building/business rates(e.g. council tax payments and building rent/mortgage repayments); building maintenance (e.g. cleaning, painting and repairs); storage fees for assets; building improvements including amended office layouts, extensions etc.; garbage removal; utility charges (e.g. water, telephone, etc.); testing of office equipment for health and safety, including maintenance contracts; information technology upgrades (software only) and cloud computing fees; environmental certification; facilities management or corporate body payments (sinking funds) for maintenance; security alarm contract fee and/or static guards; building contents and fire insurance; professional subscriptions; audit fees; solicitors/legal fees for business advice; stationary purchase and printing costs when not a project expense; clothing purchases and any hired plant items when not for a project; and the cost for any debentures or security bonds that are not project specific.

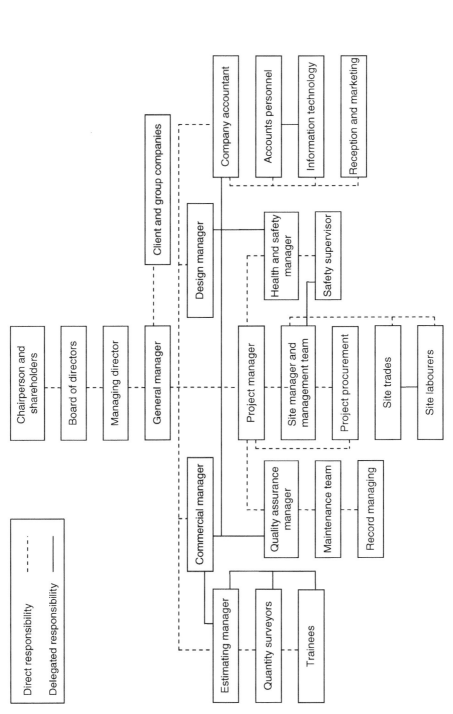

Figure 3.1 Organisation chart.

- *Vehicles*: vehicle insurance, tax and registration; the price of vehicle leasing and fuel consumption if not fringe staff costs or onsite charge; and yard and workshop maintenance of vehicles and plant if not an onsite charge.
- *Write-downs and write-offs*: depreciation of office furniture, including computer hardware; depreciation of owned cars and/or vans including fuel consumption; depreciation of purchased site engineering equipment; loss of financial interest on overdue debts; unrecoverable fees (e.g. fines; write-offs through client insolvency or lost claims); legal costs incurred for disputed sums considered high risk; and tender costs for lost projects (e.g. design modelling and charges for outsourced BQ preparation.
- *Promotional*: business promotions including travel fairs and roadshows; marketing costs (e.g. press and media advertisements, portfolios, etc.); photography and sponsorships; copyright purchasing for designs; and corporate signs if not specific to a project.

These sums can be mitigated by income received from the sale or rental of assets, offsets from tax allowances and rules under VAT legislation as determined by the company accountant. Once an amount is ascertained and mitigating circumstances considered, the offsite overheads amount is expressed as a percentage of the turnover for reference. To demonstrate, let us say a company's offsite overheads are assessed at £2.5 million and income received from projects at the end of the financial year is £50 million. Assuming income for the following year is to remain stable, the percentage to apply to the base cost as a charge out to customers is calculated as follows:

$$\pounds \frac{50}{50-2.5} \text{million} = 105.3\%.$$

The combination of offsite overheads plus an allowance for profit is known as the contractor's margin, and is the percentage addition a contractor adds to the base cost estimate when submitting tenders for projects. In other words (and in the example given here), for every £1 spent as a purchase for a project, the project must return £1.053 as a contribution for business overheads before making a profit.

Fixed assets are items a contractor owns or is purchasing through a loan that gives economic value to the business with their running expenses shown on balance sheets by the accountant as either onsite (preliminaries) overheads or offsite overheads. In order to keep track of the whereabouts of assets, it is necessary for the contractor to create records with the simplest type being an inventory, which is a document used to record small asset items such as scaffolding tubes, boards, etc. The inventory records the quantities and type of assets and their location while they remain under the ownership of the contractor. With larger items such as mechanical plant an Asset Register is used, and is likely to be managed by a plant manager who is responsible for each assets' insurance, repair and maintenance and coordinating their delivery and collection from site. Quantity surveyors and estimators need to be aware of these registers and the use of each asset with construction activities as well as the methods in place for recovering their running costs. For this reason, senior management of a contracting organisation may require the estimator to give advice regarding the viability of a purchase, requiring a financial appraisal. In carrying out this task, answers to the following questions will help steer an assessment.

- What is the purchase price of the asset under consideration?
- What is the estimated useful life of the asset once it is purchased?
- What are the units of production during the useful life and how are they to be measured?
- What will the disposal cost be?

To demonstrate, let us consider a scenario where a contractor is considering purchasing a forklift truck for transporting materials around site instead of hiring. A characteristic of plant machinery is that it depreciates in value over time which must be considered in a financial appraisal, and a method for assessing a write-down depreciation of a forklift truck during its working life, together with items of expense for its operation, is shown in Table 3.1.

From reviewing the table, the contractor will need to yield £1,883 per month over the term of the investment to break even. Income could be derived from a project's preliminaries charged out to a client, providing of course there is a project requirement for the forklift and a preserved amount included in the preliminaries. When considering this

Table 3.1 Asset purchase appraisal

	£
Purchase price	**100,000**
Investment period (say 5 years)	
Depreciation year 0–1 (35%)	−35,000
Depreciation year 1–2 (15%)	−9,750
Depreciation year 2–3 (15%)	−8,288
Depreciation year 3–4 (15%)	−7,044
Depreciation year 4–5 (15%)	−6,000
Total depreciation	**−66,082**
Net worth at resale/disposal value	33,918
Running costs	
1. A site labourer with a license will operate and manage the forklift. No costs to be incurred for the labourer as wages and employment costs are included in the preliminaries as project specific.	–
2. Operating life 12 months × 5 years = 60 months. Service every 6 months = 10 Nr @ £750	7,500
3. Replacement of worn parts: year 0–1 Nil; year 1–2 £1,500; years 2–5: £7,500	9,000
4. Loss of interest on investment, say 3% per annum	15,000
5. Fuel, average of 30 litres/week × 48 operating weeks per year × 5 years at £1.50 per litre (allowing for times when the forklift will be idle)	10,800
6. Insurance at 1.5% of value: year 0–1: £1,500; year 1–2: £975; year 2–3: £828; year 3–4: £704; year 4–5: £600	4,607
Total of running costs	**46,907**
Total expense (depreciation of £66,082 plus running costs of £46,907)	**112,989**
Monthly cost over 5 years = total expense/(5 × 12)	**1,883**

type of asset as a purchase it is worth considering the alternative, which is to hire from companies in the equipment hire market. These companies offer attractive hire rates that are competitive with sliding scale charges depending on the duration of hire. The drawback is that rates may attract insurance cover as hire companies are often reluctant to rely on a contractor's insurance in the event of damage and prefer dealing with their own broker or insurance company in the event of a claim.

3.2   Management Systems

The management systems of a contractor's business are the backbone of its internal functions and operations. Tier-1 and -2-sized contractors usually produce a policies and procedures manual in soft and hard copy for distribution to each department, describing the culture of the company, the standards expected from employees and how the business operates. In line with usual corporate policy, the manual is updated from time to time to comply with changes in a company's internal procedures and legislation affecting the business. The structure of a building company's business is usually regulated internally, and a set of policies and procedures promotes standards for business efficacy while sustaining the reputation it has earned from clients. The manual addresses business activities applicable to office procedures and may include supplements applicable to construction sites. A typical manual includes the following.

- *Business structure*: a flowchart showing the management organisation structure; office plan layout, including locations of emergency escape routes; management structure of each department, including functions and job titles; management review procedure for changing policies and procedures and the methods of notifying; and recruitment and employment selection methods.
- *Client focus policy with mission statements outlining objectives and intentions regarding*: quality assurance; health and safety; the environment; and equal opportunities and diversity.
- *In-house policies*: procedures regarding the use of information technology, computerised management systems and the reporting of faults; electronic filing and document control procedures; use of the internet; coding system for project identification; employees' responsibilities with their contracts of employment; periods for salary review; personnel/people management (e.g. sick leave, expense forms, discipline, grievance procedures); alcohol and drugs policy; motor vehicles policy and the use of company assets; behaviour and codes of conduct between employees; confidentiality of projects with third parties and the media; archiving procedures for completed projects; methods for business promotion; training procedures; and evacuation procedures in the event of fire etc., applicable to the office. Procedures for construction sites are site specific and do not usually form part of a generic manual.

Employees should be encouraged by the contractor's management to refer to the manual from time to time to become familiar with the contents, and pay particular attention to the issue of any changes.

3.2.1 Health and Safety Management

Health and safety legislation requires a contractor to act responsibly with regards to health and safety management, an example being the endorsement as principal contractor by a project client to comply with the requirements of the Health and Safety Executive (HSE) when a project falls under the rules of CDM 2015. The HSE produces statistics on the number of fatalities as well as injuries reported by employers and self-reported injuries, with data collected to comply with the Reporting of Injuries, Diseases and Dangerous Occurrences Regulations 2013 (as amended) (RIDDOR). This piece of legislation makes it mandatory to report deaths, injuries, diseases and 'dangerous occurrences', including near-misses that take place at work or in connection with work. Since implementation, figures have recorded a steady decline in the number of fatalities reported to the HSE, which is important as it emphasises the need for contractors to remain vigilant regarding the planning of works in order to minimise risk. This is possible with a health and safety management manual that demonstrates a contractor's commitment to safe working practices as part of its corporate policy, and applies to the contractor's corporate office and any site project it undertakes. The growth in this field of management, and more so since the introduction of CDM, has seen the cultivation of management tools, mannerisms and industrial characteristics all with the goal of promoting a safe working environment. Methods of CDM compliance and the internal matters of a contractor's business that impact health and safety standards are included in the manual, which is usually created by the contractor's Health and Safety Manager who issues information to the contractor's various departments and each operating site. The use of a health and safety management manual is separate to the requirements of CDM 2015, where the principal contractor must produce a construction phase health and safety file that can include relevant extracts from the manual which must be suitable for the project.

3.2.2 Environmental Management

A building contractor must acknowledge the legal requirements written into environmental legislation when carrying out construction works. In addition, a contractor will need to make a conscious decision regarding the attitude it wishes to take towards safeguarding the environment, which is usually demonstrated in an environmental management plan. By creating this plan, a contractor is demonstrating a commitment to legislation and to protecting the environment. A typical plan is usually divided into sections comprising of: waste management planning, including demolition and methods of dealing with contaminated waste; air quality management; storm and wastewater management; and noise reduction management.

Each section usually commences with an explanation of why the section is included in the plan and a description of the active measures the contractor will implement to control construction processes in the interest of protecting the environment. For example, a section on waste management planning may state the plan is to minimise waste generated from works under construction, with further aims of promoting the use of recycled materials for the manufacture of new products for installation in buildings. The contents to include in an environmental management plan may be influenced by the contractor's client's policies that can occur when a client issues a mission statement or environmental management plan as part of an invitation to tender. When doing so, a client may make it a condition of tender that any appointed contractor must reciprocate the intent and, if involving expense to the contractor, is to be included in the tendered price. In this scenario, a fortunate contractor may already have taken the initiative by implementing a policy and creating its own environmental management plan in order to promote a corporate image which in general meets the same standards of the clients' policy. Once a plan is active, it is necessary for the contractor's office-based personnel and site-based project team tasked with delivering a scheme to be cognisant with the policies it promotes.

3.2.3   Quality Management

A quality management system is of benefit to a contractor, as it implements procedures to ensure services on offer are those eventually delivered. To be effective, a system should include a quality management policy and a statement endorsed by a managing director stating that the system complies with ISO 9001-2015 and is certified by an independent authorised body. ISO is an international standard which focuses on meeting the needs of clients and delivering satisfaction to the quality provided. The ISO 9001 concept evaluates a quality management system to ensure it is appropriate and effective, while committing a contractor to identifying and implementing improvements. ISO 9001-2015 is the first revision since ISO 9001-2008 and grants organisations a three-year transition period after the revision to migrate their existing quality management system to the new edition. Where the system is in place and procedures are active, a contractor may receive the following benefits:

- better job performance that enhances the portfolio of projects;
- improves morale of staff for a 'job well done';
- creates opportunities for securing new clients as the system is accredited;
- sets a benchmark to retain clients;
- reduces building defects;
- meets regulatory, contractual and legal requirements; and
- helps to avoid disputes and claims.

Moreover, ISO 9001 assists in achieving consistent results by continually improving the quality control process. The standard does not guarantee to define the actual quality of a final product or the service it is intended to provide. However, it has the aims of mitigating the likelihood of flaws and, for construction purposes, applies to the design, installation and operating processes.

For quantity surveyors and estimators, the inclusion of a quality management plan within a contracting organisation is one that embraces good services which comes at a cost because by default, the high quality of anything system driven incurs additional expense. This can affect the price of a product and service to a client because of the way a contractor manages its production and can apply to the running of the corporate office and works carried out on a construction site. For example, an estimating manager located in an office will deal with an invitation to tender commencing from receipt of the design and documentation through to the date of submitting the tender. The process of dealing with this linear arrangement can be managed with the use of a quality management system that could include the following:

- a register recording the date of receipt of the consignment as well as date(s) of any revisions and date the tender is to be submitted;
- a flowchart defining the methods of obtaining trade, material and plant hire prices;

- a register delegating roles and responsibilities for the take-off procedure and preparation of builder's quantities;
- a register delegating the roles and responsibilities for rating builder's quantities; and
- a time frame for checking prices to achieve the deadline for submitting a tender.

Outside the office, a contractor may implement a quality management system for use with construction activities that can include the use of a quality control checklist to assess the quality of work carried out by site trades; refer to Table 3.2 as an example.

Another member of the ISO family is ISO 9004-2009, which focuses on long-term economic survival for a business. It states that an organisation must develop a strategy which is supported by other parties to sustain success, for example stakeholders. To achieve this, a contractor must have a systems approach which means developing a system that integrates the processes. An example is the use of a template for a construction site such as that demonstrated in Table 3.2, which is monitored and updated by a competent person who may be either the contractor's site or quality control manager. To be effective, the system must undergo a validation process using evidence to confirm that the requirements of the intended use are met. This is one part of a design and development system that uses objective evidence (such as templates) to confirm that the end result meets the requirements. This verification leads to vision and value that a contractor can set as a target which aids progression of the business and helps to understand how it is perceived by clients, hopefully creating repeat business in the process.

Table 3.2 Quality control checklist

Quality control checklist						
Project nr	A1234	Title: Construction of new industrial unit Subject: Substructure works	Week ending:		Revision: 2	
Item	Reference	Work activity	Complete (yes/no)	Assessor (initial)	Date	Comments
1	Excavations	Foundation excavations complete	Yes			Formation 1200 mm below existing ground level of 22.000
2	Excavations	Excavated materials stockpiled	Yes			Stockpiled along site boundary in two heaps
3	Excavations	Arrangement for disposal of excavated material off site	No			The larger of two heaps to be disposed; other kept for filling
4	Excavations	Foundation excavation inspected for follow-on trades	Yes			Dry formation – no presence of water

3.3 Marketing for Contracts

As with any business, a contracting organisation will only remain solvent if payments received enable it to pay creditors and cover business overheads with anything more being a profit. To remain in business, a contractor must secure profitable work from project clients, which in main contracting is obtained from receipt of awards won by tender or through negotiation, the latter of which is usually for repeat work stemming from an earlier award won by competitive tender. A way a contractor can compete for work is in an open market where clients or their agents advertise projects in the press, online or by referral, and invite tenders for works. Here, an invitee (the contractor) is issued the tender information with little or no background checking. This has an advantage to a contractor because it is pricing works in the hope of receiving an award which is better than pricing nothing. However, a disadvantage is the possible number of competing contractors involved which can mean the competition is fierce, especially during an economic downturn when markets are suppressed. Furthermore, a condition of an invitation to tender may be that each contractor submits details of their business functions with their tender (e.g. number of employees; statement of current and past projects including financial value; copy of endorsed quality management plan; completion of questionnaire on health and safety record, etc.), which can make the tender process a time-consuming exercise.

An alternative to open market tendering is selective tendering, where a list of contractors is selected by a client/client's agent because of their suitability for a project. The prerequisites for selection include the suitability of a contractor's financial turnover in relation to the value of the works to be tendered and the relative experience and sector(s) of the industry the contractor operates in. The inclusion of a contractor on a selective tender list may be achieved by either the satisfactory performance of current and past projects with existing clients, or branching out and seeking new, which must be endorsed by the client's procurement officer to enable the release of the tender consignment. To find new business opportunities, it is necessary for a contractor to instigate a business development plan; to be effective, this requires sources for potential leads. This includes creating a contact list and dividing them into two categories: (1) those who have an awareness of the contractor's business; and (2) those without knowledge of the contractor's operating capabilities. There are a number of avenues worth exploring when seeking new contacts, including:

- reviewing existing client's accounts within the contractor's database that are no more than 12 months old, and making contact;
- existing client referrals for issue to potential clients;
- attending trade shows and looking through magazines and journals;
- attending social and business functions;
- proper use of social media such as LinkedIn;
- communicating with competitors who decline offers of work, to avoid over-commitment;
- attending business seminar invitations;
- trade and professional affiliations and their recommendations;
- being aware of media advertisements on television, radio, websites and billboards;
- involvement with charities; and
- registering with online tender notices, including projects undergoing planning approval.

By using some of these leads, a contractor can place itself in a good position to engage in business pursuits in accordance with a business plan. The public sector can be bureaucratic in 'breaking the ice', for placement on a selective tender list that will take time and effort, and a higher success rate may be possible with the private sector. However, abuse of the tendering procedure in the private sector is not unheard of, and a contractor would be wise to avoid overburdening itself with an overwhelming number of invitations to tender if the works are not imminent, or clients or their agents are too zealous in their attitudes. Signs to look for with this scenario include the following.

- An excessive number of competing contractors issued with invitations to tender in a selective tender environment.
- Where a phase of an existing scheme is in progress and the contractors' performance is noticeably good it is likely subsequent phases would be negotiated, meaning any invitation to tender may be procedural.
- Conflicts of interest.
- The commencement date of the works is too distant.
- The design and documentation is underdeveloped where a fixed price is required.
- Funding for the project is unresolved.
- The client's preferred project duration is considered too short.
- Planning permission has not been granted.
- Abnormal conditions of tender meaning a hard bargain is being sought, for example a request for a site visit involving the opening-up of existing works affected by the new works to understand the physical condition, with the contractor deemed to be aware of the status when tendering. This may be acceptable if the risk is limited (e.g. identifying the size of existing drainage pipes for connecting). It may not be suitable where a site visit is required to ascertain the extent of finishes on a benchmarked project, such as a completed hotel in one city with the terms of the invitation to tender for the contractor to price the design and works to the same standard for a new hotel in another city.
- Expressed condition of tender that no advanced payment will be made to the contractor upon the signing of a contract.
- Lengthy payment terms.
- Where new works are involved and a fixed price is sought, a site investigation (SI) should be supplied.
- Draft form of contract, including terms that transfer an unusual amount of risk to the contractor.

Receipt of an invitation to tender with unfavourable terms can be returned, which to be effective involves formally declining to submit an offer that must be given in writing within as short a timeframe as possible after receipt of the consignment and certainly before the date the tender is due. This may be the best thing to do if the tendering procedure is considered as being abused. However, if an invitation to tender indicates or expresses an award as imminent after the tender closure date with the contractor electing to submit an offer, the offer should comply with the terms and conditions of the invitation to tender. Alternatively, if the terms and conditions appear biased or pose a high risk to the contractor, a submitted tender can qualify certain conditions for negotiation post-tender. Here, it is far better for a contractor to price, negotiate and win the works to the satisfaction of the parties to a contract instead of obtaining one that would be a burden to the contractor, which in the long run serves no purpose to the contractor or client.

3.4 Procurement

Procurement is the method of obtaining something and, with reference to the construction industry, it is the chosen route from a range of systematic pathways that provides a client with the most suitable outcome for the successful delivery of the design, construction and commissioning of a project.

3.4.1 The Client's Brief

The client's brief is a document usually prepared by a consultant for a project client that outlines the objectives required for implementing a construction project and to determine whether the final product will meet the client's expectations. There are no set rules for the format of the brief, as this depends on the client's type of business and type of project with the brief developed over time and issued in stages following client and stakeholder input. As part of this development, answers to questions as well as statements and considerations within the brief usually revolve around the following:

- aims of the project, that is, how it is to be developed and how it will be accomplished;
- design and construction process, including appointments procedures and priority of any early engagements (e.g. architect for design, quantity surveyor for cost advice, etc.);
- impact of the design and construction process on a client's usual business activity;
- client's requirements at the time of preparing the brief and those for the future, and whether the completed building can accommodate predicted changes or upgrades with ease once operational;
- end-product function (i.e. building type and height, number and size of rooms, locality, floor areas and spaces);
- perceived constraints (e.g. planning permission, location of existing utility services and any restricted access for construction purposes);
- budget in terms of design and construction costs, which may have been determined as part of a feasibility study to appraise the worth of the scheme;
- maintenance aspects of the final product;
- consideration of renovating an existing building as an option to a proposed new scheme;
- the time it will take to develop a design and construct the building;
- market value of the completed scheme and future potential;
- any specific requirements (e.g. sectional completion as opposed to a single handover, if some works are to be carried out by client-engaged contractors, etc.);
- end-user requirements who are to provide input with development of the brief if not the client (e.g. tenants or owners);
- aspects of any completed schemes that a client would like to see in the new scheme; and
- client involvement with the design and construction processes and procurement selection.

At a suitable time after a decision is made to commence a scheme, a client may delegate responsibility for developing the brief and managing the project to an intermediary who would become the client-side project manager (or client's agent), that is, the eyes and ears for progression of the project in the client's interest. Client-side project management is not a procurement system in itself, and should be considered a supervisory role that can apply to any systematic pathway for procuring a construction project.

3.4.2 Traditional Pathways

This procurement route is suitable where a project client seeks price certainty and control of the design, and is prepared to devote time and resources to accomplish the objectives. There a number of variations of the traditional pathway, all of which are linked to time, cost and design responsibilities with selection usually stemming from the client's brief. At the earliest opportunity, this involves a client appointing an architect to prepare a concept design and a cost consultant/quantity surveyor to issue cost advice. Thereafter, a client-side project manager is ideally appointed and usually makes recommendations for the appointment of further design consultants (e.g. structural engineer, building services engineers, etc.) that become the design team. With this procurement method, design performance obligations rest with the design team and construction obligations with the contractor that is tasked to carry out the works under a construct-only contract. For the process to be effective, contractual links as service agreements are formed between the client and each design team member; an architect will therefore have a contractual link with the client as will the cost consultant/quantity surveyor, yet the architect and cost consultant/quantity surveyor will have no contractual link. Likewise, the client-side project manager and contractor will have no contractual link with any stakeholder except the client. However, a control link exists by default as a management style between stakeholders; in general this is not enforceable, and relies on team collaboration to be instrumental. If there is a failure in the control link that requires remedy by contractual means, it can only be remedied under the respective agreement with the project client.

3.4.2.1 Lump-Sum Fixed Price

Under this arrangement, competing contractors are issued the design and documentation in an invitation to tender that may include a bill of firm or approximate quantities. Alternatively, a bill of firm or approximate quantities may not be supplied, which would transfer the risk for scheduling works and quantities to competing contractors. The tendering procedure may be open or selective, and is usually a single-stage process with the intention to obtain *bona fide* (in good faith) tenders from contractors that agree to carry out the works in accordance with the design and documentation provided. Following receipt of tenders, it is usual for the client/client's project manager to enter into negotiations with one or more favourable contractors to discuss the contents of their tenders and address any new issues that may arise in the process. The aim of this type of tendering system is for a client to obtain a lump-sum fixed priced for the works with the fewest tender clarifications, qualifications and conditions as possible. This is to enable the client and contractor to enter into a binding agreement based upon price certainty and project duration, and for the parties to understand their rights, risks and obligations regarding the contract they have entered into.

3.4.2.2 Two-Stage Tendering

This method is based on the design being procured by the client and the appointment of a contractor to carry out early works for a projects' first stage (e.g. site clearance and substructure works), won by competitive tender or the most economically advantageous tender regarding price and deliverables. Thereafter, the client retains the contractor's services for a second and final stage, which involves the contractor assisting with the design process.

The first stage involves inviting tenders for early works based on concept or detailed drawings and documentation, which may include a bill of quantities. Contractors then submit tenders in competition, and for best value are required to state their commitment to the project regarding collaboration and integration with the design team and agree to a programme it can manage. Thereafter, the contractor enters into a contract based upon a lump-sum price or schedule of rates or the terms stated in the invitation to tender. Selection at this stage does not guarantee the contractor will be awarded a second stage as the appointment for a second and final stage requires negotiation, which is linked to performance on the first stage as well as the contractor's commitment to time and price for the next.

The second stage aims at converting the contractor's services and commitment into a full service agreement with the contractor, possibly taking over the development of the design and documentation needed to complete the project. If doing so, this second stage would effectively create a design-and-build contract between the client and contractor (see Section 3.4.3 below for details on design-and-build contracts). For this to happen, the client would need to plan ahead and appoint design consultants for a first stage with the proviso each would be prepared to novate their partially completed design to the contractor for completion on the second stage. However, a wise client may be aware that consultants could resist an appointment for a first stage only, and might request the successful contractor to design discrete parts for the second stage. Using this approach, each design consultant's appointment is secured for both stages with the client.

The two-stage arrangement is beneficial to client and contractor on large projects as it permits early works to commence while the design undergoes development, which theoretically mitigates any delay on site. This early start also reduces the combined length of the design development and construction periods, meaning the project can be completed sooner than it would under a single-stage competitive tender.

3.4.2.3 Cost-Plus Contracts

This is a cost reimbursement arrangement where the contractor is reimbursed the base cost incurred for carrying out the works, plus a margin to cover offsite overheads and profit. It is suitable for emergency works such as insurance repairs, where a rapid start on site is required in the absence of a full design when time is considered priority. It can also be used where time is of the essence and a client wishes to appoint a contractor to carry out works under a construct-only agreement which is to commence within a stated timeframe after the design becomes available. A normal method of calculating the 'plus' amount is to apply an agreed percentage to the base cost; this means the greater the value, the higher the margin paid to the contractor. The form of contract with this procurement method must be clear regarding what constitutes cost, and is usually restricted to direct expense incurred pertaining to the project including supervision, labour supply, plant hire, material purchase and subcontracted works. Normally indirect involvement does not constitute cost, for example, the purchase of a computer for contract administration, as the computer would be a business asset.

With this arrangement there is no tendered amount, and for cost control a client may initiate a management system to avoid uncontrollable expense. Here, an architect managing the design process and acting as an agent of the client would be a suitable interface with the contractor. One method of cost control is for the contractor to obtain a stated number of fixed-price quotations for each trade package (a suggestion would be three) for approval by the client (or agent) prior to the works commencing with the contractor engaging the subcontractors on its own terms. An advantage of this arrangement to a contractor is that it should never lose out financially on a scheme as compensation is made in full through timely payments, with the client retaining the largest share of financial risk. A disadvantage to the contractor is if the contract has a termination by convenience clause, meaning the contractor may only complete portions of the works with the client free to terminate at any time. Furthermore, administrative involvement for preparing a request for payment is prolonged because of the number of invoices, time sheets and records required by the client (or agent) for auditing purposes to certify each payment.

3.4.2.4 Target Cost
This is a variance to the cost-plus arrangement and applies where the project client awards financial incentives to a contractor for earlier completion than the agreed completion date, usually to a set rate per week. The inverse may also apply where the date of completion arrives and the works are incomplete, with a financial penalty applied as a disincentive. When adopting this strategy, the requirement must be included within the contract as it is not possible to introduce the idea as an afterthought during the construction phase, and if requested by the client can be voided by the contractor as there is no obligation to accept such a condition.

3.4.2.5 Competitive Negotiation
Here, the design team prepares a set of preliminary working drawings and documentation defining the scope of works for tendering purposes that may also include a provisional bill of quantities to assist the tender process. Suitably vetted contractors are then invited to submit tenders for the project preliminaries and requested to include a statement of time required to carry out the works including a comprehensive programme of activities. In their tenders, contractors state a percentage they require to the base cost of trade packages they agree to manage as a contribution to offsite overheads and profit, plus an amount for builder's work in connection for trades requiring specific attendances. Following a recommendation from the client's agent, the successful contractor joins the design team as a building consultant and adopts a proactive duty, aiding the development of the design and providing advice on construction matters, while serving in the capacity of main and principal contractor for the purpose of CDM 2015. During the construction phase, the contractor invites tenders for trade work prices as the information becomes available and, once approved by the client's agent, awards packages for the works.

The system is advantageous to a contractor as it is managing its own preliminaries and takes little risk with the monetary value of trade packages; it is also suitable for repeat projects or projects that are carried out in phases. In addition, the experience of the contractor is valued by the project client, especially if a record of accomplishment on past projects shows schemes as completed on time and within budget.

3.4.2.6 Termed or Scheduled Contracts

This type of procurement is adopted by local authorities for the maintenance and repair of tenanted properties. Here, contractor selection is driven by local authority policy-making and funding available over a fixed term, meaning the expense must be committed before the end of a financial tax year as budgets may vary between fiscal years. The tendering procedure involves inviting contractors to submit a schedule of rates for a number of trade works that may have a target or a maximum quantity stated in a bill of quantities. Regarding preliminaries, bids are based on the scope of the project with a commitment to time and money as well as profit. Upon receipt of tenders, the assessor ascertains a provisional project value using the rates and approximate quantities from either the bills or as part of an internal cost plan. This helps to assess an approximate value and scope to include in a contract that can be adjusted to meet the maximum affordable capital expenditure. For example, a scheme of a few hundred tenanted properties may seek to replace windows and external doors; with a schedule of rates, it is possible to calculate a contractor's projected final account derived from the borough surveyor's inspections. If the projected final account exceeds the expenditure allowance, it is possible to scale back the scope and only replace doors and windows considered in need of urgent replacement. An advantage of this arrangement is that rates are fixed with the quantities subject to change, permitting the scope to be adjusted to suit the maximum expenditure. A disadvantage is the re-measuring process both during the works and after completion that can be time-consuming and requires close attention to administrative management.

3.4.2.7 Serial Tendering

This is a variance to termed or scheduled contracts and is a commercial test of prices where a client requests tenders for a number of projects or phases of a project. Here, the client commits to a fixed scope and creates a priced schedule/bill of quantities, creating a predetermined budget and value to the worth of a scheme. Competing contractors are issued the priced schedule/bill of quantities for accepting or adjusting and, if electing to adjust, is carried out by inserting plus or minus amounts (or percentages) to each item in the schedule/bill of quantities, creating a price that may be higher or lower than the predetermined budget. The successful contractor can be awarded a number of concurrent schemes that may run for a long period, providing work continuity in the process. This acts as an incentive for contractors to provide competitive rates as continuity is attractive, especially in times of a downturn when work is scarce.

3.4.3 Design and Build

This term, often referred to as design and construct, is a procurement route founded on risk transfer and exists where a contractor agrees to initiate or inherit a design from a project client, complete it for working purposes, and carry out the construction and commissioning works. The process commences with the project client (or client's agent) issuing an invitation to tender to competing contractors cable of carrying out a design-and-build service. A prerequisite of the selection process usually requires each competing contractor to submit proposals in their tender which is driven from criteria stated in the client's brief. At this stage however, contractors may only be prepared to offer a preliminary order of cost estimate, possibly by building element, that would be subject to negotiation and agreement prior to entering into a contract. A sticking point on the negotiation could be clarification regarding the contents of the client's brief and the quality of the design provided when a fixed price is being sought that poses risk to a contractor. Moreover, the willingness of client-engaged design consultants prepared to assign their services and participate in a design-and-construct contract requires understanding. When carried out successfully, this assignment becomes the legal transfer of contractual rights under an agreement with another party. This links in isolation to novation, a term used to describe the transfer of obligations and benefits to a third party (the design-and-build contractor) with the consent of the original party. To aid a smooth transfer, a clause outlining assignment must be stated in each design consultant's agreement with the client instead of being an afterthought; if not included, the client would be in breach of contract by making such a request. If assignment is included in the agreement(s) and carried out, each design consultant's contract of services is reinstated by the design-and-build contractor. The contractor then takes responsibility to coordinate the process and monitor progression of the design until the final production stage, and retains this responsibility throughout the construction phase. With novation, the contractor appoints the original design consultants under a series of agreements, with fresh contracts brought into existence between the contractor and each consultant. This is usually under the same terms and conditions of the consultant's appointment that applied before, the idea being for each consultant engaged by the contractor to be in the same position they were in before the novation. As an alternative, the design may be novated for completion by others subject to agreement and copyright protection, thus discharging the original consultant's involvement with the project.

Prior to electing for the design-and-build option, a client needs to consider any drawbacks it may encounter during the process, as development of the design is delegated to the contractor, meaning the client discharges prior involvement and must rely on the contractor's proposals. A criticism of this procurement route is that the final design may be streamlined in favour of the contractor, especially if the client's brief and preliminary design is limited in scope. For this reason, a client may seek the advice of a project manager as client's agent to oversee the design and construction process to avoid any surprises with the final building. A project manager with experience of this type of procurement would be of benefit to a client, as the appointment has the aims of ensuring expectations are met in the final product. However, it is not unusual for variations to arise during the construction phase; for this reason the supply of a statement of requirements, including room data sheets, should be agreed by the contracting parties and appended to the construction contract to help avoid any dispute.

An advantage to a contractor with this procurement route is entrusted delegated control that a suitably experienced design-and-construct contractor relies on for business. A disadvantage is the contractor's risk with handling the design process and ascertaining the price of variations due to a priced bill of quantities being absent that must be developed by the contractor, possibly stemming from the cost plan created for the purpose of preparing the tender. Moreover, disbandment of the design team from the client's grasp requires the contractor to foster new working relationships that may take time to develop. Despite the risks, this pathway is popular in both the public and private sectors and, once tested, can mean project clients and contractors are able to harvest working relationships for future work with a risk-sharing mechanism each party is willing to accept.

3.4.3.1 Turnkey Project

A variation of the design-and-build option is a turnkey project, sometimes called a 'package deal', and applies where a contractor provides a client with one of its standard designs, often referred to as off-the-shelf designs. Here, a client purchases a prototype or system-type building to suit their immediate needs. The arrangement has an advantage over other procurement systems as the design is readily available due to the contractor being in the business of providing repeat designs, saving time in the process. The blueprint to be purchased may also be modified to suit the client's brief subject to approval by the contractor. An example of this type of procurement is found with residential construction when a client is a housing association in need of affordable housing stock, who decides to form a contract with a house builder that designs and builds a standard range of house types (e.g. two-bedroom semi-detached). Here, it is simply a matter for the client to examine the range available and select a design suitable for the purpose, budget and location of the scheme, which may be part of a larger development by the house builder or on land owned by the housing association. The cost of each standard house type superstructure is usually fixed, with the substructure, external works, site conditions, infrastructure and preliminaries assessed on a case-by-case basis.

3.4.3.2 Develop and Construct

Another variation of the design-and-build concept is to develop and construct, and is an arrangement whereby a client procures a concept design and passes design management responsibilities to the contractor while retaining the services of the design consultants. The design consultants develop the design to a detailed stage. It is then submitted to the client/client's agent who usually seeks validation from the contractor that when validated, and in turn endorsed by the client/client's agent, allows the contractor to prepare shop drawings. Subject to the conditions of contract the process can be timely and repetitive, especially if the contractor is compelled to validate, check and give advice on any time and cost impact with the latter possibly applying if the client has changed something beyond natural design development that is abnormal to the design-and-build concept. The develop-and-construct process helps avoid conflict regarding the client's perception of a design in comparison with the final product, and dispels the concept of 'design creep', a term used to describe the enhancing of a design in a design-and-build contract beyond what the contractor considers should be acceptable because of client influence. Natural design development such as determining the cross section dimensions of concrete beams and columns would normally be deemed acceptable at validation stage and have no cost and time impact on the project. However, changes that clearly deviate from the contract such as a change from carpet floor finishes to stone flooring would be 'design creep' if not advised to the contractor beforehand; if seeking validation from the contractor, the impact on cost and the programme must be advised in writing as part of the process.

3.4.4 Construction Management Schemes

Construction management schemes became popular in the 1980s and continue to be adopted for projects where the bidding and construction phases are accelerated. This permits early works such as site clearance and substructure operations to commence before later designs become available, and is controlled by management schemes that seek to obtain best value with price and delivery.

3.4.4.1 Construction Management

This service is provided by a construction management company that becomes the contractor for a project. The management company charges the client a lump-sum fee for managing the project in addition to providing the project's preliminaries and acting as site consultant. Management companies differ in style and culture regarding their fee structure, which in general includes:

- pre-construction advisory services for planning and programming site accommodation;
- preliminaries for site staff charges as project running costs;
- initiation and running a collaborative information technology system, including staff training;
- mobilising and demobilising site accommodation, including maintenance and cleaning;
- fixed charges for specified mechanical plant (e.g. delivery, erection, dismantle and removal from site applicable to cranes, hoists, etc.);
- fixed charges for specified non-mechanical plant (e.g. skips, including delivery and collection);
- time-related charges for the hire of specified plant, usually on a daily schedules of rates;
- incidental charges for operating specified mechanical plant (e.g. insurance, maintenance and labour); and
- contribution to the contractor's offsite overheads and profit.

This procurement route is suitable for complex projects where the manager has experience of the arrangement with the scheme expected to run a number of years, possibly in phases, and where it is unlikely a lump-sum fixed price could be obtained even if a full design is available because of the project's duration. Appointment of a construction manager is via competitive tender where the client (either directly, or through the client's agent) issues an invitation to tender to companies selected for their capabilities. The invitation usually provides the names of the design team (or design consultant disciplines if appointments are incomplete) together with a list of objectives required from the construction manager. Objectives usually focus on the client's expectations, and include a series of questions for answering in a tender, such as:

- methods of achieving the project budget in accordance with the cost plan;
- how maximum value is to be obtained from trade contractors;
- the construction manager's ability to gain respect from the design team;
- the construction manager's policies regarding health, safety and the environment;
- demonstration of experience in achieving the completion date by referring to past projects; and
- demonstration of leadership capabilities appropriate to the project, and methods of mitigating impact and disruption caused by the works to the existing built environment.

Under this procurement arrangement, the client appoints a design team without the construction manager committing to trade prices or involvement in the design process. Subject to the terms of the appointment, a construction manager might provide a cost advisory service, which means committing to carrying out a cost analysis of site and trade works for budgeting purposes. This is for advice only with the construction manager expected to tender trade works as and when designs become available for working purposes, and provide feedback to the client (or client's agent) with a tender analysis and recommendation for each award. At the request of the client, the client's quantity surveyor may produce a trade bill of quantities for issue with each trade invitation to tender that, when priced, may form part of a binding agreement with the appropriate trade contractor. Final decision for the placement of each award rests with the client, or an agent of the client, with trade contractors in receipt of awards responsible to the construction manager for their performance on site. A characteristic of this procurement route is that each trade package is awarded by the client with the construction management company having no contractual link or responsibility for payment to the trade contractor. As part of managing the process, the construction manager's scope of services may involve reviewing each trade contractor's request for payment and recommending a payment amount.

3.4.4.2 Management Contracting

Management contracting operates in a similar fashion to construction management, with the exception that the client does not create a contractual link with trade contractors. In order to procure works, the management contracting company invites tenders for the works, makes recommendations and awards trade packages after authorisation by the client/client's agent. These companies become works package contractors, and in the process create contractual links with the management contractor and agree to provide warranties for the client. The expression 'works package contractors' can also refer to trade packages awarded by the client under construction management schemes as discussed earlier.

An award to a management contractor is based upon a fixed lump sum for a range of services including the project preliminaries, generally as per the construction management concept. The management fee is agreed as either a percentage of the project budget or as a fixed price, with the choice stated in the invitation to tender or subsequent agreement. From a works package contractor's viewpoint, the arrangement has little or no difference from a subcontract award under a traditional procurement, as the terms and conditions of subcontract regarding payment terms, etc. are the same. From a management contractors' perspective, there is little risk with the value of each trade package award versus the budget allowance as works packages are tendered upon release of the design with the client absorbing any difference between the budget and value of each award.

Management contracting has advantages to a client because it provides relief from the administrative and contractual responsibilities associated with the award of packages found with construction management. A disadvantage is the level of financial risk because the final price is unknown with only the budget to rely on, which can be exceeded if cost management is not enforced. Furthermore, the buoyancy of a programme that depends on design development can be a risk to a client, which must be monitored and controlled by the management contractor to prevent time delays and cost overruns.

3.4.4.3 Guaranteed Maximum Price (GMP)

A GMP may be offered by a construction management company to comply with the conditions of an invitation to tender, or may be offered in an attempt to fend off competition where not a condition. When accepted by the parties in an agreement, the construction manager commits to a price limit of the award of trade or works packages through effective management. This provides the client with price certainty which, to be effective, requires a commitment from the client to ensure there is no revision of the scope once the design is complete.

The GMP ideology can also apply to other procurement routes when the contractor is considered the most suitable party to accommodate the risk. Here, the contractor accepts the responsibility for unforeseen conditions/risks associated with, for example, bad weather, unresolved ground conditions, state of industrial relations, design information, the accuracy of the design, etc. When accepting a GMP for inclusion in an agreement, the client only reimburses additional amounts for changes in works that the contractor agrees to carry out, and which must be confirmed in an instruction from the client/client's agent.

The GMP ideology is not restricted to a project client/main contractor arrangement and a main contractor with a responsibility to procure a supply chain can adopt the idea for complex portions of works such as civil works, structural engineering or building services. Here, a contractor could employ a specialist manager on a fee basis to tender and award specific subcontract packages on their behalf. This approach may also be the deciding factor for a contractor, who may elect to offer a GMP to a client if they are confident they can manage the concept effectively.

3.4.5 Private Finance Initiative (PFI) and Public–Private Partnerships (PPPs)

A PFI scheme can be described as the basis of a risk-sharing partnership between public and private sector organisations that in combination initiate a project to create an end-product that also brings a desirable public policy outcome. A driving force behind the selection of a PFI scheme is the need to satisfy public demand for the end-product that would otherwise be unavailable from the public sector. When seeking this type of partnership, the public sector seeks a private sector investor to 'come on board' with the investor agreeing to design, construct and finance a project through the use of mortgages, grants or subsidies, and to operate the completed facility under an *ad hoc* arrangement that includes facilities management, usually for a minimum period of 25 years. This way, each member of the partnership reaps the benefits over the long term, that is, the public sector obtains the product it otherwise cannot afford and the private sector secures a long-term business investment. The PFI concept creates a public facility which minimises expense from the public purse, and can apply to a range of works including infrastructure and public buildings such as hospitals and prisons.

3.4.5.1 Traditional PFI

To initiate a scheme a public sector client will need to identify their needs, which usually forms part of the client's brief. The brief may include reference to the PFI concept as well as a decision to invite private sector businesses to explore the arrangement, which is usually influenced by public departmental procurement policy, the funding available and the functional requirement of the completed facility (e.g. the construction of a new hospital to set criteria such as the number of beds, theatres and consultation rooms). Once the need is identified, the next step is for the public awarding authority to appoint skilled advisors to provide advice on the relevance of a PFI to the scheme. Advisors may either be a department of the public sector's business or sourced independently, and must be considered competent and experienced enough to provide suitable advice. Part of the process involves the advisors adopting the use of a Public Sector Comparator (PSC), a management tool used to compare the viability of a proposed PFI arrangement and the cost to the public sector with other types of procurement such as design and build or construction management. The PSC has the following characteristics:

- it is project specific and demonstrates the cost to the public sector to the same standards if procured from the private sector using an alternative to PFI;
- it is expressed as net present cost, excluding adjustments for rise and fall in prices over time;
- it is risk adjusted;
- it includes design and construction costs as well as maintenance charges and the life-cycle cost of assets to the standards prescribed by the PFI arrangement for the scheme;
- it recognises risks and gains to the public sector if opting out of the PFI arrangement; and
- it recognises risks and benefits regarding long-term involvement of the public and private sectors committing to a scheme.

Once a PFI scheme is selected as the preferred option, bids (as opposed to tenders) are sought from a shortlist of private sector contractors. These contractors must be deemed competent to design and build a scheme, as well as being capable of funding the project while under construction and proficient in managing and maintaining the facility for a specified duration. This is a huge task for bidders that, in the process, usually introduce other parties to the scheme with a view to create a consortium specifically for the project. A consortium may comprise one or more companies already in business, or have an understanding they would create a new business specifically for the scheme with the required skills to design, construct and manage the facility, which usually involves banks and insurance companies to provide equity. In each bid, the business efficacy needs to be demonstrated to the public sector client to show how financial risk is to be managed. (Banks tend to view these schemes as lucrative because of public sector involvement and the lower risk of stakeholder insolvency associated with the project, which provides stability over the long term.)

The combination of the consortium and equity providers is the key provider of the integrity of each bidder's response, reviewed by the public sector client and merited on best value (not necessarily linked to lowest construction price). The selection process may commence with up to six bidders, which is reduced by negotiation and competitive dialogue to create a shortlist. From a bidders' perspective, the cost in time and money invested in preparing a bid can be substantial. This is especially so if involving numerous rounds of negotiation; for this reason, bidders are usually appreciative of their position with their competitors and may elect to decline further involvement when the time is appropriate. Conclusion of the dialogue leaves a preferred bidder to obtain full planning permission and demonstrate facilities management arrangements while finalising the legal aspects of the bid prior to negotiating a contract. At this stage, the intentions of the parties to the imminent partnership are clear, taking the following format.

- The scope of the design and construction of the works, including equipping the facility for handing over to the public client. An important factor here is the procurement method for constructing the project which must be decided (e.g. design and build, construction management, etc.).
- Normal operational and business matters pertinent to the function of the building are under the control of the end-user (e.g. hospital for health care, schools for teaching etc.).
- The level of services to be sold to the public sector as part of the PFI scheme are clarified (e.g. a private sector firm selling operational equipment to a new hospital or leased accommodation for day-to-day care for the elderly in a retirement home).
- A statement of assets for the equipped building defining levels of responsibility for their maintenance and replacement during the occupational phase is materialised. This is usually partitioned into three parts: hard facilities management applicable to the building fabric, structure, finishes and range of building services; soft facilities management, not limited to catering and cleaning; and assets under the permanent ownership of the public sector that are salvaged, donated or procured by the public sector during the construction phase.
- The creation of the stipulated lease period, including expiry date when full control is transferred to the public client.

A traditional PFI structure is shown in Figure 3.2.

3.4.5.2 Joint-Venture PFI

This is a variation to the traditional PFI arrangement, where central government (as opposed to a local authority public client under traditional PFI) partners with the private sector to partially fund a project or series of projects, leaving overall control weighted towards the private sector. Government involvement includes the supply of grants or loans and the transfer of equity assets to aid urgent programmes. The creation of these projects are in the interest of the public and are generally more fast tracked than traditional PFI projects as the negotiations are less involved, with the public sector

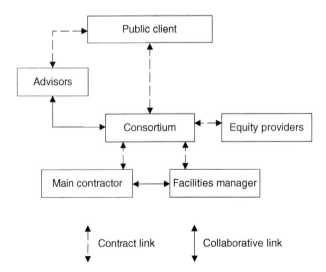

Figure 3.2 Traditional PFI structure.

accepting a larger portion of risk. With this type of PFI, the government receives advice on the potential of a joint venture scheme from a number of sources, such as HM Treasury, and when electing for the pathway invites bids from competent contractors, similar to the traditional PFI arrangement. The phrase 'joint venture' can apply to the merging of one or more contractors to carry out building or infrastructure works on a grand scale and can apply to any procurement route. This is not to be confused with a joint venture PFI which is a method of financing a scheme that is separate to the risk-sharing methods adopted for designing, managing and constructing the works.

3.4.5.3   Free-Standing PFI

A characteristic of this PFI arrangement is that the public sector client reimburses the private sector with regular unitary payments for ongoing services such as maintenance, facilities management and the life-cycle replacement of assets during the lease-back period. On some PFI projects, the consortium may consider the lease-back period and unitary payments as contributing factors to perceiving a project as being viable. Here, the consensus of a consortium at the bidding stage may be to expect little or no profit from the construction works, and view the scheme as a long-term investment that will recoup profits during the lease-back period because of the unitary payments arrangement. The longer the lease-back period, the stronger are the chances that a sustainable profit will be derived.

3.4.5.4   Advantages, Disadvantages and Criticisms

With these types of projects, the time and money invested in their initiation, design, construction, running and operation can be vast, with the advantages and disadvantages often contentious. A PFI scheme has obvious benefit to the public sector because it is procured without full burden on public spending with a degree of risk transferred to the consortium. On the flip side, and applicable to bidders, a lost scheme is of no benefit as the devotion to time, expense and innovation is a salient matter for demonstrating the opportunities a project is to provide while also making it an attractive bid. This goes hand in hand with the risks involved in creating a successful consortium, especially if the consortium's strategy is to provide the construction works at or near cost and rely on facilities management operations to produce a profit. Here, the consortium takes risk with debt to equity ratios that may take years to level off to produce the envisaged margin, which is a risk with this type of bidding.

The consortium carries additional risk during the occupational phase because of abatements that may be a condition of the project deed, the formal instrument of agreement between the consortium and awarding authority. Abatements are a series of detrimental measures an awarding authority may lodge against a consortium or party to the consortium responsible for the management of the facility whose performance fails to meet the standards benchmarked against key performance indicators (KPIs). KPIs measure the level of services provided in comparison with the services expected, as per determined factors in the project deed, and includes a mechanism to apply abatements as financial disincentives based upon the failed results of the KPIs, including:

- meeting deadlines for the timely issue of reports to the awarding authority;
- meeting the timely placing of awards for certain routine maintenance works through facilities management agreements;

- variances between recommended and actual intervals of maintenance works that may reduce the efficiency of operational plant, resulting in higher energy consumption; and
- the actual time taken to respond and attend to any breakdowns and carry out repairs in comparison with benchmarked times required by the KPIs.

The triggering of abatements can be seen as a reflection of the performance of a facilities management company for the maintenance of a building/facility. This may be considered drastic in practice as the amounts involved may not reflect any actual loss to the public sector that could be considered a penalty. However, in a situation where there is a failure to respond to a breakdown or carry out repairs in a timely manner, the application of an abatement may become contentious, especially if the public sector is not fulfilling its obligations (e.g. improper use of an asset or not following the guidelines of an operating manual for general maintenance). Furthermore, and from an administrative point of view, the process is expensive to initiate and run as it involves the public sector incurring additional expense to administer the process, which may be considered counterproductive.

Criticisms of PFI projects vary and, for clarity and balance, it is necessary to examine the facts to understand what is considered best value. Some experts on the schemes consider that buildings are generally more expensive to run under the private sector than they are with the public sector, with criticisms of the size of the profits made by some facilities management contractors. This argument revolves around the subject of public sector overspending to the benefit of the private sector, which is not possible to control because of the long-term agreement in place. Moreover, spending on public services may change over the term which is generally of no concern to the private sector, yet the public sector must meet its obligations to the private sector because of the agreement. In the case of a health care trust, central government may decide to cut grants in any fiscal year with the trust forced to reduce expenditure; the trust is however not empowered to amend facilities management scope or reduce payments from the consortium, and to meet budgets may have to deprive the public of some public service benefits. A further criticism comes with the use of Design, Construct, Manage and Finance (DCMF) projects, a variance of traditional PFI. Under this arrangement, the private sector has an additional role during the occupational phase which involves managing part of the public sector. With this arrangement, key staff such as wardens in prisons, doctors in hospitals, etc., have their services retained under employment contracts with the public sector while other staff are offered new employment contracts from the private sector, thus creating a two-tier system. Where private sector employers have control of DCMF schemes, it requires personnel to negotiate the terms and conditions of their employment or to accept trust assurances they are not accustomed to. This creates a shift in culture that can literally occur overnight, involving a change of employer often without consultation which can lead to resettling issues and disputes. There is however a general opinion that the private sector is the best party to exercise facilities management because the public sector has primary interests in managing public services only, and would otherwise neglect the management needs of a completed facility. Furthermore, and without private sector investment, the creation of a facility (and therefore benefit to community) may be delayed for years or decades if funding is only available from the public sector rather than under a PFI arrangement. Whatever the opinion of best value, PFI initiatives are initiated by the public sector for private sector integration. In doing so, projects are created with long-term strategies that have a positive impact on the construction industry which also benefits other industries in the process and the public at large.

3.4.5.5 Public–Private Partnership (PPP)

Although appearing a contradictory term, PPP is a generic reference to the creation of a government service obtained with private sector involvement where a building(s) is financed through a consortium. In many ways a PPP arrangement is the same as a PFI scheme and is the alternative term adopted by countries outside the UK. From a British perspective however, PPPs cover a broader context than PFI schemes which are often on a smaller scale in terms of scope. PPPs can also provide better value as the government debt is less with the risk mitigated because it is transferred to a consortium. There are a number of PPP arrangements in place in the UK with some examples discussed in the following sections.

3.4.5.6 Arms' Length Management Organisation (ALMO)

This organisation is a not-for profit company created in 2002 that helps to manage, improve and repair housing stock properties on behalf of local authorities, with stock levels at 2017 numbering approximately half a million across 37 local authorities. Under the arrangement each local authority has an agreement with an ALMO organisation to promote participation from the community, with central government providing subsidies on the proviso the local authority separates its management functions from strategic policies. This allows ALMOs to provide a role in the community that includes the collection of rent and a say in the maintenance of properties; the local authority has a smaller role, for example limited to administration and policy making for rent strategies. The ALMO concept is run by members who may be part of a community, with each having the benefit of being independently managed and separate from local authority input.

3.4.5.7 Local Partnerships (Formerly Public–Private Partnerships Programme, or 4Ps)

Local Partnerships is a joint venture between Partnerships UK and the Local Government Association (LGA), incorporating the former 4Ps ideology and services. The entity is involved with partnering arrangements and services provided by local authorities, and has the aim of strengthening the public sector to help deliver a more effective system while giving value for money to the taxpayer and public sector customer. Local Partnerships is owned by HM Treasury and the LGA, and derives some of its income from grants given by the LGA as well as fees derived from central government departments and local authorities that engage them for their services. In the interests of improving the delivery of public services and infrastructure, Local Partnerships provides professional support to peers, including: identifying funding and savings in PFI schemes (without having contractual involvement with the public sector/consortium for delivering a scheme); advising how to create and nurture effective partnerships; and issuing advice to local authorities regarding procurement strategies, negotiation and contract management.

3.4.5.8 Build Operate Transfer (BOT) Schemes

BOT is a project delivery method where a private sector entity is granted a concession by a public sector sponsor to design, finance, construct and operate a facility for a specified period which is transferred to the host government upon expiry. There are a variety of BOT schemes available for an array of project types which are distinguished by their risk-sharing mechanisms. In prioritised order of high to low risk to the public sector, these schemes include:

- DBM: design, build and maintain;
- DBOM: design, build, operate, maintain;
- DBO: design, build, operate;
- BTO: build, transfer, operate;
- BRT: build, rent, transfer;
- BOOT: build, own, operate, transfer;
- BOLT: build, operate, lease, transfer;
- BOL: build, operate, lease; and
- BOO: build, own, operate.

The PFI, PPP (international) and BOT concept is a method of financially underpinning a scheme and is not a procurement system in itself, as the methods of obtaining a building/facility as part of an arrangement can be derived from a choice of procurement routes (e.g. design and build). However, a building contractor's direct involvement with a consortium adopting the concept for a particular scheme is one method of participating in such a scheme, especially if in a position to offer different procurement options for the consideration of the consortium.

3.4.6 Prime Contracting

This is a long-term contracting relationship used for the procurement of new buildings and the maintenance of existing, and is adopted by the public sector where a prime contractor designs, constructs and operates a built asset for a defined (or open-ended) period. It is a suitable contracting arrangement for managing estates where new build, maintenance and refurbishments are required on a regular basis, with the UK's Ministry of Defence a staunch user of the system. A core culture of the arrangement is the effective use of strong supply chains and partnering principles that integrate to provide an efficient service for the client and end-user. These supply chains comprise a network of integrated businesses that design and plan operations with the objective of providing leverage and value to a client. Supply chain participation is not restricted to the delivery of goods only and embraces a range of services including design, subcontractors and manufacturers, which are sourced locally, nationally and internationally.

A prime contractor is selected for its expertise and proactive use of supply chain management; for this reason, it must have sufficient business contacts. Selected contractors are usually large construction businesses comprising at least one division that focuses on the prime contracting concept, which usually includes financiers and asset/facilities managers that have an understanding of the use of supply chain management. The values of supply chain management and characteristics for use with this procurement route include:

- the clustering of project elements to ensure an effective supply chain exists within the parameters of each element (e.g. mechanical and electrical trades as part of a building services cluster);
- making designers aware of lean value in terms of a building's function without diversifying the details that could cost time and money;
- including the supply chain with cost planning, value management, target costing and risk identification;
- motivating the supply chain to design and construct buildings to a cost limit (as opposed to costing the design and construction works, that can create budget blowouts), which is possible to mitigate with standard designs and prototype structures;
- developing and nurturing contractor/supply chain relationships over the long term by participating in regular meetings for updates on future business needs, seeking commitments to the guarantee of work, negotiating fixed prices and ensuring payments are made on time; and
- promoting leadership with collaboration, training and incentives; long-term projects require commitment from the supply chain, and to avoid apathy the initiation of incentives such as introducing time-related bonuses, learning new skills and creating workshops to deal with industrial trends will help reduce any monotony.

Figure 3.3 demonstrates a prime contracting arrangement.

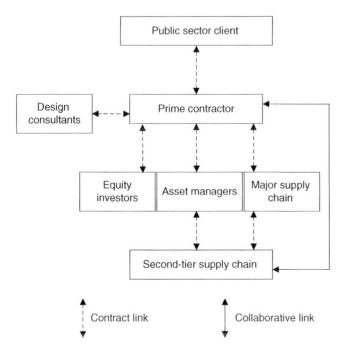

Figure 3.3 Prime contracting arrangement.

3.4.7 Partnering and Strategic Alliances

In its various forms, partnering is a structured non-contractual arrangement between business organisations that has a theme based upon trust, cooperation and the understanding of objectives to help improve the overall performance of a project, while at the same time setting the pathway for repeat work. The theme of partnering is based upon:

- forming an *ad hoc* management arrangement that defines the responsibilities of the client, design team, contractor and facilities management company (if applicable), allowing each to have an understanding of the roles required for a successful project;
- recognising methods that deals with risk and conflict avoidance (e.g. with the use of workshops to identify the needs of a project and possible pitfalls and how to deal with them beforehand);
- maintaining a flexible approach with procurement selection by considering contractor involvement (e.g. design and build);
- creating trust with potential partners without emphasis on competition (e.g. limiting the number of design-and-build contractors tendering for work);
- seeking commitment from potential partners to ensure they are capable of contributing to a scheme and are prepared to be collaborative;
- outlining the form of contract(s) to create lasting working relations with nothing more or less expected (i.e. no surprise clauses once a contract is issued for execution); and
- adopting innovative measures to overcome obstacles.

For partnering to be effective, ideologists must plan ahead and identify key project matters using specific criteria (e.g. by preparing a client's brief). Once these items are identified, the next step is to select and invite potential partners to participate in a scheme prior to asking for bids or tenders. This is possible by compiling a pre-qualification questionnaire to measure the suitability of a potential partner for a scheme, both culturally and commercially.

A strategic alliance is different from partnering and is specific to the creation of two or more businesses that agree to collaborate and obtain a mutual goal. A key characteristic with this arrangement is that each business maintains its identity while gaining the benefits of the alliance. Where a client seeks to transfer risk in a project, an alliance may prove to be the strongest and best party to accommodate risk that involves the supply of a full range of services which a single business could not normally provide. In turn, this benefits alliance members with new business opportunities and fulfils the needs for the project client. The creation of these alliances may cross boundaries in the construction and engineering industries and take individual organisations to a business level they have not experienced before. For this reason, risk assessments are required to gain insights into the impact the alliance may have on each partners' business activity. Here, each partner usually has a percentage stake in the name of the created alliance that involves investment in capital and the possible involvement of banks and insurance companies to provide bonds as surety.

From a project client's perspective, the concept of strategic alliancing and partnering is viewed as positive, as it fosters teamwork, maintains healthy competition and can cut costs. For example, two developers may each purchase a subdivision of land adjacent to each other with the intention of constructing residential properties, and may share the infrastructure costs in a joint venture. In this scenario, a theme of trust could be created where one developer acts in the capacity of project client and appoints a contractor to construct a network of fully serviced estate roads and a shared spine road with the cost split proportionally, thus providing both developers with a benefit at half the price.

3.4.8 Project Alliances

Project alliances differ from strategic alliances formed from business pursuits, and exist where a party (usually the project client) merges with a minimum of one services provider such as a designer, contractor or facilities management company to create an alliance board for a specific project. A characteristic of this board is a culture of binding the alliance as a single entity to share levels of good or bad fortunes if they arise on a project they wish to undertake. The contractor carrying out the works may agree to carry out a scope of works to a maximum sum in order to limit risk with overspending a cost target which, if exceeded, becomes a floated sum that is normally absorbed by the board. However, this risk to the project board can be mitigated if the contractor appoints key subcontractors that agree to be part of a sub-alliance and part of the alliance board. For this to be effective, a carefully written agreement will outline the burden of any loss as a maximum sum or percentage of the project value. This may be proportional to each member's stake, including sub-alliances, and applies inversely to any financial gains that would normally be to capped amounts.

Project alliances revolve around complex projects of significant value, and were introduced in the early 1990s when British Petroleum (BP) found it uneconomical to exploit oil reserves in the North Sea. BP drew up contracts to generate trust and teamwork between the players, and wrote a contract based on a mutual pain-share and gain-share outcome. Following this concept, and wherever this type of agreement exists today, the procurement route retains the following characteristics.

- Obligations with the board are contractually collective, e.g. the wording 'The Contractor will carry out the following' is replaced with 'The Alliance Partner will carry out the following'.
- Where a fixed lump sum is not used for parts of the works, reimbursement is via an open book arrangement where partners prove their incurred expense plus an agreed lump sum as a contribution to business overheads and profit.
- There is an equity-sharing mechanism in place to identify the spirit of the alliance as win or lose, applicable to all members at the same time.
- There is an expressed will of the alliance to resolve disputes without the need for litigation by choosing alternative dispute resolution methods.
- Participants share a common goal of achieving specific objectives relevant to the project and not for individual gain.
- Principles for the creation of the alliance are founded on collective responsibilities as well as a no-blame culture, collaborative mannerisms and equal say opportunities.
- Project board members have no interest in reaping rewards or suffering losses, as there is no private/public investor as found with PPP/PFI schemes.

Due to the specialist nature of this procurement route, the arrangement is best administered by those experienced in the concept. In addition, and because of the risk and expense involved, an experienced lawyer is usually appointed in the early stages to provide recommendations on risk and prepare draft agreements for the stakeholder's consideration. The risk-sharing strategies of these alliances are complex, and a project client may concede to execute an agreement with a project alliance board based upon a cost target and not a lump-sum fixed price. This is because of the high value involved, uncertainty with the project's duration and possible scarcity of contractors willing to commit to a fixed or guaranteed maximum price.

3.4.9 Framework Arrangements

A framework arrangement is a term referring to a buyer's dealings with a supplier of goods or goods and services where a purchase is made through a set of predetermined terms and conditions. The terms and conditions include obligations of sale as well as price, deliverables, quantity and period of the supply. To be reliable, a framework agreement must be created in a spirit of good faith. This is usually demonstrated by the purchaser who expresses an intention to a supplier of a desire to enter into a contract once the need for the supply of specified goods or goods and services materialises. However, a usual precondition of the arrangement is that the requirement is a preliminary agreement only and not a guarantee of purchase, as the purchaser will be relying on third-party needs and may have to adjust or cancel the arrangement at any time should circumstances change beyond the purchaser's control. Prior to entering into a framework agreement, a purchaser will need to consider if the arrangement is suitable for a project which can be determined by reviewing the effect on existing procurement policies. In order to overturn any existing policies and create new policies, it is necessary to have answers to the following questions.

- Is the existing procurement scheme suitable and is there a real need to change?
- The framework concept may take time to organise and will involve vetting new service providers that will also take time; is the time available or is there a possibility the project lead time will be delayed because of the change?
- The framework will provide opportunities for competition; will this be utilised to full potential, and is lowest price a priority or are there other matters equally important, for example delivery periods and user training of product use?
- How many frameworks are needed, and is staff training required to manage them?

Framework arrangements have advantages to purchasers and suppliers as they are flexible and can act as a catalyst for creating partnering and long-term agreements with the supply chain. Furthermore, they can create a network for further business opportunities gained from the trust and respect of prior dealings once the parties to an agreement become aware of their customs and terms of trade. To ensure that mutual trust and respect is maintained, a purchaser will need to consider their relationship with the supply chain and should view framework arrangements as asset tools that acknowledge the spirit of long-term partnerships, and not see them solely as bargaining tools to obtain the lowest price.

3.4.10 Public Procurement and European Legislation

European Union Directives 2014/23/EU, 2014/24/EU and 2014/25/EU, applying to concession contracts, public construction works and utilities works, respectively, make it mandatory for member states to raise public awareness of contract announcements where the request is for supplies, services or works when the value exceeds certain financial thresholds. The British response to the directives is legislated in England, Wales and Northern Ireland by the Public Contracts Regulations 2015 and The Utilities Contracts Regulations 2016, and in Scotland is via a coordinated response in the Procurement Reform (Scotland) Contract Act 2014. As at January 2016, thresholds (excluding value added tax) stand at €5,225,000 for public construction and utility works and €135,000 for supplies and/or design services for central government clients. This rises to €209,000 when the client is not a central government body, which is doubled to €418,000 for utility services. If the value of an award is to exceed these thresholds, it is mandatory for awarding authorities to advertise their required works, services or supply requirements online through the Official Journal of the European Union (OJEU). OJEU announces invitations to tender under a list of supplements applicable to all European Union states once the requirement to advertise is initiated.

Following the referendum result in June 2016 which favoured BREXIT (i.e. Britain's decision to cease being a member state), the OJEU requirement will remain at least until Britain extracts itself from the EU. After that, it will depend on the trade agreements negotiated as to whether this is still a requirement as well as any consequential changes to British legislation. Should the UK be the fifth state to join the European Free Trade Association (EFTA), an intergovernmental trade organisation and free trade area, it may participate and conduct trade in the European Economic Area (EEA) which comprises EU and EFTA members, that in the process would maintain the status quo.

3.4.10.1 Prior Information Notice (PIN)

This is the notification of a pending project where planning permission has been granted with the PIN issued as advice to the tendering community to prepare them in advance for tender participation. The PIN gives an approximate financial value of the project, as well as the anticipated commencement date so tendering companies are aware of the timeframe and scope. Changes to a PIN by a project client are not mandatory as the details are for information only meaning issue of a PIN is not a guarantee the works will proceed strictly as advertised.

3.4.10.2 Main Competitive Procedures

There are four main competitive procedures, the selection of which rests with the awarding authority, and are influenced by the type of project in terms of functionality, duration, financial value and complexity.

1) *Open procedure*: This is a single-stage process where the contracting/awarding authority (or client) advertises the project online and issues the design, documentation and draft form of contract to an unrestricted number of participants. However, the issuing of this information is usually subject to each participant meeting the contracting authority's terms and conditions. This can create a large number of responses that the advertiser may intentionally seek because of a project's locality, which may be remote, or if the requirements are of a specialist nature. Each respondent must demonstrate their capability to service the scheme as well as an amount to tender. Tenders are assessed on the lowest price or most advantageous proposal with the contracting authority reserving the right to not enter into negotiations once tenders are received. However, the terms of the advertisement may describe the tender vetting process and the procedure for issuing an award.

2) *Restricted procedure*: This is similar to the open procedure with the addition of an earlier stage, which requires participants to submit expressions of interest and participate in a pre-qualification round of questions and answers. This is to permit the contracting authority to be sure those expressing an interest are suitable contenders. Once participants are suitably vetted, they are issued the design, documentation and draft form of contract, thus limiting the number of tender receipts. As with the open procedure, negotiations are not entered into post tender which is usually advised during the vetting process or might be stated in the invitation to tender documents.

3) *Competitive dialogue*: A feature of competitive dialogue is the contracting authority's intention to enter into dialogue with bidders to discuss their proposals and solutions for a scheme prior to the issue of an award. The process commences with an advertisement in the OJEU stating the inclusion of a two-stage process comprising an initial invitation to tender and a later round of competitive dialogue. For the first stage, interested participants are asked to complete a pre-qualification questionnaire for vetting purposes, similar to that found with the restrictive procedure, and for the parties to understand the intention to enter into a competitive dialogue phase. Upon receipt of suitable bids and commencement of the competitive dialogue phase, the parties identify solutions for the scheme that usually involves meetings and negotiations prior to the number of tendering companies being reduced. Once favourable solutions are considered viable by the contracting authority, final or revised bids are submitted for evaluating. This can culminate in a possible further round of dialogue to clarify key points and fine-tune the requirements to understand the basis of each offer. The final round is not usually as intense as earlier rounds as solutions have usually crystallised, providing clarity in the process with the round the home stretch towards selection of a preferred bid. Under this procedure, tenders are evaluated on the basis of them being economically advantageous, which does not mean lowest price, and can mean other aspects such as time, quality and risk transfer or retention that is acceptable to the parties.

4) *Negotiated procedure with prior publication of a notice*: Similar to competitive dialogue, this is also a two-stage process where the contracting authority advertises the project with prior notification and invites participants to enter into the pre-qualification and selection rounds. Thereafter, the design, documentation and a draft form of contract is issued to selected participants in order to receive tenders. Upon receipt of tenders, the contracting authority enters into negotiations with a short list of tendering companies to discuss the proposals. This procedure permits tenders to be assessed with an award placed on the lowest value or considered as being the most economically advantageous (which does not necessarily mean that the lowest price is accepted).

3.4.10.3 Contract Notices

These are mandatory notices that publish advice on the status of received tenders and award criteria via Tenders Electronic Daily (TED), a database supplement of the OJEU providing procurement notices of contract announcements. The notices provide advice regarding the outcome of an award together with a brief explanation of the status, which is elaborated if a choice is not driven by accepting the lowest price. The award procedure requires a careful balance between a 'most economical advantageous tender'

(MEAT) and what is considered a suitable tender, as the analysis of tenders including risk identification and the party considered the best to shoulder risk can influence a contracting authority's decision with regards best value for determining an award.

3.4.10.4   Online Auctions

Online auctioning (or e-auctioning) is a process of obtaining goods and services via the internet in a competitive market. It is a transparent and efficient method for obtaining goods only as well as goods and services, and is used extensively by the private and public sectors. The process involves a contracting authority advertising a notice online specifying the needs, quality, scope and quantity (where the supply of goods only is required) and contract duration (where the supply of goods and services is required). A characteristic of the arrangement is that the notice can be open or restricted with the aim of advising bidders of a MEAT arrangement, where best offer is considered best value not restricted to the lowest price. This is reverse-style auctioning (as opposed to normal auctioning, which is bidding to create the highest offer to a buyer as found with e.g. e-Bay), where sellers or bidders submit offers in competition with one another up to and including a predetermined date as stated in the auction criteria, after which the auction ceases. Each bidder submits a price with any incentives, such as an extended warranty expressed as a value, even though a competitor may wish to include the incentive free of charge. In this scenario, the bidder must apply a charge and reduce the cost of other components to maintain a lump-sum bid, as the theme of this type of auctioning is to not enter into negotiations once an auction is closed. At the end of the auction, initial bids are adjusted in line with any changes made during the auction period by the contracting authority to identify the lowest or preferable bid. The system is advantageous to bidders as they are made aware of other bids and may reduce their prices if wishing to do so, with their identity remaining anonymous except to the contracting authority.

3.5   Estimating and the Contractor's Quantity Surveyor

The contractor's quantity surveyor is most likely to be involved with commercial activities of a construction project during the post-contract period, carrying out project and contract administration to assist in the delivery of a scheme. Large-sized contractors may also engage quantity surveyors for their measurement skills to assist with the estimating function of the business outside the confinements of existing projects. This is when a contractor offers tenders for works based upon specifications and sets of drawings, or a design-and-build service when responsible for producing quantities. Alternatively, the quantity surveyor may assist with pricing a bill of quantities provided with an invitation to tender.

   A quantity surveyor has the traits and skills to become an estimator as the qualifications and training overlap, with some estimators possibly already trained as quantity surveyors. Estimators may also be from a commercial management background or tradesmen who have diversified their careers into management. Small-sized contractors and subcontractors often employ estimators from a trade background because of their hands-on experience gained over the years while working 'on the tools'. This can be highly regarded by a company, as their experience is real and valuable for preparing prices to include in a tender. However, the larger-sized contractors involved with competitive bidding schemes worth millions of pounds usually make no preference and leave the complexity of the tender process and tasks to professionals deemed competent enough to carry out a tendering service.

A postgraduate could find working in an estimating department of a contractors' organisation a valuable experience, as it provides insight into the working practices of the business operations and an understanding of the process for submitting a tender. This part of the chapter focuses on the role a quantity surveyor may play as part of an estimating team in a medium- or large-sized contractor's office when preparing a tender worth upwards of £5 million under a traditional procurement arrangement.

3.5.1 Activity on Receipt of Tender Documents

The date stated on an invitation to tender until the stated date the tender is to be submitted is known as the tender period. The invitation to tender is usually accompanied with a cover letter from the client's agent stating the contents of the consignment, which as a minimum includes drawings and specifications, with the letter usually advising the method for delivering the tender (e.g. tender box, uploaded to a host website, etc.). The date on the cover letter and the date the contents are received by the contractor may differ, and if there is an abnormally long period between, the client's agent should be notified as, in effect, failure to notify would be understood to mean acceptance of the situation, where in reality it is a shortened tender period for the contractor. A reduced tender period would naturally put the contractor at a disadvantage and an extension to the date of submitting a tender should be requested if the original is considered hindered, providing of course there is no error on the contractor's part. Influencing factors that determine the length of a tender period include:

- the client or client agent's choosing;
- size of building and scope of works, in terms of floor area and number of storeys;
- financial value and complexity of the project;
- quality of the design and documentation provided;
- contractor's required input with the design;
- start date on site;
- procurement route; and/or
- whether or not a bill of quantities is provided.

A tender period for traditional procurement where a bill of quantities is provided is usually 4 weeks, which can be extended to 8 weeks for design and build due to contractor involvement. During the tender period, it is important for the contractor to create a plan of activities commencing from the date of receiving the consignment until such time the tender is due. A usual first task for the manager in charge is to review the contents of the consignment and delegate duties and responsibilities to quantity surveyors and other team members. Effective management requires good communication as well as adequate contributions from team members; Figure 3.4 shows a participative and delegation style of management applicable to a tender period.

Each participative task must have a commencement date and duration; this is necessary because one task may need completing before the next can commence. For example (and as shown in Figure 3.4), the last responsibility of the quantity surveyor is to

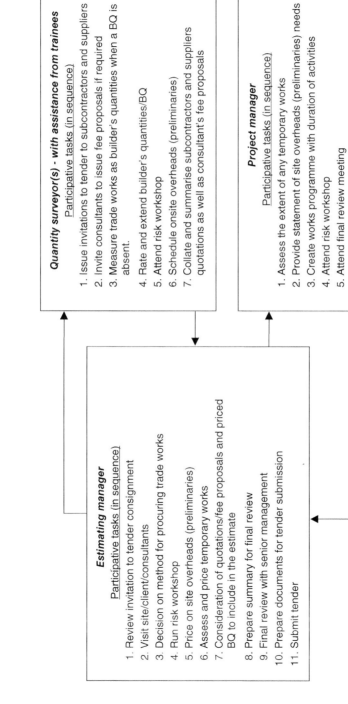

Estimating manager

Participative tasks (in sequence)

1. Review invitation to tender consignment
2. Visit site/client/consultants
3. Decision on method for procuring trade works
4. Run risk workshop
5. Price on site overheads (preliminaries)
6. Assess and price temporary works
7. Consideration of quotations/fee proposals and priced BQ to include in the estimate
8. Prepare summary for final review
9. Final review with senior management
10. Prepare documents for tender submission
11. Submit tender

Quantity surveyor(s) - with assistance from trainees

Participative tasks (in sequence)

1. Issue invitations to tender to subcontractors and suppliers
2. Invite consultants to issue fee proposals if required
3. Measure trade works as builder's quantities when a BQ is absent.
4. Rate and extend builder's quantities/BQ
5. Attend risk workshop
6. Schedule onsite overheads (preliminaries)
7. Collate and summarise subcontractors and suppliers quotations as well as consultant's fee proposals

Project manager

Participative tasks (in sequence)

1. Assess the extent of any temporary works
2. Provide statement of site overheads (preliminaries) needs
3. Create works programme with duration of activities
4. Attend risk workshop
5. Attend final review meeting

Figure 3.4 Tender management process.

collate and summarise quotations received from subcontractors and suppliers, which is necessary to allow the estimator to consider a price to include in the estimate. It is therefore important for the team to adhere to stated durations to prevent delays, as a rush towards the end of a tender period can lead to errors, incorrect judgment and parts of a tender being under- or overstated.

3.5.2   Trade Contractor Pricing

During the tender period, it will be necessary to obtain estimates or quotations from trade contractors that would normally carry out works on a project should the main contractor's tender be successful. The supply of trade contractors' services on offer to the construction industry means they either supply labour only as a workforce, provide plant and labour only, or offer a full package comprising labour, plant and materials and become subcontractors to the main contractor once the parties bind themselves to an agreement. Subcontractors carry out most of the works on a construction project and, if wishing to be competitive with a tender, a main contractor will need to decide which trade works would be subcontracted and test current market prices as opposed to 'guestimating' an approximate price instead. Once decided, invitations to tender to potential subcontracting organizations can be issued. The term 'domestic' suggests that the scope of trade contractors works is limited to residential projects. This is not the case however, as the term is used to describe a contractual arrangement between a main contractor and subcontractor for any type of project not contractually involving the project client. The selection of these potential subcontractors is carried out under any of the following arrangements:

- an open tender approach, where a contractor has no preference to the selection of companies and seeks competition with all trades; this might apply when a contractor is in short supply of certain trades;
- subcontractors that have existing partnering agreements in place with the contractor;
- subcontractors already engaged on a phase of a project with further phases imminent;
- those that offer a partial design-and-construct service to the contractor;
- those capable of servicing a project with adequate labour, material and plant resources;
- those that are financially stable and reliable;
- the contractor's experience of subcontractor's performance on current and past projects; and
- companies named in the tender documents.

As there may be numerous domestic subcontract trades required for a project, the issuing of invitations to tender to three or four companies for each trade could result in scores of issues that will absorb time in the process. The number of invitations to issue must be managed effectively, and as a minimum needs to include trades the contractor would not carry out themselves if successful in winning the project. This applies to the likes of the building service trades, civil engineering, piling operations, etc., where a trade contractor supplies and fixes materials and provides necessary plant to carry out the operations as well as testing, commissioning and certifying of the works. When one or more trade contractors are named in the tender documents as preferred companies, the quantity surveyor must acknowledge the requirement by ensuring they also receive an invitation to tender as it would do the contractor no favour to overlook the request.

Prior to issuing invitations, it would be wise for the quantity surveyor to contact selected trade contractors to discuss the project to see if they are willing to participate in the tender process and, where willing, commit to submitting a price within a stated timeframe. A prior discussion with competing trade contractors will avoid later disappointment if the design and documentation is issued with the trade contractor(s) declining to provide a price because of existing commitments that could have been made known prior to sending the information. The time of a response must be within the tender period, thus creating a series of mini-tender periods in the process; these must obviously be of a shorter duration than the overall tender period, as anything longer would defeat the purpose.

If a contractor is pricing a number of projects concurrently with tender submission dates around the same time, it would be unwise to burden each trade contractor with a request to price all projects. Here, a suitable approach would be to submit multiple enquiries to an abundance of trade contractors to increase the chance of receiving a number of prices for each trade. Each invitation to tender must be formal and can be formatted from templates the contractor already uses or purpose-made, to include:

- a cover letter or email request addressed to the person who confirmed they would accept the invitation and provide a response in a timely manner;
- a list of design drawings and documents and a date by which the tender is to be provided;
- a draft form of the agreement the trade contractor would be prepared to enter into with the main contractor if successful with an award; and
- the appropriate trade section of a bill of quantities if provided as a tender document.

Where a bill of quantities is not a tender document and the contractor has produced builder's quantities/schedules, it is discretional if the document is provided and should only be issued as a last resort (e.g. if there is a limited chance of receiving a price). Where provided, the contractor should advise competing trade contractors that supply of the information is for advice only and the document is not intended to form part of any contract. If a trade contractor is required to measure and quantify their own works, they should be given adequate time to carry out the assessment and provide a submission.

Once invitations to tender are issued to trade contractors, a register listing the trades, company names, contact details and dates tenders are due should be created to keep track of activities. Notwithstanding this, it would be wise to remind tendering companies a few days before their tenders are due of their commitment to the contractor. This is to ensure prices are submitted on time, as the system can become lapsed if prices are allowed to trickle through as and when they become available, which does not help the contractor's position.

After tender periods are closed and tenders are received, the quantity surveyor may compile a tender return analysis of each trade for comparison. This can be created on a computerised spreadsheet to check arithmetic and scope, which involves entering works descriptions, quantities and rates in the worksheet that form the basis of each priced submission. When a contract bill of quantities is issued, the checking is a straightforward process as trade contractors are bound to the stated quantities. Where trade contractors are responsible for preparing their own quantities, or builder's quantities/

schedules has been provided instead, the quantified and rated schedules should be provided for checking. However, some trade contractors might be reluctant to provide this information by stating they are providing a lump-sum price based upon the terms and conditions of the invitation to tender. Where applying, this decision must be respected because the scope and quantities on a scheduled document would not usually form part of any future contract. Nevertheless, some type of a breakdown is required where a lump-sum price is offered in order to assess the competition on an equal basis; trade contractors would normally supply this on request, especially if works are of a high value. The supply of such information is necessary for the quantity surveyor to confirm to the estimator that received tenders comply with the scope of works. To demonstrate, Table 3.3 comprises a tender return analysis from civil engineering contractors for the construction of an estate road complete with drainage to serve residential properties, which is offered as a lump-sum fixed price where the civil contractors have provided schedules of the works including quantities and rates.

With building service trades, the various consultants involved with preparing the design and documentation may create a tender return form which may be found within the appropriate trade specification. Where provided, this form usually includes a breakdown of the service trade(s) into parts. For example, in the case of electrical works these might be: (a) electrical distribution boards; (b) sub mains; (c) temporary installations, etc. The form usually requires each tendering company to insert lump-sum prices for each item to arrive at a total price, which should be completed, signed, dated and returned with each tender. The completion of this form may not be an option for the main contractor and could be a condition of the project under tender, requiring the return of at least one completed form for review by the client's agent who would delegate responsibility for vetting to the appropriate design consultants following receipt of main contractor's tenders.

When comparing prices on any trade contractor's tender, it is important for the quantity surveyor to observe if it is an estimate or quotation, which may be stated on the written response. The difference between the two is that an estimate is a statement of the approximate cost that may change, and is not considered a firm price. An estimate may be prepared as an interim assessment because of a lack of time on the part of the tenderer, or if the design and documentation lack sufficient criteria with the tendering company issuing the estimate for advice, possibly describing it as being for budgeting purposes only. On the other hand, a quotation usually means the price is firm and a bona fide offer to carry out the works for the stated price including any conditions for the contractor's acceptance.

An estimate or quotation may include qualifications, which are conditional notes intentionally appended to a selling price. These notes usually comprise a list of items stated as being specifically included or excluded from the price that may also include a list of assumptions by the tendering company to clarify vague, ambiguous and unmentioned matters in the invitation to tender to remove any doubt of the responsibilities of the parties, and the risks and intentions of what to expect for the stated price. Examples include: 'the removal of rock encountered below ground level is excluded'; 'vertical transportation of goods using hoists and cranes to the required floor level for their permanent installation in the building is to be by the main contractor'; and 'horizontal transportation of goods to their permanent position within the building is to be by the trade contractor'. Tender qualifications made by a trade contractor can be perceived by

Table 3.3 Tender return analysis

Tender return analysis

Project nr 1600

Description	Company 1			Company 2			Company 3		
Trade: Civil Engineering	Qty	Rate	Total	Qty	Rate	Total	Qty	Rate	Total
Site mobilisation and demobilisation	Item	Incl	Incl	Item	2,500.00	2,500	Item	Incl	Incl
Preliminaries	Item	Incl	Incl	Item	22,500.00	22,500	Item	Incl	Incl
Site clearance: topsoil strip	$650\,m^2$	2.00	1,300	$620\,m^2$	1.75	1,085	$670\,m^2$	2.00	1,340
Roads: excavation	$395\,m^3$	10.00	3,950	$375\,m^3$	9.00	3,375	$400\,m^3$	11.00	4,400
Roads: stone sub-base, 250 mm th	$680\,m^2$	13.00	8,840	$630\,m^2$	11.00	6,930	$690\,m^2$	13.00	8,970
Roads: temporary kerbs	$260\,m$	25.00	6,500	$240\,m$	22.00	5,280	$270\,m$	26.00	7,020
Roads: bitumen base 100 mm th; base course 50 mm th	$590\,m^2$	35.00	20,650	$550\,m^2$	32.00	17,600	$600\,m^2$	35.00	21,000
Roads: asphalt surface 40 mm th	$590\,m^2$	27.00	15,930	$550\,m^2$	22.00	12,100	$600\,m^2$	28.00	16,800
Black water: trench B1–B2 with imported backfill	$90\,m^3$	100.00	9,000	$90\,m^3$	90.00	8,100	$95\,m^3$	100.00	9,500
Black water: trench B2–B3 with imported backfill	$75\,m^3$	100.00	7,500	$75\,m^3$	90.00	6,750	$80\,m^3$	100.00	8,000
Black water: trench B3 - existg. with imported backfill	$75\,m^3$	100.00	7,500	$75\,m^3$	90.00	6,750	$80\,m^3$	100.00	8,000
Black water: drainage pipe B1–B2	$45\,m$	112.00	4,950	$45\,m$	100.00	4,500	$45\,m$	115.00	5,175
Black water: drainage pipe B2–B3	$35\,m$	112.00	5,040	$35\,m$	100.00	3,500	$35\,m$	115.00	4,025
Black water: drainage pipe B3 - existing	$35\,m$	112.00	3,920	$35\,m$	100.00	3,500	$35\,m$	115.00	4,025
Storm water: trench S1–S2 with imported backfill	$55\,m^3$	78.00	4,290	$55\,m^3$	70.00	3,850	$58\,m^3$	80.00	4,640
Storm water: trench S2–S3 with imported backfill	$55\,m^3$	78.00	4,290	$55\,m^3$	70.00	3,850	$58\,m^3$	80.00	4,640
Storm water: trench S3 - existg. with imported backfill	$35\,m^3$	78.00	2,730	$35\,m^3$	70.00	2,450	$38\,m^3$	80.00	4,640

(Continued)

Table 3.3 (Continued)

Tender return analysis

Project nr 1600

Description	Company 1			Company 2			Company 3		
Trade: Civil Engineering	Qty	Rate	Total	Qty	Rate	Total	Qty	Rate	Total
Storm water: drainage pipe S1–S2	40 m	70.00	2,800	40 m	65.00	2,600	42 m	72.00	3,040
Storm water: drainage pipe S2–S3	40 m	70.00	2,800	40 m	65.00	2,600	42 m	72.00	3,024
Storm water: drainage pipe S3 - existing	30 m	70.00	2,100	30 m	65.00	1,950	30 m	72.00	2,160
Black water: inspection chambers 2–3 m deep	3 Nr	3,150.00	9,450	3 Nr	2,750.00	8,250	3 Nr	3,300.00	9,900
Storm water: inspection chambers 1–2 m deep	3 Nr	1,900.00	5,700	3 Nr	1,700.00	5,100	3 Nr	1,950.00	5,850
Road gully: connections into storm water drainage	45 m	105.00	4,725	43 m	92.00	3,956	48 m	110.00	5,280
Road gully: frame and cover	9 Nr	450.00	4,050	9 Nr	400.00	3,600	9 Nr	450.00	4,050
Disposal of excavated spoil off site	790 m³	55.00	43,450	800 m³	50.00	40,000	810 m³	55.00	44,550
TOTAL lump sum (£, excluding VAT)			180,435			182,676			188,413

the main contractor as a lack of intention to enter into an agreement based on the terms of the invitation to tender. This might not be the case however, and the conduct must not be seen as negative, especially if the theme of the qualification(s) is consistent with other tenders. This can occur if the main contractor is seeking to drive too hard a bargain either by policy or by passing on the terms and conditions of the project under tender (e.g. unfavourable credit terms or high commercial risk, or no advance payment to be made).

A trade contractor may change the reference quotation on a letterhead and call it a tender or offer. In effect this means the same and, if also including a set of terms and conditions, is a price to carry out works subject to those terms and conditions which may be qualifications, albeit under an alternative guise. If left to chance and the offer is accepted unconditionally by the main contractor, it would create a contract with the terms and conditions/qualifications effective, which can pose risk to the contractor.

A main contractor may look favourably on trade contractors tenders, yet may wish to avoid the scenario of accepting certain terms and conditions/qualifications should the parties create an agreement, as there may be hidden costs or unforeseen risks that may only come to light after accepting the offer. The goal at this stage of course is to ensure submissions comply with the works under tender and for the contractor to understand what is included in each price. To answer this, the following helps to clarify matters:

- all works should be priced in accordance with the documents provided and a priced bill of quantities will verify the inclusion;
- where a bill of quantities is not part of the enquiry, the scope on the priced submission should reflect the requirements;
- if a price appears too low or scope is missing, the estimator requires alerting to the matter to make due allowances or should contact the provider of the priced submission to clarify, with any amendments confirmed in writing;
- trade contractors' submitted prices should comply with the contractual requirements and not depart from any clauses of the draft form of agreement;
- any site-specific and/or client's requirements should be acknowledged as being inclusive;
- prices need to be realistic, of the correct currency and reflect market conditions;
- prices should state if value added tax is included, excluded or applicable;
- works excluded from a trade contractor's price that require input from other trades or the main contractor as builders work in connection requires determining, as this will add to the base cost;
- prices should be fixed for the duration of the works or in accordance with a fluctuation clause in the contract; and
- where trade contractors are reluctant to fix prices and no fluctuation clause exists, the estimator will have to make allowances.

Once priced submissions are summarised, the estimator makes due consideration for risk to assess the anticipated price of the works. The assessment of risk to the contractor is not usually a responsibility of the quantity surveyor. However, the role played in collating the information aids the estimator to arrive at a decision for including suitable allowances.

3.5.3 Works Schedules

Where a contractor elects to carry out certain trade works that would not be outsourced as domestic subcontract trade packages, this involves the contractor purchasing materials for installation by labour-only subcontractors. This arrangement has an advantage to a contractor, as the buying power can mean the percentage addition applied to the base cost charged by domestic subcontractors for overheads and profit and employment costs is cancelled, making the overall price cheaper; this may be a deciding factor for the contractor when preparing a tender in order to fend off competition. Trades to procure as separate labour and material purchases usually apply to building works, primarily because of the associated manpower skills and commercial advantages under the direct control of main contracting organisations such as:

- brick and block wall construction;
- steel mesh and bar reinforcement as part of structural components;
- carpentry works including timber joists, roof timbers, timber stud partitioning, doors, skirting boards, etc., but excluding joinery items such as high-quality benches and cabinets;
- in-situ concrete; and
- certain types of formwork, except on large-sized projects or if a patent product is to be used.

A decision on which trades would be procured in this manner is usually made upon receipt of the invitation to tender consignment, possibly by the contractor's estimator. Alternatively, the choice may be influenced by company policy or advice received from the project manager responsible for the deliverables if the tender is successful. When electing for this strategy, the contractor must be in a position to accept the risk involved and have access to reputable prices. The quantity surveyor will be of assistance here, and would need to schedule the required trade works as builder's quantities and split the chargeable rates into labour, materials and plant. Rates can be sourced from current and past projects or through market research with the constituents expressed as split percentages (e.g. the rate for the placement of mesh to slabs could be divided as 20% labour, 75% materials and 5% plant for offloading the materials once delivered to site).

The format and presentation of builder's quantities/schedules for pricing by the contractor can be created with spreadsheets, which require minimal training, or alternatively with purpose-designed software that involves a prior period of training. The Tier-1 and -2-sized contractors usually elect to use industrial software because of the limited time available for preparing tenders, assisted by an efficient estimating system. See Chapter 2, Section 2.8 for a further discussion on this subject.

The scheduling process should be considered flexible, and be applicable to those works that would be procured as split trade packages if the contractor wins the award. Trades to be procured as combined labour and material packages can have a schedule prepared in brief without the need to quantify and fully describe the works, as the contractor will be relying on receipt of tenders (the process of which is discussed in Section 3.5.2 above). However, the quantity surveyor should remain flexible in this matter, as there may be a need to produce builders quantities/schedules for price checking if receipt of trade contractor's quotations have excessive prices or show considerable variances. For example, a situation can occur when three lump-sum prices are received as tenders for £500,000, £800,000 and £1,000,000 for the supply, pour, pump and place-ment of concrete, including testing and surface finishes to foundations, floors, columns and beams, absent of any obvious errors and appearing consistent with scope. In such a scenario there is obviously uncertainty regarding the price and amount the contractor should include in the tender. Here, the creation of builder's quantities/schedules charged at marketable rates for the concrete works would provide certainty with the price to include in a tender. Trade contractor's submitted prices may vary for a variety of reasons as follows.

- Current market conditions might be dictating a demand for certain trades, with low supply meaning trade contractors are overstretched and not pricing competitively.
- A trade contractor may be seeking to shift from one sector of the industry to another, for example residential to commercial, and only pricing residential works at a premium.
- Inadequate design and specification criteria or a prolonged period of sustaining a fixed price may result, with trade contractors adding risk to their prices to comply with the terms and conditions of the invitation to tender.
- The existence of cartels where businesses agree not to compete with each other.

Cartels exist where a number of sellers of goods and/or goods and services mutually agree to fix selling prices, apply the same discounts and/or fix credit terms with buyers. It may also be seen as an attempt to controls prices and exclude the entry of new com-petition in a market, and is known as bid rigging. Article 101 of the Treaty on the Functioning of the European Union prohibits cartels and other agreements that could disrupt free competition between member states. In the UK, the Enterprise Act 2002 empowers the Office of Fair Trading to address cartels. In general, this act comprises a range of measures to replace any monopoly scheme by strengthening the competitive legal framework and introducing criminal sanctions for individuals who involve them-selves with hard-core cartels.

During a tender period, if a cartel appears to exist (e.g. abnormal commercial condi-tions applying, such as requests for unusually high advanced payments (over 50%) or excessively priced tenders) commercial and estimating managers may already be aware of the situation, and to ensure competitiveness may recommend the creation of builder's quantities/schedules. This is to ensure they are suitably rated by the contractor to create a reasonable price that includes risk, which is not influenced by cartel prices. This action by respective managers is to be commended as artificially inflated prices may be temporary and, if included in a tender, could jeopardise the contractor's chances of providing a viable offer and possibly missing out on an award of contract.

3.5.4   Resources Costing

A core skill of a qualified quantity surveyor is an understanding of the constituents of a price in a bill of quantities/schedule. Determining the quantity is one part of the equa-tion; the other is the unit rate to apply, which in combination drives the price. The unit rate is derived from an assessment of resources costing, comprising the cost to employ labour, the purchase price of materials including consumables and the cost of hired mechanical plant where applicable. Contractors require this information if wishing to carry out certain works themselves or for assessing trade contractor's prices for assurance if there are anomalies with received quotations.

3.5.4.1 Labour

This refers to the human resources required to produce, install, test and commission works that involves working, cutting, fixing, placing, handling and the installation of materials. To asses an amount to charge, it is necessary to calculate or be advised of an hourly rate for employing labour. The basic hourly rate payable by a contractor for labour is known as the 'flat rate' and the cost to charge out for works payable to the contractor the 'all-in rate'; the all-in rate is the flat rate plus the cost to employ and a contribution to offsite overheads and profit. These rates vary due to the different skills of operatives, and apply to the trade labour component included within the chargeable rates in a bill of quantities/builder's bill and general labour assistance to aid skilled operations which forms part of the contractor's preliminaries. The basis of calculating a rate of return varies with companies as businesses operate differently, and is calculated on the number of productive hours worked per annum charged against the calculated all-in rate. The flat rate payable in the construction industry is regulated via a Working Rule Agreement with rates published annually by the Building and Allied Trades Joint Industrial Council (BATJIC). The formation of BATJIC stems from an alliance between the Federation of Master Builders and Trade and General Workers Union, which provides a valuable forum to help and maintain good industrial relations in the workplace. A contractor can log each of these rates and allowances and build up a library of charge-out hourly rates for a number of skilled operatives, which can be used for estimating the price of works. Table 3.4 shows a format for calculating a labour rate which recognises incentives, and options an employer may pay in addition to the statutory requirements. This example excludes site-specific allowances such lodging and site-conditional pay, as well as other working rule agreements that would be considered on an *ad hoc* basis and specific for a project. The base rate used in the example is for a skilled operative qualified with an S/NVQ3: City and Guilds London Institute Advanced Craft pay award for 2016/2017.

3.5.4.2 Materials

It would assist a contractor if the purchase prices of materials for works under tender are available in a database and updated regularly, as such information would keep the contractor abreast of prices that are not usually shown in trade contractor's quotations, which includes the supply and installation of materials. However, with a vast array of products available for use on a construction project, the full database concept may be impractical to implement and manage. Time would therefore be well spent seeking quotations and storing data for material prices that could be procured by the contractor if successful with the tender, or if in the habit of purchasing frequently for any project.

The process of selecting companies to be issued with requests for material quotations is no different than the selection of trade contractors, with the addition of adhering to the specification where certain suppliers or manufacturers are specified. However, the process might be accelerated in scenarios where material suppliers have existing agreements in place with a contractor because of repeat business and agree to fix prices to certain dates, which are subject to negotiation thereafter. Where these agreements exist, with goods applicable for a project under tender, stated rates/prices should be collated for the estimate as they are reliable and competitive. Where price agreements are not in place, the quantity surveyor can assist the estimating process by issuing enquiries; if a bill of quantities/builder's schedule is available, it could be included as part of an enquiry. Rarely

Table 3.4 Labour charge-out rate

Productive hours	Calculation	Notes
Total days per year:	365	
Less weekends	−104	Saturday and Sunday are not
Total weekdays	261	normal workdays
LESS Annual holidays	−22 days	BATJIC Working Rule 13, 30
Bank holidays	−8 days	days holiday incl statutory days
Sick leave with pay (estimate)	−10 days	Loss of production
Wet days with pay (estimate)	−10 days	Loss of production
	−50 days	
Total productive days:	−50	
Production is based on a 39 hour week for 5 working days. Allowing a lunch break of 30 minutes, equates to 7.50 hours per day. However, allowing 0.25 hour (15 minutes) break, twice daily, = 7.00 hours productive time	211	BATJIC Working Rule 1, 39 hour week
	7.00 hr × 211 days = **1477 productive hours**	

Labour costing	Calculation	Notes
Wages per day		
Hourly rate	£11.79	BATJIC Working Rule 1
Daily productive hours	7.00	
Daily wage before allowances	**£82.53**	
Plus site allowance hourly rate	NIL	Calculated as site specific
Daily travel (Allowance for 25 km each way)	£6.56	BATJIC 1–50 km
Lodging allowance	NIL	Calculated as site specific
Site conditional pay	NIL	Calculated as site specific
Tool money: contractor's incentive (daily)	£9.00	Other incentives are included
Others	NIL	here
Total	**£98.09**	This is a contractor's risk
Allow non-productive overtime, 1 hour per day @ time and a half rate = 1.5 × £11.79	£17.69	allowance for overtime not payable by the client
Daily pay including non-productive time	**£115.78**	
Wages per annum		
Daily pay × total working days £115.78 × 211	£ 24,429.58	
Plus lost productive days, 10 × £98.09	£ 980.90	Sick leave and wet days payable
Total wages per annum	**£25,410.48**	to employee

Annual statutory and overheads costs	Calculation	Notes
Statutory costs		
1. Employer's National Insurance Contributions; 13.8% of total wages per annum applicable after first £156 per week/£8,112 (£25,410.48−£8,112.00)	£ 2,387.19	
2. BATJIC holidays with pay 234 hours @ £11.79	£ 2,758.86	BATJIC, 30 days @ 7.80 hours

(Continued)

Table 3.4 (Continued)

Annual statutory and overheads costs	Calculation	Notes
Statutory costs		
3. Retirement benefit scheme; 1% of £25,410.48 + £2,758.86 = £28,169.34	£ 281.69	Applicable to basic + holiday pay
4. BATJIC Stakeholder scheme £6.75/week	£ 351.00	
5. CITB Training levy 0.5% of £28,169.34	£ 140.85	Applicable to basic + holiday pay
6. BATJIC Death benefit scheme £16.13/month	£ 193.56	BATJIC Working Rule 13
Overhead costs		
1. Severance pay 261 days × £5.00	£ 1,305.00	Employer's policy cost
2. Workers insurances 0.8% of £28,169.34	£ 225.35	Workers compensation
Total statutory and overheads costs	**£ 7,643.50**	insurance scheme

Total labour cost and hourly rates	Calculation	Notes
Totals		
Total wages per annum	£25,410.48	
Total statutory and overhead costs	£ 7,643.50	
Total cost to employ	£33,053.98	
Summary	£ **22.38**	This rate excludes offsite overheads and profit.
Hourly rate to charge to projects as an on-site cost: £33,053.98/1,477 productive hours		

will a material supplier issue a firmly priced quotation based upon the quantities in a schedule, as suppliers are generally not responsible for damage, waste and the security of goods once delivered to site. However, by providing a quantified schedule there are advantages to the contractor, as it gives suppliers an indication of approximate quantities; this can attract handsome discounts with suppliers possibly submitting quotations based upon a schedule of rates or approximate value based upon the schedule/bill of quantities provided. Where works are to be phased, the contractor should advise accordingly.

Material prices must comply with the specification, as the contractor cannot usually depart from the requirement during a tender period. In response to an enquiry, a supplier may wish to offer an alternative product, possibly because it is one of their manufactured brands, and may describe the product as 'equal and approved'. This means there is an assumption by the supplier that their product is equal to the specified choice and is seeking approval from the contractor to alter the selection. Indeed, the alternative could be less expensive with the quality on par with the specified product. However, the supplier must be made aware by the contractor that it has no authority to amend the specification without the expressed permission of the client's agent. An exception would be if a specified product is obsolete, where the contractor should alert the client's agent and seek a tender amendment. Any deviation from the documents without obtaining prior approval from the client's agent can create havoc for the quantity surveyor and estimator, and the golden rule is to adhere to the specification unless instructed otherwise. If a supplier (or trade contractor) is keen to amend a product specification, there is no harm in directing them to the client's agent to request a change, as this is the entity that will have the final say. With a design-and-build project, any referral should be to the estimating manager as the contractor is effectively driving the quality of the specification.

When assessing the price of materials to include in an estimate, it is necessary for the contractor to make due allowance for wastage. Usually, quantified works as components/items in a trade bill of quantities/schedule are measured 'net as fixed in position'. This is also endorsed by the RICS in the new rules of measurement (nrm2) section 3.3, 'Quantities' meaning stated quantities make no allowance for product waste. A quantity surveyor will be expected to understand material wastage allowances to include in a chargeable unit rate and apply logic to the purchase cost which varies with the type of material. Waste allowances are included for surplus or consumed materials used in the site production processes, and examples include:

- cutting and trimming timber to length from standard stock sizes with unusable off-cuts;
- laps for working or as specified, applicable to mesh reinforcement, membrane sheeting, etc.;
- sand and hardcore purchased and stockpiled in bulk that is compacted to a lesser volume;
- transporting and disposing of excavated material that bulks in size after excavation;
- mixing water with sand, cement and additives which consolidates to produce mortar;
- loss of bricks and blocks due to cutting which are disposed of; and
- residual in-situ concrete from truck deliveries that is not required.

The term wastage also encompasses allowances which an estimator must consider for site breakages, material damage beyond repair which is not recoverable from an insurance claim, vandalism and theft. Allowances in an estimate for product waste are expressed as a percentage of the purchase rate/cost and vary with the type of material; they can vary from as low as 1% for singular items such as bathroom sanitary ware, to 15% for bricks and 50% for the compaction of hardcore, including waste that may be left over requiring disposal. It can therefore be seen that with the ranges so extreme, waste poses risk to the contractor, and feedback from current and past projects will act as an indicator of realistic allowances to include in an estimate.

Some materials are purchased by units of measurement that differ from the method of measurement in a bill of quantities/schedule. For example, bricks are sold per thousand where they are billed in m^2, and sand and stone is sold per tonne and billed as either m^2 stating the thickness or m^3 when in bulk. For this reason, the quantity surveyor needs to be aware of conversion factors to convert the purchase units of sale to units of quantified measurement. For example, the dimensions of a standard brick are 215 mm (length) × 65 mm (height) × 102.5 mm (depth or wall thickness) which, with a 10 mm mortar bed and surround, become 225 mm × 75 mm × 102.5 mm. Assuming that bricks are to be laid as stretcher/running course, to convert wall area to the number of bricks for a wall thickness of 102.5 mm, the calculation is:

$$\text{Nr of bricks per square metre of wall (brick and mortar bed) area} = \frac{1}{0.225 \times 0.075} = 59.$$

Similarly, when dealing with products purchased by weight and rated in a bill by volume, it is necessary to use their relative density, that is, weight in metric tonnes per cubic metre:

$$\text{Relative density} = \text{material weight in kilograms per } m^3 / 1,000.$$

For example, the weight of dry sand is 1,600 kg/m³ which, from the above equation, is converted to 1.60 tonnes/m³. Furthermore, the weight of dry course aggregate sized 25–40 mm is 1,800 kg/m³ = 1.80 tonnes/m³.

Conversion factors exclude wastage which is added as a nominal percentage. To demonstrate, let us say a bill of quantities has a requirement for a 50 mm thickness bed of sand blinding over an area of 1,000 m², and a supplier issues a quotation for the supply and delivery of dry sand to site at £20.00 per tonne. The supply rate before overheads and profit is calculated as follows:

$$\text{Volume required} = 1,000 \, \text{m}^2 \times 0.050 \, \text{m} = 50 \, \text{m}^3$$

$$\text{Weight required} = \text{dry density} \times \text{volume required} = 1.60 \, \text{tonnes} / \text{m}^3 \times 50 \, \text{m}^3 = 80 \, \text{tonnes}$$

$$\text{Weight required plus 50\% for compaction and waste} = 80 \times 1.5 = 120 \, \text{tonnes}$$

$$\text{Cost} = 120 \, \text{tonnes} \, @ \, £20 \, \text{per tonne} = £2,400$$

$$\text{Supple rate for area required} = £2,400 / \left(1,000 \, \text{m}^2\right) = £2.40 \, \text{per m}^2.$$

Upon receipt of supplier's quotations, information requires collating on a tender return analysis and should include reference to any items in the quotations that affect the base price for the estimator's consideration. Suppliers will rarely visit a site for quoting purposes, and may apply generic or specific qualifications to a quotation. Factors to consider for including in the estimate, whether qualified or not, include:

- the cost of any samples;
- period of acceptance of the quotation;
- indications of impending price increases in the short term;
- trade discounts and payment terms;
- prompt payment discounts on offer (e.g. within 7 or 14 days after receiving an invoice);
- warranties and guarantees;
- inclusion or exclusion of value added tax;
- testing prices (as with concrete);
- specific attendances required by the contractor such as a forklift truck to offload materials;
- delivery charges;
- special charges; and
- part-load charges (e.g. a quotation may state it is for a single delivery that may not be suitable).

The quantity surveyor's assessment should include calculations showing conversion factors and waste together with items not quantifiable but noted on received quotations as items of extra cost, which are added as a percentage to cover risk. The quantity surveyor does not need to show interest in wastage factors for materials included in trade contractor's prices that offer a supply and installation service, as the allowance is included in the stated price.

3.5.4.3 Plant

A construction project will require mechanical and non-mechanical plant to service site operations, especially where new works are involved. Whatever the type of plant required, items are hired and charged to the contractor by time or as a fixed cost, and there are four ways of obtaining the charges:

- by obtaining hourly/daily/weekly rates from plant hire companies;
- by obtaining fixed unit rates from plant hire companies;
- by obtaining trade contractor quotations that include the hire and use of plant; or
- by seeking advice from a company accountant when a contractor has owned assets that will be used on a construction project.

Time-related and fixed-cost items hired directly by the contractor are usually required to fulfil the obligations of the main contract preliminaries (e.g. skips for collecting waste, fork-lift trucks to transport materials, site perimeter security fencing, etc.). These are arranged by the contractor's site management for installation and/or operation by labour-only operatives under the direct control of the main contractor. Heavy items of mechanical plant requiring skilled operation will require maintenance and insurance which the main contractor does not usually provide, and includes items required to carry out the likes of demolition work, site excavation, piling, tower cranes and scaffolding. For estimating pur-poses, these items are usually treated as trade contract packages that would be market tested by issuing invitations to tender. Where items of mechanical and non-mechanical plant are not to be provided by trade contractors, the quantity surveyor can assist the estimating process by obtaining schedules of rates from the plant hire equipment market, including delivery and collection from site, that would be procured by the contractor.

Instead of hiring either directly from the equipment market or via subcontractors, a contractor may elect to purchase certain mechanical plant for productive work for utilising on site with the use of skilled operatives that in the process would become a contractor's asset. This is different than the purchase of non-productive plant such as a forklift truck as demonstrated in Table 3.1 used to assist with multiple trade operations. The decision to purchase productive plant is strategic, and can influence the procure-ment of trade contract packages because in effect, the contractor is substituting some trade works scope to carry out the works itself. To demonstrate, let us consider a sce-nario where a contractor wishes to purchase a two-tonne mini-excavator to excavate trenches for laying service ducts and has resources to supply skilled labour needed to operate the excavator. To assess the worth of the investment, the contractor would need to calculate an hourly rate to charge out on projects, which is shown in Table 3.5.

From observing the table, it can be seen the excavator will need to be operating for *c.* 1500 hours per year over the investment period charged at the stated hourly rate for the purchase to be considered viable.

3.5.4.4 Consultants

When a project is to be procured with contractor's design (e.g. design and build), the contractor will need to include the design consultant's costs in a tender submission to cover the cost of preparing the design and documentation. This is difficult to quantify and clarify during a tender period, as it depends on the level of detail prepared by the initial design consultants and whether their services will be novated to the successful contractor. However, consultants engaged by a project client to prepare a concept design may provide competing contractors with a fee proposal to continue the process of developing the design and documentation for working purposes in addition to advi-sory services during the construction phase. With this procurement arrangement, an

Table 3.5 Plant charge-out rate

	£
Purchase price of 2.0-tonne mini-excavator	**30,000**
Investment period (say 3 years)	
Depreciation year 0–1 (50%)	−15,000
Depreciation year 1–2 (15%)	−2,250
Depreciation year 2–3 (15%)	−1,912
Total depreciation	**−19,238**
Total depreciation per year = £6,413	
Net worth at resale/disposal price	10,762
Plant depreciation and running costs	
1. Depreciation cost.	4.22
Operating life per year = 365 − 104 (weekends) − 8 (bank holidays) = 253 days = 2,024 hours for 8 hour working days.	
Less 25% for down time (maintenance, wet weather) = > annual productive hours = 1,518 hours.	
Annual depreciation cost per productive hour = £6,413/1,518	
2. Hourly cost of loss of interest on investment, say 3% per annum.	0.19
[£30,000/(3 years × 3%)]/1,518	
3. Maintenance repairs and insurance.	1.00
Allow sum of £1,500 per annum for replacement parts as there are no warranties and materials for servicing, plus cost of insurance policy:	
£1,500/1,518	
4. Fuel consumption.	2.50
Allow 40 litres per week × 52 operating weeks per year = 2,080 litres	
Total fuel cost per hour = (2,080 × £1.50 per litre)/1,518 = £2.06	
Round up to allow for lubricants, consumables and oil	
5. Plant operator	19.00
Hourly rate including BATJIC/Working Rule Agreement for plant operator. Rate calculated following format on Table 3.4.	
Total hourly rate for production charges (excluding overheads and profit)	**26.91**

allowance must be included in the tender to cover consultant's fees, which are usually expressed as a percentage of the contract price that varies with each discipline. A leaner method is to include the price of any submitted fee proposals, or seek advice from a commercial or other manager involved with design management who can suggest a suitable allowance to include in a tender. Advice may also be sought from the same manager(s) if the procurement route is traditional with a requirement for the contractor to provide some design input, the extent of which should be stated in the invitation to tender. The anticipated expense for design services may form part of the preliminaries under the contract conditions, or as a separate fee provision within the estimate. This is in addition to the cost of any other consultant fees not involving design that would be outsourced (e.g. project planning).

3.5.5   Unit Rate Calculations

The resource cost of labour, materials and plant form the basis of a unit rate calculation that, when charged against a quantified description of works, creates a price. The contractor's quantity surveyor must understand this estimating concept, as it forms an integral part of preparing a price from scratch, a skill that may be required on a project under construction if the works change and there are no rates available. The calculation of a unit rate from scratch is referred to as being from 'first principles', and involves applying output for each item of productive work known as a labour constant (or sometimes referred to as labour output) charged against a monetary rate. A labour constant is the estimated length of time it takes an operative to commence and finish a task that, when charged against a monetary rate, creates the labour component of a unit rate calculation which has a relationship with the unit of measurement in a bill of quantities/schedule. By default, a labour constant is linked with any item of operational plant required to assist. For example:

- an excavator with an operator will excavate foundation trenches to 2000 mm deep in stable clay ground at an output of 15 minutes or 0.25 of an hour per cubic metre;
- a bricklayer will place mortar and flush point 45 facing bricks every 1.25 hours to one side of a cavity wall, 102.5 mm thick; and
- a carpenter/joiner will cut, place and nail 10 m of 13 mm thick × 120 mm high softwood skirting through plasterboard to stud partitions in 30 minutes, or 0.50 of an hour (0.05 hours per metre).

In addition to the labour and plant assessment, the cost of materials including wastage is added to create a built-up unit rate. To demonstrate, Table 3.6 shows a number of unit rate calculations for a range of work descriptions that may be included in a bill of quantities/schedule. All rates exclude overheads and profit, which are added to the rates or priced in isolation.

As there are many work items for construction operations, a database of output for each trade description would aid the estimating process. However, output for generic descriptions captured in a database should not be seen as suitable for every project as there may be variables that influence production. Productive outputs are influenced by 'buildability', a broad concept that includes site conditions, any restricted access to areas to carry out tasks, working environment, working at height, out of hours working, etc., meaning the quantity surveyor and estimator need to be open minded with their approach to labour constants. Here, an understanding of construction technology and appreciation of the time it takes to commence and complete a task will assist. For example, a contractor may have information in a database stating a carpenter's output as 0.05 hours per metre for nail fixing softwood skirting up to 20 mm thick and 200 mm high to walls and partitions. So, what happens on a project under tender if the request is for hardwood skirting of the same thickness range and height drilled and screwed to block walls? A fair assessment is that the process will take longer, and the allowance should be 0.10 of an hour per metre with the cost of the materials reassessed.

If a contractor creates a database of trade descriptions which captures unit rate calculations, the rates will alter over time. This is because of the increases in payable labour, changes to the cost of materials and plant hire as well as any amendments to waste allowances, and adjustments of labour constants gained from experience on current

Table 3.6 Unit rate calculations

Excavating and filling (nrm2 Section 5) — Foundation excavation not exceeding 2.0m deep, (150 m³)

	LABOUR £5.00	PLANT £8.00	MATERIALS £0	TOTAL UNIT RATE £13.00 m³

Description	Quant	Unit	Rate	Total	Comments
LABOUR					
Banksman	0.25	Hrs	20.00	5.00	A banksman is a labouring assistant to the operations.
Total labour rate				**£5.00**	
PLANT					The cost of mobilising and demobilising plant is included in preliminaries. The excavator is assessed to carry out detailed excavations at an output of 0.25 hrs/m³ using a rate including the operator.
Mobilisation to site		item		0	
Excavate trench for foundations	0.25	hrs	32.00	8.00	
Demobilisation from site		item		0	
Total plant rate				**£8.00**	
MATERIALS				£0	
Total materials rate					

Excavating and filling (nrm2 Section 5) — Hardcore beds over 50mm th, not exceeding 500 mm deep, 175 mm th, below paviors (100 m³)

	LABOUR £0	PLANT £14.50	MATERIALS £40.50	TOTAL UNIT RATE £55.00 m³

Description	Quant	Unit	Rate	Total	Comments
LABOUR					Labour is assessed as a 'gang rate', a term used for numerous operatives and is included in the plant rate.
Total labour rate				£0	
PLANT					Hardcore is excavated from the delivered stockpile and filled by an excavator. A 'gang rate' is used with an output of 0.25 m³/hr and a vibrating roller will be hired to compact the material in a single layer with the operatives included in the rate.
Excavate from stockpile and fill	0.25	hrs	52.00	13.00	
Mobilisation to site		item		0	
4 tonne vibrating roller	0.15	hrs	10.00	1.50	
Demobilisation from site		item		0	
Total plant rate				**£14.50**	
MATERIALS					The supply rate is per tonne based upon 20 tonne loads delivered and tipped on site. Shrinkage and waste is included in the waste factor.
Hardcore: convert as 1,800 kg/m³	1.80	t	15.00	27.00	
Compaction and waste + 50%	0.90	t	15.00	13.50	
Total materials rate				**£40.50**	

Table 3.6 (Continued)

General joinery (nrm2 - Section 22)	Softwood skirting 13 x 120mm nailed to plasterboard and stud partition (550 m)	LABOUR £1.17			PLANT £0		MATERIALS £3.33	TOTAL UNIT RATE £4.50 m
	Description	**Quant**	**Unit**	**Rate**	**Total**		**Comments**	
LABOUR	Carpenter to fix	0.05	hrs	23.43	1.17			
Total labour rate	Total				**£1.17**			
PLANT								
Total plant rate						**£0**		
MATERIALS	Skirting board supplied/delivered	1	m	3.00		3.00	Nails can be assessed as 2.5 kg/m^3 of timber. So, for 100 lm, $100 \times 0.02 \times 0.12 \times 2.5 = 0.60/100 = 0.006$ kg/m.	
	Waste from cutting +10%	0.10	m	3.00		0.30		
	Nails	0.006	kg/m	5.00		0.03		
Total materials rate	Total					**£3.33**		

and completed projects. To manage such a database may be unrealistic; a compromise is to log labour constants as a benchmark, which can be adjusted as necessary for the project under tender. An estimating team will only need to calculate unit rates for those trades a contractor will procure as labour only if they purchase materials themselves. The contractor will also need to rely on unit rates provided by trade contractors inclusive of their overheads and profit if a priced bill of quantities is to be submitted with a tender.

An alternative to creating a database is to seek rates through published building price books authored by advisors who collect data continuously. These publications are primarily referred to by cost consultancies and provide rates suitable for floor areas of a building by function (e.g. schools, hotels, etc.), as well as elements of construction and detailed construction pricing suitable for cost planning. A prominent source of cost data in the UK can be obtained from the BCIS (Build Cost Information Service) which has information available for subscribers available online. In addition, Spons and Laxtons publish pricing books that provide information for assessing the cost of constructing buildings; in Australia Rawlinsons and Cordell publish reputable data. Whatever the source of information, the application of unit rates must have appropriateness and relevance to the project under tender. The use of building price books must therefore be considered as indicative to possibly cost-check the prices of received quotations if there is doubt with the actual worth.

Advantages of these publications are that they are: a reputable source of information that has been tried and tested with time; inexpensive to purchase; and updated regularly. Disadvantages include the facts that: they are generic and not site-specific; they are not suitable as a market testing source for a tender; and rates are geographically based by region with any stated locality factor as an adjustment possibly considered too general.

The assessment of a unit rate calculation can be an interesting exercise, and to be confident with the result, there is need for the estimating team to be collaborative and open to suggestions for improvement. Team members may have different experiences gained from past and current projects and may be able to share knowledge if there is doubt with the extent of any works that impacts the price. When creating unit rate calculations, those involved may develop a competitive edge as the goal of course is for the contractor to secure an award. However, a wise team will not stray away from reality by pricing unrealistically, and will wish to avoid any surprises that could come to light when a contractor is questioned about their offer or, worse, secures the project with the price committed. The Chartered Institute of Building provides an authoritative guide to essential principles and good practices for estimating by construction contractors in its publication CIOB Code of Estimating Practice (COEP). The principles and guidelines produced in the COEP are theoretical, as is the information provided in this book. However, the aims are to ensure those involved with estimating understand the skills involved when pricing construction projects in order to provide a comprehensive estimate.

3.5.6 Operational Estimating

A unit rate charged against a stated quantity in a bill of quantities/schedule is a traditional approach widely recognised in the industry as a reputable method for pricing construction works. However, it is the drawings, specifications and site-specific conditions that influence the scope and duration of trade works for a project which, if not assessed and priced suitably, can pose risk to the contractor. The assessment of a price to tender for a project is different for each case, with market rates and unit rates considered resourceful and an accurate guide to the worth of trade works. However, a market rate is for reference only and may not take into account specific site conditions; likewise a calculated unit rate may be generic and possibly not consider abnormal/site-specific requirements. For example, construction of block walls in a basement core filled with concrete with restricted site access will mean long distances between concrete truck deliveries and the walls requiring prolonged pumping and equipment hire. Where a bill of quantities is part of an invitation to tender and subsequent contract, the client is usually protected from risk in this type of scenario by a preamble note that may say, for example, 'during the tender period the contractor shall become fully acquainted with the site conditions, drawings, specifications and bills of quantities with prices reflecting the contents therein...'. Where applying, and if a suitably prepared BQ is provided or the contractor prepares builder's quantities, the contractor will have no excuse for misunderstanding the requirement by not pricing accordingly. The solution here is for the contractor to estimate the cost of the works as working operations, that is, the amount of labour, plant and materials required to start and complete each trade works, which can be cost estimated with or without a bill of quantities. Moreover, the approach is suitable for design–and-build estimating when the design is at a stage where the scope of works can be ascertained. Trade contractors providing a service to a main contractor often estimate their cost on this basis, and an experienced estimator from a main contracting organisation may also be proficient in the concept and elect to price the works in the same manner for trades it would procure as split labour and materials packages. There are pros and cons with this operational style of estimating to a contractor:

Advantages of this style are that:

- it gives the contractor more control with pricing as it is an in-house arrangement;
- it allows the contractor to clearly identify risks;
- works can be awarded as operational packages to trade contractors;
- it is suitable for continuous work;
- it allows the contractor to schedule works and materials by splitting operations into phases to suit the master programme; and
- it ensures items are scheduled for pricing that would normally be 'deemed included' in a unit rate alongside a BQ description often without mention (e.g. supply of shop drawings).

Disadvantages of this style of estimating are that:

- it is possible to duplicate with preliminaries items (e.g. access equipment) which, if included, can distort a price and may need crediting which is difficult to calculate;
- the contractor is responsible for quantities and takes the risk;
- where works are staged/phased, the process is more complex to calculate;
- rates can appear excessive in comparison with market rates;
- it takes considerable time to calculate, which may not be available during a tender period;
- difficult to interpret as unit rates in a BQ for returning to the client/client's agent, as the descriptions and formats are different;
- client-side quantity surveyors/cost managers may not be familiar with the concept and will rely on unit rate calculations that are comparable with market rates standing

the test of time and their education, with persuasion by the contractor to consider otherwise often difficult; and

• the application of operational estimating is unique to contractors and not a usual part of technical education or curriculums taught in universities.

To demonstrate, Figure 3.5 shows an operational estimate for the construction of block walls scheduled for a single operation. This theoretical example is based on a wall constructor's price including preliminaries requirements that would normally be provided under the main contract (e.g. scaffolding and mechanical plant). In order to make for a price to be competitive for a project under tender, the contractor may elect to include the priced preliminaries within the masonry/block work package price. They would then become works package contract preliminaries, and while doing so must credit a proportion of the main contract preliminaries.

3.5.7 Preliminaries Pricing

The cost of the main contract preliminaries to the contractor generally run at 8–15% of the overall price of a project. However, this general rule is not suitable to commit to a tender as the risk is too great for the main contractor. Costing must therefore be carried out with accuracy during the tender period, which can be achieved by creating a definitive list of items for pricing. Established contractors may elect to create a standard list based upon the RICS new rules of measurement (nrm2), which includes a schedule of items for a project of any type. Alternatively, a contractor may use in-house templates which are considered equally suitable. Where labour, plant and materials are required for an item it is usual to invite tenders from appropriate companies; this applies to the likes of scaffolding, temporary works, tower cranes and hoists. It is important to obtain *bona fide* prices for these works as their cost is a considerable proportion of the preliminaries, and should not be assessed by the contractor without input. In addition to creating a standard list, it is necessary to append specific items as stated in the invitation to tender for including in the estimate (e.g. employer's requirements for accommodation). The quantity surveyor must have an appreciation of the constituents of preliminaries, and during the tender period will aid the process by scheduling the scope for rating and pricing.

To assess preliminaries expenditure, the driving force for the contractor's requirements may be determined in a risk workshop instigated by the estimating manager at the commencement of the tender period and subsequent recommendations of the project manager. The project manager usually advises on project duration and site staff management requirements, and provides a site layout of the site accommodation and a list of mechanical plant necessary to assist with trade operations. It is not a usual requirement for the project manager to schedule and rate specific items for pricing as he/she will be preoccupied with existing projects, leaving this responsibility to the devices of the estimator/quantity surveyor. Established contractors usually adopt an accounting and contract administration system using a coding structure for budget preparation, which also acts as a checklist to ensure relevant items are included in the estimate. Figure 3.6 breaks down the preliminaries using a coding structure together with guidance notes on how to assess the constituents of expense for a project to run for one year. After experiencing the requirements for at least one long-term project, a quantity surveyor should be able to gain an understanding of the process and be able assess the anticipated expense.

Project: XYZ Retail development		Trade Contractor: 123 Block Layers Ltd	
BQ Item 3.1 (a–d)	*Double block wall 350 mm thick overall (200 mm: 50 mm cavity: 100 mm) including accessories as per Drawing AE-5001 'E'*	*Wall area 1250 m²*	Rate m²

LABOUR

200 mm block wall:

Output	1.50 man hours / m²*	
Total man hours 1,250 m² × 1.50 man hours/m²	1,875 hours	
Number of gangs (2 block layers; 1 hod carrier)	6	
Number of productive working hours per day	8	
Area of wall constructed per day = 6 × 8/1.50	32 m²	
Duration = 1,250/32	39 days × daily gang rate of £472.00** × 6 gangs/3 gang members = £36,816/1,250 m²	29.45

100 mm block wall:

Output	0.80 man hours / m²*	
Total man hours 1,250 m² × 0.80 man hours/m²	1,000 hours	
Number of gangs (2 block layers; 1 hod carrier)	6	
Number of productive working hours per day	8	
Area of wall constructed per day = 6 × 8/0.80	60 m²	
Duration = 1,250/60	21 days × daily gang rate £472.00** × 6 gangs/3 gang members = £19,824/1,250 m²	15.86

Forming cavity:

Installation of wall ties (included in output of walls)		
Installation of insulation board (included in output of walls)		NIL

Damp proof course:

(a) Laying of damp proof course, flexible, Xtra-Load, 7.24 kg/m, not exceeding 500 mm wide

Output, including lapping/cutting	0.33 man hours/m	
Total man hours - each wall is 416.5 m = 833 m × 0.33	275 hours	
Number of productive working hours per day	8	
Length of DPC laid per day 8/0.33	24 m	
Duration = 833/24	35 days @ £472.00 × 37.5% = £6,195/1,250 m²	4.96

Figure 3.5 Operational estimate.

(b) Laying of damp proof course as cavity trays, flexible, Xtra-Load, 19.30 kg/m, over 500 mm wide

Door width	Ends 300 mm	Nr of doors	Door cav tray	Window width	Ends 300 mm	Nr windows	Wdw cav tray	
900 mm	600 mm	20	10.80 m²	2100 mm	600 mm	15	24.30 m²	

Output, including lapping/cutting, forming weep holes/open perps and inserting stop ends 1.66 man hours/m²
Total man hours (10.80 + 24.30) = 35 m² × 1.66   58 hours
Number of productive working hours per day   8
Length of DPC laid per day 8/1.66   4.82 m²
Duration = 35/4.82   7 days @ £472.00 × 37.5% = £1,239/1,250 m²   **0.99**

Lintels:
Proprietary heavy duty SS lintels, 200 × 233, SWL 36Kn, built in as works proceed - main contractor to supply as free issue, i.e.
20 nr, 1500 mm long; 15 nr 2700 mm long

Output   1.00 man hours/each
Total man hours 35nr × 1.00   35 hours
Number of productive working hours per day 35/8   8
Nr of lintels installed per day 35/4.38   4.38 nr
Duration = 35/4.38   8 days @ £472.00 × 37.5% = £1,416/1,250 m²   **1.13**

Sundries:
(a) Forming opening in walls for doors/windows including building in cavity closures - main contractor to supply closers as free issue

Total duration of works = 60 gang days, allow 10% for pointing, cutting, chasing and making good:
60 days @ £472.00 × 10% = £2,832/1,250 m²   **2.27**

(b) General operative assistants 1 nr per 3 gangs, i.e. 2 nr = 60 days × 2 × 8 hrs @ £17.50/1,250 m²   **13.44**

(c) Acid clean wall areas on completion of works, 1250 m² @ £2.00 × 2 sides = £5,000/1,250 m²   **4.00**

* Based on experience of past projects and considered suitable for this project. Includes for the following: Setting out works - assumes datums and a minimum of 4 gridlines provided by main contractor; risk for lost time associated with bad weather; forming cavities; installing partial fill cavity wall insulation board up to 50 mm thick; hand tools; PPE; flush pointing to 1 side of wall; protection of works after each working shift; daily cleaning; rubbish removal and placing into skips provided by main contractor; and maintenance of datums, gridlines and equipment.
** Based on 2:1 gang × 8 hrs day @ gang rate of £59.00 per hour (split as 2 × 37.5%; 25%) including supervision.

LABOUR - TOTAL	**72.10**

Figure 3.5 (Continued)

MATERIALS	Rate m²

200 mm block wall:

£0.50 Each × 11.50 nr/m² × 1,250 m² = £7,187.50 +10% waste = £718.75 … Total £7,906.25/1,250 m² — 6.33

Ready spread mortar, from past projects, 0.028 m³/m² @ £150.00/m³ — 4.20

100 mm block wall:

£0.35 Each × 11.50 nr/m² × 1,250 m² = £5,031.25 +10% waste = £503.12 … Total £5,534.38/1,250 m² — 4.43

Ready spread mortar, from past projects, 0.014 m³/m² @ £150.00/m³ — 2.10

Forming cavity:

Wall area	Wall length	Wall height	Ties - vertical	Ties - horiz.	Nr - vert	Nr - horiz	Total ties
1,250 m²	416.500 m	3.000 m	Every 750 mm	Every 1200 mm	4	350	1,400

Wall ties, Ancon ST1, Type 1, 200 mm long - 1,400 nr @ 0.25 = £350.00 +20% waste = £70.00…. Total £420.00/1,250 m² — 0.34

Insulation board, Kingspan Kooltherm K8, 0.021 W/m.k partial fill cavity boards, 1200 × 450 mm boards (0.54 m²)

Wall area 1,250 m²/0.54 = 2,315 nr @ £2.00 incl. accessories = £4,630.00 +10% waste = £463.00… Total £5,093/1,250 m² — 4.07

Damp proof course:

(a) Damp proof course, flexible, Xtra-Load, 7.24 kg/m

Specified DPC is in 8 m lengths, available 225 mm wide

833 m/8 = 104 nr rolls @ £17.50 = £1,820.00 +10% waste = £182.00…..Total £2,002/1,250 m² — 1.60

(b) Damp proof course as cavity trays, flexible, Xtra-Load, 19.30 kg/m

Specified DPC is in 8 m lengths, available 225 mm wide, or 1.80 m²

35m²/1.80 = 19 nr rolls @ £47.50 = £902.50

Stop ends 70 nr @ 0.30   £21.00

Total   £923.50 +10% waste = £ 92.35…..Total £1,015.85/1,250 m² — 0.81

Lintels:

Main contractor to supply as free issue — NIL

Figure 3.5 (Continued)

	Rate m²
Programme:	
Preparation and issue of 1nr programme of works prepared from Microsoft Project for issue to the main contractor	
Item cost £750.00/1,250 m²	0.60
Shop drawings and as-built information:	
2 draftsmen × 1.50 months @ £4,500.00 = £13,500.00, allow +3% for prints, folders and consumables = £13,905/1,250 m²	11.12
MATERIALS-TOTAL	35.60

PLANT	Rate m²
Scaffolding:	
Both sides of cavity wall, based on 1575 mm wide (7 boards, 225 mm wide)	
1,250 m² × 2 = 2,500 m² @ £9.00 = £22,500.00/1,250 m²	18.00
Transportation of materials:	
Main contractor to place materials in central location for retrieval by 123 Block Layers Ltd.	
Forklift including driver and fuel (storage area for fuel to be provided free of charge by main contractor)	
60 days × 8 hrs = 480 hrs @ £28.50 = £13,680.00/1,250 m²	10.94
Mechanical plant:	
Grinder and cutter, 60 days × 15% applicable to works = 9 days @ £35.00/day = £315.00/1,250 m²	0.25
Silo mixing plant for mortar by main contractor	
Task lighting for works by main contractor	
Water supply for works by main contractor	
Power supply for works by main contractor	
PLANT-TOTAL	29.19

Figure 3.5 (Continued)

Preliminaries - cost assessment

Project title: New medical centre - Salford   Project duration: 52 weeks

Date:

P100 Pre-commencement (6 weeks)	*Time-related expense*				*Fixed expense*				
Description	*Quant*	*Unit*	*Rate*	*Total*	*Quant*	*Unit*	*Rate*	*Total*	*Comments*
Project manager	6	wks							This represents the time and work involved between the date of the agreement with the project client through to the date of establishing the contractor's presence on site. This period is lead in time and does not form part of the construction phase.
Site manager	6	wks							
Health and safety manager (50%)	3	wks							
Quantity surveyor (50%)	3	wks							
Student quantity surveyor (50%)	3	wks							
Design manager	6	wks							
Construction programme						item			
Consultant fee for site plan details						Item			
Consultant fee for BIM/IT training						item			
Total									

P101 Management and staff	*Time-related expense*				*Fixed expense*				
Description	*Quant*	*Unit*	*Rate*	*Total*	*Quant*	*Unit*	*Rate*	*Total*	*Comments*
Site manager (100%)	52	wks							Weekly rates for site staff charges are assessed on a cost to employ basis, usually provided by the contractor's accountant.
Structural foreman (50%)	26	wks							
Finishing foreman (50%)	26	wks							
Quantity surveyor (100%)	52	wks							
Student quantity surveyor (25%)	13	wks							
Project manager - visitor	6	wks							
Health and safety manager - visitor	6	wks							
Planning manager - visitor	2	wks							
Construction programme - updates					4	nr			
Site labourer - Nr 1 (100%)	52	wks							Rate as per cost to employ - See Table 3.4. Apprentices included where not considered chargeable as measured/productive work.
Site labourer - Nr 2 (50%)	26	wks							
Apprentice carpenter (25%)	13	wks							
Total									

P102 Site establishment and main contractor's presence	*Time-related expense*				*Fixed expense*				
Description	*Quant*	*Unit*	*Rate*	*Total*	*Quant*	*Unit*	*Rate*	*Total*	*Comments*
Mobilisation*						item			Previous projects or current quotations will provide a source of cost.
Demobilisation*						item			
* For listing, see Main contractor's running costs P103									
Compound perimeter hoarding							m		Allow for posts, plywood sheeting and painting or other type of screening system.
Access gate					1	nr			
Temp service connection - water						item			Consider distance from the source, and request a price from the service trades that price the contract works. Also include connection fees by statutory authorities. IT requirements provided by IT specialist or contractors' in-house resources.
Temp service connection - electric						item			
Temp service connection - IT/other						item			
IT commission/decommission						item			
Drainage						m			
Disconnection of temp services						item			
Excavate/stone bases/car park						m³			
Excavate & remove stoned areas						m³			
Total									

P103 Main contractor's running costs	*Time-related expense*				*Fixed expense*				
Description	*Quant*	*Unit*	*Rate*	*Total*	*Quant*	*Unit*	*Rate*	*Total*	*Comments*
Site cabin & canteen	52	wks							Consider hire rates from suppliers or, if a contractor's asset, apply a weekly rate. Rates to include insurance and maintenance.
First aid room	52	wks							
Toilets	52	wks							
Storage containers (2 Nr)	52	wks							
Furniture/office equipment hire	52	wks							
Mobile phone/telephone charges	52	wks							Research past projects for average prices.
Test portable appliances (quarterly)	4	nr							
Operating supplies incl. admin	52	wks							
Total									

Figure 3.6 Preliminaries cost assessment template.

P104 Temporary services	Time-related expense				Fixed expense				Comments
Description	Quant	Unit	Rate	Total	Quant	Unit	Rate	Total	Comments
Temp service connection - water						item			Notes as per site establishment apply
Temp service connection - electric						item			(P102). Allow here for consumption by the
Temp service connection - gas						item			site establishment for 52 weeks.
Utility services consumption - low	13	wks							Varies with site operations/workforce.
Utility services consumption - med	13	wks							Refer past projects. Where potable drinking
Utility services consumption - high	26	wks							water consumed is not from a service
Temp services disconnections						item			connection, consider the litres purchased.
Total									

P105 Temporary works	Time-related expense				Fixed expense				Comments
Description	Quant	Unit	Rate	Total	Quant	Unit	Rate	Total	Comments
Excavate/stone up haulage road						m³			Form access routes and material storage
Excavate/stone up hardstand area						m³			areas. Maintain areas for 50% of the
Maintain areas (2 hrs/week x 50%)	52	hrs							contract phase.
Excavate & remove stoned areas						m³			Applicable at the end of the project.
Contractor's notice board					1	nr			
Total									

P106 Security	Time-related expense				Fixed expense				Comments
Description	Quant	Unit	Rate	Total	Quant	Unit	Rate	Total	Comments
Security guards @ weekends only	3,120	hrs							Allow as required, say 60 hrs x 52 weeks
Total									

P107 Employer's requirements	Time-related expense				Fixed expense				Comments
Description	Quant	Unit	Rate	Total	Quant	Unit	Rate	Total	Comments
Erection of employer's notice board					16	hrs			As noted in the tender documents.
Sundry materials for fixing board						Item			Employer to supply & deliver the board.
Total									

P108 Contract conditions	Time-related expense				Fixed expense				Comments
Description	Quant	Unit	Rate	Total	Quant	Unit	Rate	Total	Comments
Fixed price contract - no rise & fall									Documents request contractor to provide
Insurance for laboratory equipment						Item			an insurance policy for the value of goods
									installed by the employer's contractors
									until date of handover.
Total									

P109 Fees and charges	Time-related expense				Fixed expense				Comments
Description	Quant	Unit	Rate	Total	Quant	Unit	Rate	Total	Comments
Considerate Constructor's Scheme						Item			Allow registration fees as required by the
Construction licenses/permits									invitation to tender and industrial practices.
Total									

P110 Insurance, warranties, guarantees and bonds	Time-related expense				Fixed expense				Comments
Description	Quant	Unit	Rate	Total	Quant	Unit	Rate	Total	Comments
Contractor's 'All Risk' policy						Item			Each requirement is expressed as a % of
Bank guarantee incl bank's fees						Item			the contract sum, or rated per £1000 of the
Professional indemnity insurance						Item			estimated cost to be insured. Rates will
Advanced payment bond						Item			vary depending on the amount of policy
Total			Total					Total	excess or conditions of bond.

P111 Safety and environmental protection	Time-related expense				Fixed expense				Comments
Description	Quant	Unit	Rate	Total	Quant	Unit	Rate	Total	Comments
Audit of CDM health and safety file						Item			Research cost of past projects and express
Signage supply						Item			as % of the contract value.
First aid supplies						Item			
Safety item purchases, i.e. PPE, etc.						Item			
Total									

Figure 3.6   (Continued)

P112 Control and protection	Time-related expense				Fixed expense				
Description	*Quant*	*Unit*	*Rate*	*Total*	*Quant*	*Unit*	*Rate*	*Total*	*Comments*
Setting out building, datums/levels						Item			Surveyors quotation
Protection of surfaces						Item			Material expenses only. Site labourers to
Maintain datums/levels/surfaces		Item							install and maintain. Consider risks.
Material storage pallets 1.8 x 1.8 m					40	nr			Supplier's price.
Total									

P113 Site records	Time-related expense				Fixed expense				
Description	*Quant*	*Unit*	*Rate*	*Total*	*Quant*	*Unit*	*Rate*	*Total*	*Comments*
Supply of operating manuals						item			Manuals etc. may be part of trade prices. If
Supply of as-built info + CDM file						item			not, allow here. Include CDM file and end-
Allowance for training						item			user training of installed equipment.
Total									

P114 Cleaning	Time-related expense				Fixed expense				
Description	*Quant*	*Unit*	*Rate*	*Total*	*Quant*	*Unit*	*Rate*	*Total*	*Comments*
Site accommodation cleans	52	wks							Subcontractor costs.
Building early cleans/recleans		nr							
Building final clean						Item			
Road sweeps (1 per week)	52	nr							
Total									

P115 Non-mechanical plant	Time-related expense				Fixed expense				
Description	*Quant*	*Unit*	*Rate*	*Total*	*Quant*	*Unit*	*Rate*	*Total*	*Comments*
Perimeter scaffold - erect/dismantle						item			Subcontract quote. Consider suitability of
Additional weekly hire		wks							the hire period in the quotation.
Access equipment						item			Consider internal scaffold/access over 3 m.
Small hand tools hire/purchase	52	wks							Research past projects. Consider control of
Waste skips						nr			generated waste.
Total									

P116 Mechanical plant - major	Time-related expense				Fixed expense				
Description	*Quant*	*Unit*	*Rate*	*Total*	*Quant*	*Unit*	*Rate*	*Total*	*Comments*
Crane		wks							Not required.
Hoist		wks							When required, source quotations which
Labour hire to operate crane		hr							include maintenance, insurance and
Labour hire to operate hoist		hr							operatives hourly labour rates & consider
Mobilisation to site - all plant						item			contractor's risk, i.e. overtime. Allow for
Demobilisation from site - all plant						item			main contractor's attendances for the
Design works for temporary bases						item			length of time the plant will be on site.
Cost to construct temporary bases						item			Energy use usually included under P104.
Total									

P117 Mechanical plant -minor	Time-related expense				Fixed expense				
Description	*Quant*	*Unit*	*Rate*	*Total*	*Quant*	*Unit*	*Rate*	*Total*	*Comments*
Mobilisation						item			Consider hire rates plus an allowance for
Demobilisation						item			energy consumption if not under P104. If a
Forklift truck (75%)	39	wks							contractor's asset, charge as time - related
Mortar mixers (3 nr) x 25%	13	wks							basis including fuel using a calculated rate
Dewatering equipment	6	wks							as shown in Table 3.1.
Fuel storage container (75%)	39	wks							
Drying out dehumidifier	4	wks							
Fuel supply on hired items		litre							
Total									

P118 Post-completion requirements	Time-related expense				Fixed expense				
Description	*Quant*	*Unit*	*Rate*	*Total*	*Quant*	*Unit*	*Rate*	*Total*	*Comments*
Site manager					4	wks			Rectification period allowance
Material purchase/insurance cost						item			
Total									

Figure 3.6 (Continued)

On large-scale projects of considerable financial value that may run for a year or more, trade contractors should be asked to provide priced preliminaries for their own purposes which are in isolation from the main contract preliminaries. Where applicable, these are included in trade contractor's quotations and kept separate from trade works as the main contractor will need to verify there is no duplication with the main contract preliminaries, and that the period covered is for the duration of the trade contractor's works. The RICS new rules of measurement (nrm2) includes sections defining items to include in a bill of quantities/pricing schedule for both the main contract and 'works package contract' preliminaries. The format of each has similar requirements, which when priced clarifies the constituents of the price.

3.5.8   Cost-Planned Tenders

With traditional procurement, an estimating team estimates the cost of a project from an advanced building design. However, a contractor that carries out a design-and-build service may offer a cost-planned tender as an initial price as part of the process towards securing an award of contract. A contractor offering this service must be established with professional staff that possibly includes in-house design teams, and must have access to resources that creates cost data. Quantity surveyors experienced in cost planning or originating from a consultancy background are of benefit here, as they are aware of the importance of collecting cost data and the recording of information on current and past projects for estimating purposes. Cost-planned tenders are a risk to a contractor because the market is not tested. However, as the intention is for the design to be under the eventual control of the contractor, the ethos used for creating a tender is for the design to be completed to a predetermined cost as opposed to costing a predetermined design because in effect, the contractor is willing to commit to a cost limit. Advantages of cost-planned tenders to a contractor include:

- it provides a contractor with options to undertake a diverse range of projects;
- it permits the contractor to take more control of the process than with traditional procurement;
- it is relatively quick to calculate a price as market testing is non-existent or limited; and
- it reduces the possibility of contractual disputes with a project client.

Disadvantages include:

- the transfer of risk from the client to the contractor, meaning the contractor is more vulnerable;
- the contractor must have adequate resources to accurately capture cost data of past projects;
- possible lengthy negotiations with the client after issue of a tender (e.g. inclusions/exclusions);
- possible complex arrangements for obtaining the initial design for developing by the contractor; and
- the initial tender may need numerous updates to reflect client negotiations.

3.5.8.1   New Works and Cost-Planned Tenders

Contractors offering cost-planned tenders usually focus on the types of buildings they are accustomed to producing. In order to cost plan effectively when there is limited or no design, estimates are based upon rates charged against gross internal floor areas

(GIFA) of a building. In general, GIFA is the sum of floor areas by building level measured from the internal faces of external walls, without deductions for columns, internal projections, floor openings and voids such as stairwells, and excludes external projections that do not interfere with the internal layout such as canopies, and all types of outbuildings. Rates to apply to GIFA are obtained from current or completed projects and stored as data which record the project specifications. These specifications are retained as a source of reference for future use when a contractor considers they can be applied to a new project under tender that can be modified to meet a client's needs. In order to do this, a contractor must have certainty with the source as they become the starting point of a cost-planned tender.

The triggering of the tender process is usually initiated by the client's brief and the appointment of consultants that prepare architectural concept designs and other outline proposals for associated works including structural, civil and building services engineering criteria. At a particular stage when the client's objectives are met with the design deemed sufficient for tendering purposes by the client/client's agent, invitations to tender are issued to competing contractors to provide preliminary cost-planned tenders. Depending on the quality of the design, the submission of a tendered amount is usually subject to negotiation or, if stated in the tender documents, may be tendered as a maximum sum which becomes a target cost. Target cost plans may be provided where a guaranteed maximum price is explored on the understanding a client accepts the objective of a contractor's tender and remains inactive with progression of the design leaving the completing to the contractor. To demonstrate, let us consider a contractor that has experience of medical centre construction and is approached by a client/client's agent to provide a cost-planned tender for a new medical centre. To assess the cost, the contractor must access the cost of a medical centre project it has completed or is currently building, and break down the price into elemental parts. Each element is then *pro rata* priced to the GIFA of the building. Table 3.7 demonstrates this principle on a medical centre project already constructed with a total GIFA of $3000\,m^2$; because the data includes the price of building elements, the rate of each can be assessed as a benchmark for the project under tender.

To assess benchmarked rates, the total cost of each element is divided by the GIFA to create a cost per square metre of the element in relation to the GIFA. Once these rates are ascertained they can be applied to the new project on a *pro rata* basis. When selecting a benchmark project it must be consistent with that under tender; for example, an office or industrial building should not be used as a benchmark for a medical centre or similar health facility due to the different functional requirements and inappropriate specifications. An important factor here is that the GIFA for the new building must be provided when determining a price, which is usually included in the client's brief or available from the design.

The *pro rata* basis of estimating described above is relatively straightforward. However, calculating a price using this method is only one part of the picture, as in effect the process is supplanting the same (or similar) building of the same or a different floor area and placing it elsewhere. This might not be ideal, and to determine a price it is necessary to consider factors for the project under tender in comparison with that benchmarked, including:

- external works and finishes including parking areas, roads, paved areas and landscaping;
- changes to the prices as the rates are based upon a project that has surpassed;

Table 3.7 Cost model (completed project)

Project description and parameters			

Completed project: Site 'x', new medical centre, GIFA 3000 m²

Element	Total cost	Cost m²	%
Substructure			
Foundations and substructure	£215,864	£71.95	7.94
Total (1)	**£215,864**	**£71.95**	**7.94**
Superstructure			
Columns	£30,556	£10.19	1.12
Upper floors	£261,110	£87.04	9.61
Staircases	£21,066	£7.02	0.77
Roof	£86,175	£28.73	3.17
External walls	£277,280	£92.43	10.20
Windows	£185,175	£61.73	6.81
External doors	£20,120	£6.71	0.74
Internal walls	£29,411	£9.80	1.08
Internal screens	£78,144	£26.05	2.87
Internal doors	£65,076	£21.69	2.39
Total (2)	**£1,054,113**	**£351.39**	**38.78**
Internal finishes			
Wall finishes	£172,084	£57.36	6.33
Floor finishes	£125,008	£41.67	4.60
Ceiling finishes	£57,036	£19.01	2.10
Total (3)	**£354,128**	**£118.04**	**13.03**
Fittings and equipment			
Fitments	£322,188	£107.40	11.85
Total (4)	**£322,188**	**£107.40**	**11.85**
Services			
Sanitary fixtures and plumbing	£225,079	£75.03	8.28
Water and gas supply	£58,122	£19.37	2.14
Space heating	£172,449	£57.48	6.34
Fire protection services	£28,514	£9.51	1.05
Mechanical services	£156,028	£52.01	5.74
Electrical power and lighting	£119,006	£39.67	4.38
Other services	£0	£0	0.00
Builders work with service trades	£12,795	£4.27	0.47
Total (5)	**£771,993**	**£257.34**	**28.40**
Total amount (1, 2, 3, 4 and 5)	**£2,718,286**	**£906.12**	**100.00**

Excludes external works, design fees, preliminaries and margin

Date of agreeing contract sum: xx/xx/xxxx

- design and construction risk regarding specification, especially to quality;
- duration of the construction phase for the new project;
- any site-specific and abnormal requirements, such as demolition of existing structures;
- differences in client requirements; and
- locality of the benchmarked project versus new (e.g. the cost to build a medical centre in London is generally more expensive than in Manchester).

The timeframe for issuing an award from the date of an initial cost-planned tender can be up to six months or more, with competing contractors possibly requested to modify their offers due to further development of the client's brief. This can involve the supply of new and/or updated information, and involves post-tender reviews with contractors reassessing the design as well as attending meetings with the client/client's agent who provides items such as schedules of accommodation and other specific needs. Here, the contractor's quantity surveyor assists by measuring the floor areas or building elements from new and/or updated designs and rating them accordingly. If there are 'financial blow-outs' compared to the preliminary cost plan which increases the price, the contractor's estimator/commercial manager will advise the client of the elements that create the scenario (e.g. finishes). This is to allow the design to be amended or for the client to accept the consequences and increase the budget.

Contractors in the business of submitting cost-planned tenders are usually prepared to accept the consequences of early involvement and understand that effective working relationships are created in the process. A client usually makes no obligation to a contractor of the guarantee of an award once an initial cost-planned tender is issued. This may also be noted in the tender documents as a condition of tendering, meaning the client may at any time instruct the design team to develop the design to reflect later proposals and to retender the works, usually including the contractors that provided a preliminary cost-planned tender. However, this may only occur if the scheme changes drastically or if the price is to be significantly altered, possibly because of a change in scope or the GIFA. A contractor's tender expense in both time and money can therefore be considerable with this type of procurement.

3.5.8.2 Refurbishment Works and Cost-Planned Tenders

Contractors experienced in the refurbishment of buildings may offer solutions for a client if a scope and predetermined budget is known. For example, a client may divulge a budget for converting a number of shops into one large retail store. Here, a contractor may elect to submit a cost-planned tender to meet the budget where plausible, with the quantity surveyor assisting the process by creating a scope of works from an indicative design and pricing accordingly. However, if a client has a limited budget with an extraordinary requirement, the contractor must advise the client/client's agent if the budget is sufficient. Here, a quantity surveyor can prepare 'spot items' and rate them accordingly from first principles and provide advice regarding a scope of works that can be accomplished for the budget. On occasion there can be a pleasant surprise for a client if a budget surplus exists, meaning the scope of works can be augmented permitting the full budget to be spent. This could involve an increase in quality such as changing the type of wall and floor finishes. If the estimated cost of refurbishment cannot meet the client's budget the option may be to demolish and rebuild, subject to planning approval, with the cost of constructing a new building assessed upon GIFA and rates obtained from the contractor's past and current projects. However, common sense should prevail here; if a client has a budget that cannot pay for refurbishment, it is unlikely the construction of a new building will be viable with the exercise being a possible waste of time. On no account should information like this be directed to a client by the quantity surveyor, as it would need approval by the estimating manager. This is because there may be items of risk that could go unnoticed with both options (e.g. the unknown presence of asbestos that is expensive to remove once disturbed because of the health and safety aspects involved. Here, a consultant is engaged to issue advice on the extent of any works that could add to the demolition and refurbishment costs).

Involvement with cost-planned tenders and the estimating function of a contractor's business can be a rewarding experience for a quantity surveyor, and may enlighten individuals who prefer being involved with pre-contract activities instead of post-contract administration of a project. Being part of an estimating team is advantageous, because it provides in-depth knowledge of the nature and risks of a project before commencing the works. It also provides an understanding of the assessment of prices and strategic methods adopted by a contractor in the attempt to secure a contract.

3.5.9   Tender Submission

Time management and effective planning from the date of receiving an invitation to tender through to the date of submission is vital for the effectiveness of the tender process. Where works are to be tendered as a lump-sum fixed price, the price should crystallise a week or so before a tender is due with the financial worth and risks of the scheme becoming known. At this stage the estimator will be in control of the final costings before presenting the details for peer review as a final assessment to steer a decision on an amount to tender.

A peer review is held as a meeting, usually between the estimator and other key members involved with the tender process (including quantity surveyor), with at least one director present. This can also involve other members of senior management if the scheme is of significant financial value or a particularly complex and large project of long duration. The director(s) attending may have had minimal involvement with the tender process due to other commitments, with delegation and the reporting of activities passed to others as a normal business arrangement. However, the involvement of at least one board director with the peer review process is crucial, as the influence could be a determining factor of the amount to tender. At the meeting, methods adopted for ascertaining prices will be discussed, as will the contents of any notices received during the tender period from the client/client's agent and other pertinent matters such as cash flow, financing charges, opportunities, competition and risk. The intention of the discussions is for the director(s) to gain an understanding of how the estimated cost has been derived and impact the tender could have on the business if resulting in an award of contract. It is at this stage a decision regarding risk and its effect on the estimate is made, as well as the margin (usually expressed as a percentage of the estimate) to apply to the estimated cost as a contribution to offsite overheads and profit, which in combination determines the amount to tender.

The identification of risk will influence a decision to confirm if the tender is to be a legal offer (where it can be called a tender or offer), or refer to it as an estimate of cost subject to negotiation, which may be suitable for a cost-planned tender. Any changes made to an estimate at the peer review can be formally stated in the tendered amount and referred to as a 'director's adjustment' as identified in the RICS new rules of measurement (nrm2), 'Definitions' 1.6.3. The inclusion of a 'director's adjustment' has advantages to the contractor and client as it demonstrates transparency regarding risk, which can be reviewed during any post-tender negotiations should the client/client's agent look favourably on the tender.

The peer review is one part of the tender submission process, with another being an assurance the tender is legally compliant (e.g. by including a health and safety file to comply with CDM 2015). While the supply of a health and safety file is a legal requirement, the contractor must also pay attention to any non-legal conditions of the invitation to tender and follow the requirements exactly, which means providing nothing more and nothing less. If submitting a tender to a potentially new client, a contractor may be tempted to over-service the invitation request by supplying information that may be surplus to requirements (e.g. portfolios of current and past projects with photographs, elaborated quality control assessment procedures, a dozen client testimonials, etc.). This approach of attempting to 'woo' a client with commercial information cannot be commended, and may in fact be a waste of the contractor's and client/client's agent's time, especially if a public client, as such information would most likely be disregarded. Furthermore, it is likely the client will have received satisfactory information regarding the contractor's capabilities with the completion of a pre-qualification questionnaire, a prerequisite for issuing the invitation to tender. However, innovation of a technical nature that could benefit the project could be provided. For example, if an amount to tender exceeds the client's budget, the contractor could identify possible savings and express a desire to enter into negotiations post-tender.

The amount to include in a tender and the format in which it is to be offered is most likely stated in the invitation to tender, and is usually on a form of tender supplied with the initial consignment. Where provided, the form of tender must be completed, signed, dated and offered by the contractor as a stated price to carry out the works in accordance with the terms and conditions of the invitation to tender and supply of a priced BQ if provided. Where a form is not provided, templates such as appendices D or E as found in RICS nrm2 can be used. Subject to the tender not stating otherwise, the quoted price is considered the contractor's legal offer and, where applying, must include a note stating that it either includes or excludes VAT. As good practice, the contractor should also provide a signed and dated certificate entitled 'Certificate of bona fide tender' to demonstrate the contractor's tender is issued in good faith, and may include a statement to the affect that the tender or tendered amount has not been communicated to other parties.

A contractor should aim at issuing the tender on or before the date it is due. The supply of an early tender gives no advantage to the contractor and, if more than two working days earlier, could give an impression prices have been prepared without resourcing the market sufficiently. Submission of the tender must be to the correct address and may be hard or soft copy submitted by no later than the given date and specified hour where applicable. If submission is via the e-tendering procedure or other collaborative electronic system, time should be allowed for uploading information and for possible delays in the recipient's server, with confirmation of receipt usually provided as a matter of course.

3.5.10 Value Management

Value management (or sometimes called value engineering when a project has a salient engineering theme) is a generic term used to describe the collection and control of a set of principles and practices that produces best value for those concerned. With construction projects, the process seeks to satisfy a client's needs by ensuring all necessary functions and facilities of a building are achieved for the lowest possible cost while maximising their performance. This involves identifying aspects of a building's design in terms of the cost to construct and considering the benefits that may be gained through the life cycle. A general misperception of value management in construction projects is that it aims to reduce construction costs in line with a budget. This is when a component, element or product is substituted or modified, yet still deemed suitable for the purpose. However, a bigger picture involves value managing a project that intentionally creates or modifies a design and specification for long-term benefit. This occurs when a client considers that the tendered sum incurs an acceptable financial premium that can be offset by benefits through the life cycle. An example is the inclusion of sustainable methods, for example the cost to install photovoltaic solar panels creating renewable energy to be consumed during the occupancy of a building, resulting in lower energy running costs. If the extra cost to install solar panels to a building is £100,000, with the panels anticipated to last 20 years and produce an approximate saving of £10,000 per annum when using renewable energy, the system may start to pay for itself after the first 10 years.

A client may have a predetermined budget for a project and be open to suggestions for ways of maintaining the budget or exploring the benefits of value management. Here, stakeholders (usually limited to the client's team) hold discussions and provide opinions regarding what they consider is acceptable for a project regarding the expansion, removal, replacement, substitution or amendment of any parts that can be incorporated into the design. A method of addressing these issues is during a workshop, with objectives being to:

- bring the client and team members up to date with the status of the project;
- identify suggestions and contents of a value study if the study is available;
- foster teamwork towards achieving project goals;
- identify value-for-money solutions and cost savings that maintain the scope of works through the introduction of efficiencies;
- identify scope reductions and/or savings for the project to be within the current budget;
- schedule risks and opportunities;
- prompt the client to explore opportunities;
- challenge the client's brief;
- apply rules in order to achieve objectives;
- seek a win-win outcome (for both client and team);
- listen to others' points of view; and
- consider constraints such as time, planning approval, topography and items not subject to negotiation because the client insists on their inclusion without compromise.

From the offset, participants in a workshop should be advised by a team leader that no hierarchical attitude exists, with subjects open to discussion. An effective workshop is one that creates a relaxed atmosphere without defensive attitudes and prompts contributions from all team players, with the common goal of meeting objectives. Upon completion of the workshop, the client's quantity surveyor will usually calculate cost variances and issue the results.

When a client drives the design and appoints consultants, any considered value management criteria is usually included in the design, meaning receipt of tenders from contractors reflects the requirements. Unfortunately, if value management is required to reduce construction costs and is not addressed prior to tender, surprisingly high tenders may be submitted by contractors. This may result in a delay to the start date on site because the design goes back to the drawing board to produce cost savings, much to the dissatisfaction of all involved. Contractors tendering on construct-only projects tend to have experience of these scenarios more than other procurement routes, and may be asked by a client's agent to resubmit tenders based upon modified designs and possibly participate in any cost-saving exercises. With design-and-build projects, contractors regularly run in-house value management workshops, adopting the approach described in the bullet points above. This can also be conducted during the construction phase because the contractor has a strong influence and interest in modifying a design to meet the budget without compromising the integrity of the design-and-build contract it has undertaken. Where value management is being considered for a project, the concept should be captured in the design at the earliest opportunity as late changes to reflect new ideas could affect the works in progress and be costly and disruptive.

3.6 Construction Contracts

At their core, standard forms of contract used in the construction industry follow a set of principles that clarify each party's rights, obligations and responsibilities under an agreement for a range of building types. In this section of the chapter, a number of prominent forms in use are reviewed as well as the details of their mechanisms without examination of the clauses. This section does not make a comparison of the overall impression or suitability of a form, as the actual choice depends on the procurement route, type of project, market conditions and level of risk acceptable to the parties. It does, however, demonstrate the array of forms available and types of projects where they are most likely to be used.

3.6.1 Memorandums of Understanding

Before a contractor mobilises itself to site to commence works, it makes business sense for the contractor and project client to have a common understanding regarding the project they are about to undertake. Ideally, this should be in a written contract executed by the parties for there to be no doubt a formal agreement exists. Alternatively, the creation of a contract can occur by conduct when a contractor submits an offer that a client communicates as accepted, which is usually in writing but can be verbal. Verbal arrangements should not be encouraged however, as the basis of an agreement could be hard to prove to a suitable intervening third party if requested to settle a dispute. Here, the third party would need to recognise the existence of a contract, and could dismiss the case if matters are not evidenced in writing with the dealings between the parties possibly considered hearsay.

If a client wishes to negotiate the terms and conditions of an offer and expresses it this way, it creates a counter-offer by default which, under contract law, nullifies the original offer meaning a contract has not come into existence. For a contract to exist there must be offer and acceptance. However, due to the nature of construction operations, the negotiations may take some time to conclude that can stifle the start date despite the best intentions of the parties. To permit mobilisation or for works to commence, a project client/client's agent can communicate the matter with a Letter of Intent or a Letter of Acceptance. These letters have different meanings, and to gain an understanding of their application and use, it is best to discuss them separately.

3.6.1.1 Letter of Intent

This is an interim notice served by a client (or client's agent) to the contractor outlining the status of the party's negotiations up to and including the date of the letter. However, such letters are not restricted to project client and contractor relationships, and can apply to contractor/supply chain relationships. In a client/contractor pending working relationship, the issue of the letter signifies a client's intention to enter into a contract with the contractor, while reserving the right to negotiate and conclude certain matters beforehand. This can mean the client's agreement with any tender qualifications made by the contractor, or the acceptance by the contractor of imposed conditions by the client. The wording of the letter must be understood by the contractor which, aside from the status of negotiations, means recognising any expiry date, by which time the formal contract is expected to be executed by the parties. However, it is not uncommon to extend the date in writing, subject to the letter including such a provision stating which party will issue the extension. A poorly drafted letter can pave the way to a dispute if, during negotiations, there is failure to conclude an agreement and contract which may be by occurrence or choice (e.g. a lack of project funding or a hard commercial bargain being sought). For example, a situation could occur where a contractor is to proceed with works to a certain stage reimbursed on a schedule of rates with the remainder on a fixed price to be agreed, but subsequently the contractor cannot agree to fix this price and later suspends the works. In a situation like this, the letter should include a provision for such a scenario as well as the implications, for example:

> … the contractor will be paid on a schedule of rates applicable to the phase of works with the project client reserving the right to continue with the arrangement until such time the fixed price is agreed. If the fixed price agreement is not executed by both parties on or before xx/xx/xxxx (date), either party may terminate this Letter of Intent by giving notice in writing to the other party. In the event of this occurrence, the project client shall not be liable for consequential loss and/or indirect costs, losses or damages, including but not limited to any loss of profits suffered by the contractor as a result of either party electing not to proceed with the agreement.

This type of wording will help avoid surprises following the conduct of one or both parties after the letter is issued.

Where a party suffers financially because of a failure to conclude an agreement without a provision in the letter for remedy, the injured party can seek an award in damages through a court of law. This is demonstrated in English case law with Trollope and Colls partnership with Holland Hannen and Cubitts *versus* the consortium of Atomic Power Construction (1962). Here, the consortium issued a Letter of Intent to the partnership to carry out works prior to fully accepting all of the terms of the partnership's tender. When the terms were agreed upon, works had progressed with a variation issued. A dispute arose requiring legal intervention because the partnership believed it was free to terminate the arrangement and be paid on a fair basis for completed works. The court's ruling found with implied terms and behaviour that the partnership was bound to the consortium by a retrospective agreement that had provisions for valuing variations, and the court did not accept the partnership's argument that a contract did not exist. This case demonstrates the importance of understanding the effect of a Letter of Intent and that it can be treated as a contract.

3.6.1.2 Letter of Acceptance

Different from a Letter of Intent, this is the interim notice of an agreement pending the issue of a formal contract, and can be used in client/contractor relationships as well as contractor/supply chain agreements. Unlike a Letter of Intent, a Letter of Acceptance does not have an expiry date, and is the confirmation of the conclusion of negotiations pending an award of contract. Once issued, it is seen as an unconditional interim notice and viewed in common law as the acceptance of an offer, thus fetching a binding agreement into existence. The issue of a Letter of Acceptance precedes the issue of a contract and can be issued after a Letter of Intent, with the Letter of Acceptance referring to the conclusion of negotiations triggered by the Letter of Intent.

To avoid any misunderstanding, a contractor must receive a signed Letter of Intent or Letter of Acceptance by fax, electronically, post or hard copy delivered by courier or other hand delivery arrangement. Generally, postal rules under common law say an offer is considered accepted once an acceptance letter has been posted. Ideally, this should be registered by the sender and signed for by the receiver upon delivery. The letter should include a list of the tender (and any post-tender) information, as well as the contract sum/schedule of rates and form of contract (including confirmation of any amendments), together with a list of negotiated terms of the pending contract that will be appended to the contract. It is usual for the contractor to sign the letter to confirm receipt and return one copy to the sender, so that both parties have identically signed and copied letters.

3.6.2 The Joint Contracts Tribunal (JCT)

In the UK, the most widely used standard forms of building contract are produced by the Joint Contracts Tribunal Limited, abbreviated JCT. The JCT is a prominent tribunal in the construction industry, and comprises seven council members with industrial expertise and an interest in the creation and investment of the built environment, namely:

- British Property Federation Limited;
- Contractors Legal Group Limited;

- Local Government Association;
- National Specialist Contractors Council Limited;
- RICS;
- RIBA; and
- Scottish Building Contract Committee Limited.

JCT forms of contract are published by Thomson Reuters t/a Sweet and Maxwell as 'Contract families', and comprise of main and subcontract formal instruments of agreement that reflect a range of collaborative procurement methods. The most recent family of contracts has been overhauled as JCT Contracts 2016, the exception being the Home Owner contract that sustained an earlier version. Where applicable, these forms include separate subcontract agreements for use by main contractors with their subcontractors, and 'SubSub' agreements for use between subcontractors and their subcontractors. Where the main contract is executed under a JCT form of agreement, these subcontract forms provide a theme of consistency. Although usage is not compulsory, their adoption has advantages to the parties for identifying associated risks, rights and responsibilities.

3.6.2.1 Major Project Construction Contract (MP 2016)

The JCT Major Project Construction Contract is used by project clients (employers) that procure large-scale construction works carried out by contractors that have the experience and ability to accept greater risk than found with other JCT forms of contract. It is most suitable where the employer's requirements have been prepared and supplied to the contractor, with the contractor carrying out both the design and construction works and the employer engaging a representative to ensure the contract is enforced. The form can be used by private and public sector clients, and is best suited for parties that have established procedures in their place of business and are capable of dealing with the fundamentals of a large-scale project in terms of financial value, complexity and duration. The form includes pre-construction services agreements (general contractor and specialist), provisions for collaborative working, sustainability, third party rights and the use of framework agreements to address matters that would arise on a major project.

3.6.2.2 Standard Building Contract (SBC 2016)

As the title suggests, this form can be used with traditional and conventional procurement arrangements, and is suitable where the employer requires the services of an architect or agent to have an interface with the contractor. It is used for projects of varying value where the risk is considered high to the employer. The supply of the design and specifications is carried out by consultants appointed by the employer, with the contract also suitable where the contractor designs discrete part(s) of the works, termed 'Contractors Design Portion' (CDP). There are three types of SBC, as follows.

1) *With quantities (SBC/Q 2016)*: Applicable when a lump-sum price is required where the design, documentation and a firm bill of quantities is provided with an invitation to tender. In line with modern procurement thinking, the form has provisions that addresses collaborative working, sustainability, advanced payments, security bond for use with an advance payment, cash retention, payment for materials stored off

site, third party rights and the use of collateral warranties. Regarding subcontracting, a degree of client control exists where subcontractors can be appointed with the written permission of the architect/contract administrator or selected from a list of three names. The JCT Named Specialist Update also applies, meaning the form can name domestic subcontractors deemed suitable to carry out identifiable specialist works. The form can also be used as part of a framework agreement allowing the contractor to be paid interim payments on monthly credit terms.

2) *With approximate quantities (SBC/AQ 2016)*: This form is used when a lump-sum price cannot be provided because the design is incomplete, with works requiring an early commencement such as the first stage of a two-stage tender project. An approximate bill of quantities is provided in the invitation to tender documents, and is rated by competing contractors to create an initial fixed price which is subsequently converted once works are re-measured and valued at the schedule of rates. All other items applicable to the 'With quantities (SBC/Q)' form also apply.

3) *Without quantities (SBC/XQ 2016)*: The selection of this form is usually the result of the employer's intention to seek lump-sum prices by supplying drawings and specifications only, making no commitment to quantities which are to be determined by competing contractors. As a matter of course, a schedule of rates is normally requested from competing contractors for the assessment of variations which is appended to the contract. Separate to omitting reference to the requirement for a bill of quantities, the form has the same coverage as its sister forms that include the reference.

3.6.2.3 Minor Works Building Contract (MW 2016)

Selection of this form is suitable for low-valued works where the scope is simple in nature. The employer procures the design and specification in full or in part for issue to the contractor without a bill of quantities, with the project administered by an architect/contract administrator. It is available in two forms, either including or excluding provision for design input from the contractor.

3.6.2.4 Intermediate Building Contract (IC 2016)

Possibly seen as a gap filler between the Standard Building Contract and Minor Works Building Contract, IC 2016 is suitable where the nature of the building works are simple with trade works and building services installations uncomplicated. It is suitable for execution when the design is at an advanced stage and can be used with or without a bill of quantities where an architect/contract administrator will have an interface with the contractor. IC 2016 is available in two forms that include or exclude the provision for design input from the contractor. A characteristic of the form is the inclusion of named subcontractors, a concept stemming as a mid-way process between domestic subcontractors appointed at the contractor's discretion and those nominated by the employer. For IC 2016 to be effective, named subcontractors are involved with contractors during the tender process; in essence, an award of contract is the continuation of negotiations without the employer having a contractual link to any appointed subcontractor. The process involves contractors issuing an invitation to tender to named subcontractors using JCT form ICSub/NAM/IT, which gives details of the project and states the nature of the subcontract works. Thereafter, each subcontractor must issue any quotation on JCT form ICSub/NAM/T addressed to the employer, quoting a price for the works, with copies issued to each competing contractor including information required to comply with items T1 to T5 of the form. These items are associated with programming, contractor attendances, price fluctuations, dayworks and the installation of the subcontract works into the main contract works. The successful main contractor must appoint each named subcontractor in a timely manner, which is affirmed under clause 3.7 that compels the contractor to award such subcontract packages no later than 21 days after the contractor executes the main contract agreement.

3.6.2.5 Design-and-Build Contract (DB 2016)

With this form, the employer transfers responsibility of completing the design to the contractor who also agrees to carry out the construction works and deliver the scheme. DB 2016 can be used on projects where detailed provisions are known, with the employer having the responsibility for providing documents outlining the requirements that usually stem from the client's brief and concept/schematic design. The form has many of the mannerisms of the JCT family of contracts and elects for the contractor to be paid a lump-sum price progressively with interim stage or periodic payments, which is usual with a design-and-build arrangement.

3.6.2.6 Construction Management Contract (CM/A 2016)

This form is suitable for large-scale projects in terms of the floor area, financial value or complexity of a building, where a fixed price is not possible due to the anticipated duration of the project and the design is to be developed over time. The process involves the construction manager executing form CM/A with the employer with the manager agreeing to oversee the works until complete for a fee, during which time design appointments are arranged, usually by the employer. Under this appointment, and as the design unfolds, works are procured by the management contractor with the employer entering into separate contracts with trade contractors using form CM/TC. The creation of these agreements means there is no contractual link between the CM/A form and each CM/TC, with the construction manager's scope limited to administering the conditions of each CM/TC on behalf of the employer. The use of forms CM/A and CM/TC promotes collaborative working and sustainability as well as third party rights and the use of collateral warranties. Form CM/TC also includes provisions for advanced payments in return for security bonds, withholding of cash retention and payment for materials stored off site. As with other JCT family forms, CM/TC can be used as part of a framework agreement and elects for trade contractors to be procured with or without a bill of quantities (which would normally be prepared by the employer's quantity surveyor) on a lump-sum price broken down into parts, or as a schedule of rates for calculating interim payments.

3.6.2.7 Management Building Contract (MC 2016)

This form is similar to CM/A with the exception that the management contractor is appointed on the prime cost of the project plus a fee, and enters into individual contracts with trade contractors known as works contractors to procure the works. Works contractors are engaged on form MCWC/A and MCWC/C, that is, Articles and Conditions of a Works Contract Agreement that are only effective under an MC appointment. The Works forms of contract can also be used where the employer or works contractor provides design input, which can be in full or in part. An employer can also elect to use form MCWC/E if wishing to have a direct contractual relationship with a works contractor when carrying out services under this type of arrangement.

3.6.2.8   Measured Term Contract (MTC 2016)

The MTC form of contract is suitable where an employer is in the business of maintaining and/or improving buildings from time to time, and requires a single-stage tender process. Under this arrangement, the employer appoints design consultants independently of the contractor, and may provide a bill of approximate quantities or unquantified schedule for pricing by competing contractors where the value is ascertained as the works proceed. The form promotes sustainability and collaboration, and is used by local authorities for maintenance of their properties with the contract administered by a local authority department or other outsourced qualified professional.

3.6.2.9   Prime Cost Building Contract (PCC 2016)

This form is used for reimbursement arrangements, such as cost-plus or target contracts where the financial value is not considered high and a fully developed design is unavailable yet a quick start on site is required. In administering the contract, the contractor is reimbursed the cost of the completed works, usually on a monthly basis, plus an agreed sum (or percentage of the prime cost) as a contribution to offsite overheads and profit, and may receive incentives for early completion or for achieving the target cost. The project may be administered by an architect/contract administrator, and usually requires the services of a client-side quantity surveyor. The form can be used with a framework agreement and, as per other JCT family forms of contract, has provisions that recognise collaborative working, sustainability, third party rights and the use of collateral warranties.

3.6.2.10   Repair and Maintenance Contract (RM 2016)

This form is suitable for a single-stage tender where the scope is for the maintenance and repair of a building(s), possibly with a limited design available, and the works are a singular occurrence that does not require the services of a contract administrator. It exists in two parts, the Tender (including invitation to tender and acceptance) and the Conditions.

3.6.2.11   Constructing Excellence Contract (CE 2016)

This is an innovative form of contract addressing the industrial wake-up call to change. It was formed in conjunction with Constructing Excellence, an organisation leading the efforts of the Latham and Egan Reports, which were influential reports produced for the British government during the 1990s. The former report is a review of procurement and contractual arrangements in the construction industry, and the latter a report that paved the way for a task force to advise on opportunities for the improvement of efficiency and quality of the industry's services and products, and to make the industry more responsive to the needs of its customers.

   The Construction Excellence form of contract promotes partnering where project participants wish to adopt collaborative and integrative working. It can be used where a contractor is appointed to construct only the works or is to provide design input in addition, or can be for a complete design-and-build service awarded as either a lump-sum price or target cost. The contract has an ancillary form (CE/P), which is a project team agreement where team members enter into a multi-party agreement under a pain/gain sharing relationship where the results of risk management strategies are proportionally distributed. The form is based upon the importance of honesty and the development of trust while valuing the functionality and characteristics of a building contract. The decision-making process revolves around collective open management, which is enforceable through the ancillary form CE/P. Risk management is created as a style within the agreement that includes risk allocation schedules created prior to executing the contract. Where a risk item cannot be placed solely with a party, it becomes a 'Relief Event', and in the event of an occurrence triggering an event, the impact is shared using the pain/gain theme.

3.6.2.12   Home Owner Contracts (HO)

These contracts are based on 2005 editions, revised 2015, and are suitable for anyone seeking the benefits and protection of a building contract when wishing to appoint a consultant to design a scheme and/or contractor to execute home renovations or new residential works. The HO concept is divided into a number of forms. Form HO/B is a consumer building contract suitable for small domestic works such as building alterations where the home owner/occupier does not appoint a consultant to administer the contract on his/her behalf. If wishing to appoint a consultant, form HO/C as a building contract is used instead with the consumer entering into a separate agreement with the consultant outlining the scope of services using form HO/CA. Where the HO/B or C forms are used, the price is a lump sum with payment made upon completion of the works or with progressive payments. Where works are minor, involving small-scale repairs and/or maintenance, the Home Repair and Maintenance Contract form HO/RM is more suitable, with reimbursement made to the contractor on an agreed lump-sum price or a cost-plus basis.

3.6.3   New Engineering Contract (NEC)

The NEC suite of contracts was created by the Institution of Civil Engineering (ICE) and has distinctive themes to that of the JCT forms and are suitable for civil engineering operations as well as construction works. In contrast to JCT, NEC also has a family of contracts to facilitate the application of project management principles and practices, as well as defining proper rules for the creation of effective and binding working relationships. NEC forms are suitable for procuring a diverse range of works, services and supplies spanning from major framework projects to minor works and the purchase of supplies and goods. Launched in 1993 as the 'New Engineering Contract', the third and current edition published as NEC3 in 2005 was updated and enlarged to 39 documents in 2013. The NEC3 suite of contracts provide major benefits for projects both nationally and internationally in terms of time certainty, cost savings and improved quality. At their core, NEC3 contracts represent the traditional principles of a formal instrument of agreement regarding each party's rights, obligations and risks, plus the benefit of mannerisms that emphasise sound project management principles and practices. Fundamental styles of the forms include the following.

- *Easy to read and workable*: The forms are worded in favour of contract administration and project management and have reduced legal jargon compared to other forms of contract that use Legal English, a legal style of writing lacking punctuation that can make a contract appear more suitable to the legal profession instead of project management. Due to the wording within the NEC forms, clauses resist cross-referencing to other clauses, which could otherwise cause confusion regarding the intentions and consequences of the terms and conditions of contract.

- *Broad in application*: NEC forms are used across a spectrum of building and civil engineering works applicable to the public and private sector, for example Crossrail, (Europe's largest project), a £14.5 billion scheme of new twin-bore rail tunnels, subsurface stations and associated works. Other construction projects procured under NEC3 include Heathrow Terminal 2 and 5C and the London 2012 Olympic velodrome. On an international level, the impressive list of achievements includes the construction of Halley VI research station to monitor the Earth's atmosphere in Antarctica, Al Raha Beach waterfront development (Abu Dhabi, United Arab Emirates) and the construction of Terminal 3 at New Delhi airport, India.
- *Catalyst for effective project management*: A characteristic of the NEC3 forms is the creation of guidance notes and contract flow charts which aid project management and set the basis for effective communication and collaboration. The use of diagrams as communication tools can sometimes be better than wording as the concept can dispel misunderstandings, leaving the text confined to contractual phrases, definitions and meanings. Phrases are project-management friendly, with examples including the following.

  1) 'Early warning procedure', which places responsibility on the parties to a contract to identify any problems that could impact time, cost and the performance of the works.
  2) 'Compensation events', which address methods of reimbursement, both in time and money. This is a substitution for the word variations as found with other forms of contract.
  3) 'Activity schedules', the milestone stage breakdown of a project price for payment purposes.
  4) 'Risk register', created to identify hazards and how they are managed regarding time and cost.

3.6.3.1 Procurement Options

Suitable for use with engineering and construction works, the NEC3 Engineering and Construction Contract is the core document from which Options A–F exist. These are summarised in the following.

- Option A: Lump-sum price contract with a priced activity schedule, suitable for design-and-build procurement.
- Option B: Priced contract with a bill of quantities.
- Option C: Target cost contract with priced activity schedules, applicable to projects where activities are negotiated or tendered as cost targets and adjusted once the price becomes firm, plus an incentive mechanism payable to the contractor.
- Option D: Target cost contract with a bill of quantities. Similar to Option C, except the costs of the works are assessed with the use of bills of quantities instead of activity schedules.

- Option E: Cost reimbursable contract. Contractor is paid the cost of the works plus a fee for offsite overheads and profit
- Option F: Management Contract. A fee-based contract where a manager is appointed to supervise and place subcontracts packages upon the approval of the employer. The manager is paid the cost of the works plus a fee and is responsible for reimbursing subcontractors.

In addition, secondary options clauses can be selected from:

- X1: price adjustment for inflation;
- X2: changes in law;
- X3: multiple currencies;
- X4: parent company guarantee;
- X5: sectional completion;
- X6: bonus for early completion;
- X7: delay damages;
- X12: partnering;
- X13: performance bond;
- X14: advanced payment to the contractor;
- X15: limitation of the contractor's design liability to reasonable skill and care;
- X16: retention;
- X17: low-performance damages;
- X18: limitation of liability;
- Y (UK) 2 and 3: statutory obligations for the Housing, Grants and Construction Regeneration Act 1996 and The Contracts (Rights of Third Parties) Act 1999; and
- Z: additional conditions of contract.

Note, options X8–11 and Y (UK) 1 are not used.

NEC also produces a series of subcontract forms to maintain an NEC theme throughout a project. Furthermore, it also publishes professional services contracts for anyone providing a service–only as well as a term-service agreement and a goods-only contract, all with guidance notes. Also available is a framework contract for use with suppliers and purchasers and an adjudication contract to address a method of settling disputes.

NEC forms of contract are in their infancy compared to the plethora of long-established standard forms with their success undeniable due to their application in high-profile projects. Protagonists may be critical however as the success of such forms can only be accurately tested with the outcome of case law, for which NEC has not been prominently exposed. This is unlike the established peer forms that have weathered the storm of courts with the passage of time, meaning contract selection by experienced employers may be a case of 'better the devil you know'. However, optimists and realists tend to welcome NEC forms and praise their success in the projects they have helped to deliver, and there appears to be no stopping the popularity of this conventional suite of contracts.

3.6.4 GC/Work Contracts

GC/Works Contracts are a large family of standard government forms covering procurement arrangements for use on public projects, authored by the Property Advisers for the Civil Estate (PACE) and published by the Stationary Office. The suite of GC Works Contracts is still available, but is no longer being updated by the government (which is moving to NEC3). A summary of the forms is provided in below:

- GC/Works/1: With Quantities (1998)
- GC/Works/1: Without Quantities (1998)
- GC/Works/1: Single Stage Design and Build (1998)
- GC/Works/1: Two-Stage Design and Build Version (1999)
- GC/Works/1: With Quantities Construction Management Trade Contract (1999)
- GC/Works/1: Without Quantities Construction Management Trade Contract (1999)
- GC/Works/2: Contract for Building and Civil Engineering Minor Works (1998)
- GC/Works/3: Contract for Mechanical and Electrical Engineering Works (1998)
- GC/Works/4: Contract for Building, Civil engineering, Mechanical and Electrical Small Works (1998)
- GC/Works/5: General Conditions for the Appointment of Consultants: Framework Agreement (1998 and 1999)
- GC/Works/6: General Conditions of Contract for a Daywork Term Contract (1999)
- GC/Works/7: General Conditions of Contract for Measured Term Contracts (1999)
- GC/Works/8: General Conditions of Contract for a Specialist Term Contract for Maintenance and Equipment (1999)
- GC/Works/9: General Conditions of Contract for Operation, Repair and Maintenance and Electrical Plant, Equipment and Installation (1999)
- GC/Works/10: General Conditions of Contract for Facilities Management Contract (2000).

In addition to the above standard forms for subcontract agreements are published, together with model forms and guidance notes.

3.6.5 Association of Consultant Architects (ACA)

Founded in 1973, the ACA is a UK professional body representing architects in private practice ranging from one-person firms to large practices engaging numerous professionals and technical staff. ACA's principal aims are to encourage excellence in the quality of services that members provide, and to vigorously represent the aims and interests of architects in private practice on issues relating to the practice of architecture. ACA publishes standard forms of agreement for the engagement of architects, and also the ACA Form of Building Agreement 1982, available as a Third Edition 1998 (Revised 2003) for use as an agreement between a project client and contractor. In order to maintain a consistent theme for a project, the ACA publishes the ACA form of subcontract and ACA certificates for use with the ACA Building Agreement. ACA also produces a suite of partnering contracts prepared as a joint venture with the Association for Consultancy and Engineering (ACE), an organisation representing business interests of both its members and the consultancy and engineering industry in the United Kingdom. The range of partnering contracts is described below.

- *PPC2000 (Edition 2013, amended 2015 for CDM Regulations 2015)*: This project partnering contract is a flagship contract and the first of its kind in project partnering. Key characteristics of the form allow a client, as well as key specialists (e.g. design team, constructor (contractor), subcontractors and suppliers) to sign a single partnering contract in a non-adversarial manner. This 'one-line' approach can be extended to include addendum partners with the common goal of working collaboratively to achieve the desired results. To bind parties to the agreement, PPC 2000 includes a clause making it mandatory for a project to provide a core group of individuals that collectively agree to operate an 'early warning system' that involves undertaking regular reviews of project performance and progress. A mandatory clause also calls for the use of a Partnering Timetable to govern the contributions of partners, as well as monitoring the development of the design and the progression of supply chain procurement. The contract has been used extensively with central government projects in addition to local authority education, health care, transport, leisure and housing schemes. It is also available overseas as PPC International, omitting reference to CDM Regulations 2015 which only apply to British homeland projects.

- *SPC2000 (Edition 2008)*: This is an ancillary form to PPC2000 where the services of specialist subcontractors are required. Subcontractors are invited to join the project team culture at the earliest possible stage to evaluate risk and share the encouragement of collaborative working adopted from the philosophy of PPC2000. A Short Form Issue (2010) is also available, as well as a Short Form for SPC international use outside the UK.

- *TPC2005 (Edition 2008)*: This Term Partnering Contract is authored and published by the law firm Trowers & Hamlins LLP for ACA, and the first of its kind developed in response to the demand for a new approach to the procurement of term and service contract works.

- *STPC2005 (Edition 2010)*: The Standard Form of Specialist Contract for Term Partnering is used as a supplementary form for use with TPC2005 and later edition.

3.6.6 Institution of Civil Engineers (ICE)

The Institution of Civil Engineers (ICE) is an independent engineering institution based in the UK with members across the globe, that 'strives to promote the progress of civil engineering'. One of its acclaims is the production of a family of established forms of engineering contracts. The most dominant form is the ICE Conditions of Contract (CoC), published by Thomas Telford on behalf of the Institution of Civil Engineers (ICE), the Association for Consultancy and Engineering (ACE) and the Civil Engineering Contractors Association (CECA). The first edition was published in 1945 with the seventh and final edition published in 2001. When the seventh edition was being prepared, ICE was pressurised to withdraw its support for the CoC in favour of the established NEC forms. Subsequently, the ICE withdrew the CoC from August 2011 following the ICE council's decision to endorse the NEC3 Suite of Contracts. The CoC was rebranded in a suite of contracts as the Infrastructure Conditions of Contract (ICC), which is fully managed by the ACE and CECA following transfer of ownership from the ICE. The ICE, ACE and CECA continue to hold reference copies of the last published version of the ICE CoC, but no longer support it or offer it for sale.

3.6.7 Institution of Chemical Engineers (IChemE)

The Institution of Chemical Engineers (IChemE) is a global professional membership organisation for chemical engineering professionals and anyone involved with the processing industries. As there is a need for building and engineering works within this industry, the organisation has established a family of contracts for use in the processing industries. The contracts have a performance-based manifesto aimed at achieving the objectives of a project for civil engineering works and the construction of buildings to facilitate oil, gas, pharmaceuticals, testing, biochemical production, fluid separation, food processing and nuclear technology requirements. IChemE forms of contract are produced by a committee of senior practitioners with experience in all areas of the international processing industries, including operations and management, contracting, consultancy and construction law. For identification purposes the forms are coloured differently, and a list of current contracts within the UK for construction and civil engineering works is provided below:

- IChemE (Red) Form of Contract Lump Sum (5th Edition 2013);
- IChemE (Green) Reimbursable (4th Edition 2013);
- IChemE (Burgundy) Target Cost (2nd Edition 2013);
- IChemE (Orange) Minor Works (2nd Edition 2003);
- IChemE (Yellow) Subcontracts (4th Edition 2013); and
- IChemE (Brown) Subcontract for Civil Engineering Works (3rd Edition 2013).

In response to the international growth of the chemical industry, IChemE also produces the following contracts for use outside the UK, all as a 1st edition (2007):

- Lump Sum agreement, International Red Book;
- Reimbursable agreement, International Green Book;
- Target cost agreement, International Burgundy Book; and
- Subcontract agreement, International Yellow Book.

For use with dispute resolution seen as an alternative to litigation, IChemE produces:

- IChemE (Pink) Arbitration Rules (4th Edition 2005);
- IChemE (White) Rules for Expert Determination (4th Edition 2005);
- IChemE (Grey) Adjudication Rules (3rd Edition 2004); and
- IChemE (Beige) Rules for Dispute Boards (1st Edition 2005).

With such an array of forms available, IChemE provides training in contract law and contract management for contract users to provide insight into how the forms help deliver successful projects and develop supplier relationships.

3.6.8 Fédération Internationale des Ingénieurs-Conseils (FIDIC)

Translated from French into English as International Federation of Consulting Engineers, FIDIC produces standard forms of contract for civil engineering and construction works for use throughout the world, which are seen as forms representing international standards. Founded in 1913 by Belgian, Swiss and French authorities, with its head office currently in Geneva, FIDIC aims to represent the global consulting engineering industry in defining the conditions of contracts that involve building and civil engineering works. FIDIC carried out a major overhaul of their rainbow contracts, that is Red, Yellow, Silver and Green, in 1999, created new additional forms in the 2000s and overhauled some of the rainbow suite in 2017 with a view for the change process to continue in the future. An outline of these forms of contract for creating a project-client–contractor working relationship for the procurement of works is discussed below.

- *Conditions of Contract for Construction*, 1st Edition (1999 Red Book): This form is recommended for building and engineering works where the client has a design team and tenders the works to competing contractors in accordance with a set of tender documents in a traditional manner. The contractor may be requested to provide design input into aspects of the works such as civil engineering and mechanical, electrical and plumbing (MEP) building services. A separate ancillary form for use with the form, entitled Conditions of Subcontract for Construction 1st Edition 2011, is also available.
- *Conditions of Contract for Plant and Design-Build*, 2nd Edition (2017 Yellow Book): Suitable for large-sized projects in terms of financial value and duration, under this contract the contractor carries out a design-and-construct service for MEP services and/or civil works in accordance with the employer's requirements. In the process, the client transfers risk to the contractor for time and costs with the design consultants and MEP and/or civil contractors procured by the contractor.
- *Conditions of Contract for EPC/Turnkey Projects*, 1st Edition (1999 Silver Book): This engineering, procurement and construction contract outlines the party's obligations for creating an end-product that becomes available at the 'turn of a key'. It applies to the likes of roadworks (and other types of infrastructure works), power plants and factory construction where a single entity takes a dominant role for the development of the design, site management and the control and delivery of a fully equipped facility. Under this arrangement, the employer has a less active role, for example fulfilling basic obligations such as making interim payments in a timely manner. The term must not be confused with a turnkey project under the British system, which is the engagement of a contractor to carry out works using a contractor's prototype design with little or no design changes made.
- *Short Form of Contract*, 1st Edition (1999 Green Book): This form is suitable for both building and civil engineering works where the client or contractor is responsible for procuring the design. The form is intended for use on any project of a straightforward nature that may be modular or repetitive and is not considered high value in terms of finance and duration of the works.
- *Conditions of Contract for Design, Build and Operate Projects*, 1st Edition (2008): For use where an investor(s) designs, builds and operates an asset. The investor(s) may be a consortium or part of a joint venture agreement, with the form of contract applicable to a facility where the operational period does not exceed 20 years. Under this arrangement, the project must be new works and not a refurbishment or upgrade of an existing asset, with the contractor not accountable for the overall business success of the scheme when complying with the terms and conditions of the construction contract. The form includes a number of sample forms to enable parties to understand the implications of the agreement, and a series of flow charts addressing how parties to the contract deal with the settlement of disputes, contractor's claims, employer representative's determinations, payment provisions, design/construction/operational service periods and the timely commencement of the building commissioning.
- *MDB Harmonised Construction Contract*, 3rd Edition (2010): Multilateral Development Banks (MDBs) are institutions providing financial and professional advice to governments and project stakeholders for social and economic growth in developing countries. When participating in a project, an MDB becomes a

stakeholder which it financially underpins, and in the process is usually party to one of the FIDIC forms of contract. However, in order to make the contract workable, MDBs often require clauses to be edited, reworded or for the contract to be issued with new or appended clauses to make them acceptable for the culture of a project involving an MDB. This can hinder the administrative process, for which the solution from FIDIC was to create the Harmonised Construction Contract, drafted in a manner to reciprocate the standard requirements of projects where MDBs are stakeholders.

- *Form of Contract for Dredging and Reclamation Work*, 2nd Edition (2016): This form is prepared in consultation with the International Association of Dredging Companies (IADC) with the aim of producing a straightforward document that includes the necessary provisions for a variety of administration requirements associated with dredging, reclamation and ancillary works.

FIDIC also produces business practice information for consulting engineers on the management of risk, project sustainability, environmental issues, quality management and integrity compliance. This is in addition to dispute resolution techniques, insurance, capacity building, the transfer of technology, law and other business issues. Guides are also available focusing on quality-based selection including procurement and tendering procedures, consultant selection (White Book 2017, 5th Edition), quality of construction and other documents dealing the appointment of consulting engineers for projects.

3.6.9   Edited and Bespoke Forms of Contract

The advantages of standard forms of contract are that they are drafted by reputable sources, readily available and structured in a manner so they can address each and every occurrence while the agreement is active. Disadvantages include their generic structure, which may not provide complete solutions, and that some clauses may be irrelevant or are unworkable for the parties wishing to conclude an agreement. One solution to this is to edit the forms to the satisfaction of the parties and/or provide supplementary conditions or schedules that to be effective must be legal and not contradict legislation. For this reason, a contracts manager should consult a lawyer for legal advice on suggested changes because if unintentional illegal editing is included in an executed contract, the contract as a whole could be struck down by a court of law if a court is requested to determine the outcome of a dispute that in the process exposes the illegality. Furthermore, any changes must be agreed by the parties prior to executing the contract, which to be effective must be initialled in ink by each party. Where a clause serves no purpose it could be left in place, as in effect it is redundant and would not be referred to in any situation.

The parties to a contract may wish to change clauses of a standard form or append new so that it becomes either contractor or employer friendly, thus clarifying to whom the risk rests. This must not be confused with biased changes, as in general such changes would defeat the purpose of the spirit of an agreement, which is to create a productive working relationship. To demonstrate, below is an extract from the FIDIC Conditions of Contract for Construction 1st Edition (1999 Red Book) regarding timing for the issue of a project final account, which FIDIC refers to as a Statement at Completion. With the contractor-friendly version, the parties may in an upfront manner agree that the timeframe is too short because of the complexity and duration of the project and edit the term accordingly. By contrast, the timeframe written into an employer-friendly contract can be reduced if the construction project is of a short duration. Naturally, in each scenario, only one edit can apply.

- *Standard term* (Clause 14.10): 'Statement at Completion: Within 84 days after receiving the Taking-Over Certificate for the Works, the Contractor shall submit to the Engineer six copies of a Statement at completion with supporting documents…'
- *Contractor-friendly term* (Clause 14.10): 'Statement at Completion: Within ~~84~~ 126 days after receiving the Taking-Over Certificate for the Works, the Contractor shall submit to the Engineer six copies of a Statement at completion with supporting documents…'
- *Employer-friendly term* (Clause 14.10): 'Statement at Completion: Within ~~84~~ 42 days after receiving the Taking-Over Certificate for the Works, the Contractor shall submit to the Engineer six copies of a Statement at completion with supporting documents …'

Where the parties to a contract elect to include numerous changes leaving the original form with less prominence, or the nature of the project is unique, it may be practical to create a bespoke form of contract instead of adopting a standard form. A bespoke form of contract is a uniquely prepared formal instrument of agreement usually drafted by the project employer. For the contractor's quantity surveyor inexperienced with bespoke forms of contract, the contents may take time to understand as they may clash with education and experience gained from the use of standard forms of contract. Clauses written in bespoke forms vary from the standard forms and may include the following.

- A catch-all clause (a broad term used to describe a clause in a contract with an overarching and onerous effect) where the contractor is deemed to be satisfied with the quality of the contract design and documentation and has an understanding of the contract with any inconsistencies rectified at the contractor's own expense.
- The time barring of claims for an extension of time to the project end date by the contractor and/or the reimbursement for associated loss and expense. In the event of a delay giving rise to a claim by the contractor, an expressed term may state that the notification must be served in writing and within a stated timeframe after the event giving rise to the claim and if submitted late, the conduct may be grounds for rejection.
- A contractor's responsibility to coordinate the design on a construct-only contract and advise the architect within a stated timeframe of any inconsistencies in the design and documentation received together with associated costs. Here, a clause may state that if a stated period lapses from the date the contractor receives the information with the contractor not issuing any advice, the employer may assume there are no cost implications with issue of the information considered an instruction to proceed. The inclusion of such a clause is to ensure the contractor is proactive with design management, especially if designing discrete parts of the works.
- A responsibility of the contractor to audit the supply of design and documentation, and act upon any errors or omissions whether or not the contractor produces them. Here, a clause may express a requirement for the contractor to check 'for

construction' information, and at its own cost carry out alterations or remedial works caused by the error. This clause would permit a contractor to amend drawings without in anyway reducing the functional performance of the component.

- Expressed exclusion for the payment of unfixed materials
- In the event of litigation, the law applying in the country where the project exists is the rule of law. If a project is in country 'A' and the employer is based in country 'A, B or C' and the contractor is based in country 'A, B or C', the effective courts and their rulings will be in country 'A'.

Advantages of bespoke forms to a contractor include the following.

- They are suitable when the contractor and employer have repeat business with the form standing the test of time with their working relationship.
- They may have specific clauses suitable for a project that are otherwise vague or missing in a standard form (e.g. a construction contract part of a PFI scheme with specific clauses worded to comply with the requirements of the consortium).
- Certain forms may have been tested by the courts with clauses that are enforceable.
- The contractor does not absorb the cost of producing the form.

Disadvantages to the contractor include the following.

- They generally disadvantage the contractor because clauses favour the party requesting the form (i.e. the employer).
- The contractor may need to appoint a lawyer to review the clauses and advise on any pitfalls, which upon the contractor's objection of the form may require the employer to edit certain parts; this can take time to arrange.
- There may be illegal clauses included which the parties are unaware of until tested by the courts.
- The quantity surveyor will spend additional time reviewing the form than for a standard contract which may be more familiar.

As with a standard form of contract, if the contractor's quantity surveyor is involved in a project using a bespoke form, a review as early as possible will aid an understanding of how the contract is structured and to be administered, which will save time once the project is flourishing.

3.7 Remedies for Breach of Contract

Once a contract exists, the parties to the agreement must not lose sight of their arrangement. In the event of a breach by either party giving rise to a formal dispute, the dispute resolution process (or processes) is/are usually stated in the contract. Formal disputes in the construction industry usually revolve around time, money or both, and it is important for the dispute resolution methods stated in the contract to be appropriate to the size of the project and what is at stake in the event of a dispute. Litigation is one such method that deals with dispute resolution on legal grounds, and involves a court of law applying a remedy using legal precedent and citing statutory compliance using common law principles. The use of litigation does not require to be mentioned in a contract because access to the legal system is a statutory right, and is usually included in the contract as a matter of course. The judicial outcome with litigation is prolonged and decided on a win/lose scenario, which may be an undesirable outcome for the parties with the dispute itself possibly a waste of the legal system's time. For this reason, methods of resolving disputes other than litigation are available, known as alternative dispute resolution (ADR), with the types usually stated in the contract. Where plausible, ADR is the most suitable pathway for resolving disputes in the construction industry as each dispute is heard on technical grounds more than the legalities, with possibly the reverse in effect with litigation. ADR in a construction contract does not apply to a situation where an independent person or persons can act as a character, expert or judicial witness in a judicial or semi-judicial process such as industrial tribunals or other settlement processes. Furthermore, the term is not used with reference to a situation where a party issues a complaint to a professional body regarding the conduct of one of its members.

3.7.1 Litigation

Once a court is satisfied a contract is legal, in their proceedings (a term used to describe the invoking of power to enforce law) a courts' ruling can be made based on the terms and conditions of the contract even if appearing one-sided, unfair or unreasonable. This can involve construing certain terms and conditions that are vague or ambiguous where the court decides what it considers is the meaning or intention of such terms and conditions. A court may also make a ruling based on statutory provisions written in legislation. One such piece of legislation applicable to the construction industry in the UK is the Supply of Goods and Services Act 1982 (clause 29), which requires traders to provide services to a proper standard of workmanship. Where this statute does not apply, and depending on the case, remedy may be decided upon a case's relevance to another industry's case if there is contrast. This could be demonstrated between the construction and shipping industries if the findings relate to determined cases where there was a failure to deliver a product on time if one party in a new case is seeking damages (or financial compensation) for late delivery. For example, in a shipping industry case dealing with a scenario where a cargo ship laden with goods sets sail to travel halfway around the world but is delayed because of bad weather, the judicial outcome of that case may influence the decision of a new case involving a delayed construction project. Here, a ruling may decide if the late delivery is vital or not.

The lowest form of a civil court in England and Wales is a tribunal court, followed by the County Court. Whether these courts become involved in a dispute under a construction contract depends on the value at stake and the legal advice received. If the case is complex, hearings can be through the High Court of Justice and, at the extreme, the Supreme Court (replaced in 2009 from the Appellate Committee of the House of Lords) of the UK. Variances of the civil court arrangements applicable to England and Wales exist in Scotland and in Northern Ireland.

Advantages of litigation include:

- judges of disputes are experts of the law;
- a judge is impartial to the construction industry and will refer to case law and statute for guidance;
- jurisdiction is on a win/lose outcome which is binding unless overturned on appeal; and
- the legal system is reputable for providing a fair outcome using natural justice.

Disadvantages include:

- expense, with the legal proceedings possibly taking a long time to get to court;
- judges may not relate to technical aspects of construction issues and will need to rely on disputing parties calling on character and/or expert witnesses to aid their case and assist the court with its decision, adding time and cost to the proceedings;
- the case is public and may involve the press which the parties may wish to avoid because corporate profiles could be tarnished irrespective of the judicial outcome; and
- unlike other dispute resolution methods, it is a win or lose jurisdiction with nothing in between.

In the UK, legal proceedings involving construction cases may be referred by counsel to the Technology and Construction Court (TCC). This is a subdivision of the Queen's Bench Division and part of the High Court of Justice which expedites litigation procedures. As the title suggests, TCC litigates on technological issues involving disputes in the information technology, engineering and construction industries. The Ministry of Justice issues Civil Procedures Rules (CPR); CPR Part 60 addresses the type of claims that can be managed and run by the TCC applicable to claims by and against builders and consultants, as well as challenges to decisions made by other dispute resolution methods. The rulings apply to England and Wales with courts established in various locations. Cases determined by the TCC and other courts in relation to British and Irish case law as well as European Union case law, Law Commission reports and details of legislation can be found on the British and Irish Legal Information Institute's website (http://www.bailii.org) and, in particular with TCC decisions (http://www.bailii.org/ew/cases/EWHC/TCC).

3.7.2   Alternative Dispute Resolution (ADR)

With the exception of negotiation, all forms of ADR involve a neutral third party. A neutral third party is a person or persons skilled in the appropriate field of the selected ADR with the necessary credentials to understand the industrial complexity surrounding a dispute. The contract may nominate a body tasked with making a recommendation (e.g. RICS, RIBA, etc.) for a neutral third party and type of ADR. Where a body is not nominated, the parties should select a neutral third party with a solid reputation in the industry that is impartial, acceptable to each party and has no conflict of interest with the parties.

When ADR is included in a contract it may state more than one method, which is usually sequential in the event that the parties fail to reach an amicable outcome with a first round of ADR. Should this occur with further intervention necessary, the parties may commence a further round using another type of ADR. By adopting the sequence of the contract, the parties must restrict themselves to the methods stated and not any other because seeking another and making a subsequent appointment would amount to breach of contract.

Advantages of ADR to litigation include:

- it is generally less expensive;
- the process and outcome can be rapid, although this depends on the dispute and type of ADR;
- the arbiter is usually experienced in construction matters and will relate to the facts;
- the hearing is private; and
- there are options available towards the outcome that seek win/win and win/lose rulings.

Disadvantages include:

- the outcome may overlook common law matters; and
- if a win/lose outcome is sourced, choices of ADR are restricted and are usually the most expensive.

Appointment of a neutral third party is with a third party agreement that outlines the scope of services, including maintaining confidentiality throughout the process and the outcome. It is preferably arranged as a tripartite arrangement and executed before the process commences. A method of reimbursing the cost of the service needs clarifying in the agreement, which may be separate because of affordability and the amount in question, or may be equally shared.

3.7.3 Negotiation

This is the only ADR method that excludes the involvement of a neutral third party, and is a low-cost initial process and usually the first method of ADR where disputing parties are encouraged to resolve the situation themselves. Negotiation is considered most suitable when the parties wish to have an ongoing business relationship beyond the dispute, where each party has similar bargaining capacity to deal with the process and outcome. The triggering of negotiation involves one disputing party serving notice on the other, advising of an intention to appoint its senior executives to negotiate the dispute, and invites the other party to do the same. The notice may seek a timeframe for a response, and a date or length of time thereafter for the negotiations to commence and where they are to be held. The intention of the timeframe is to permit senior executives of each disputing party to become acquainted with the facts surrounding the dispute, ensuring they are fully aware of these when negotiations commence. The duration of the negotiations is not usually stated in the notice however and, as with any dispute, the objective should be to make this period as short as possible in the interest of keeping the working relationship intact. Should the negotiations fail or be inconclusive, the parties may agree at some point to abandon the process and appoint another neutral third party to follow a different type of ADR as required by the contract.

3.7.4 Mediation

Mediation is widely adopted by standard forms of contract and suitable when disputing parties consider intervention will aid the resolution process. It is an arrangement where disputing parties appoint a mediator to seek a win/win situation that may stem from negotiation or initially after a dispute arises. The intention of the process is for the parties to:

- stop, look and listen;
- reflect on the dispute and recognise the harm it may have already done to their business relationship;
- consider what an unresolved dispute may do to their future business relationship; and
- seek a win/win scenario.

Mediation has probably more human influence than any other type of dispute resolution method involving intervention because it focuses on issues the party's value from each other, and what may be lost if progressing to a more intrusive method of ADR.

The process is developed by a mediator who must first understand each party's position in an attempt to clearly identify the dispute before seeking to create an amicable agreement. The mediator first holds separate meetings with the parties and expands the dialogue to combined meetings once the facts become exposed and a level of trust has developed. If money is involved, the parties may agree on a preliminary settlement based upon the later performance of one or both parties. If the dispute involves a project under construction with both parties required to perform under the contract, the mediator will advise on the implications of any agreement. If the agreement includes the recommendation of a set of terms and conditions, it may require one party's lawyer to draft a preliminary agreement to include the terms and conditions. In summary, a mediator provides a judgment which is not binding and will advise the parties on pitfalls they may encounter should they wish to proceed to another type of ADR.

Mediation is a non-binding process with any written communication towards an agreement cited in the documents as being issued 'without prejudice'. This means any subsequent dealings involving another neutral third party that seeks to make a determination will not be presented with the earlier negotiations for consideration. If one party issues a copy document at a subsequent determination process that includes the reference it will not be treated as evidence, and an experienced neutral third party will dismiss its relevance because of the reference.

Mediation has advantages to other types of ADR as it is of medium cost, not usually prolonged and suitable for uncomplicated disputes. A disadvantage is the loss of some legal and contractual provisions that recognise the basis of contract law and the putting aside of the formal contract that may have taken considerable time to negotiate and conclude.

3.7.5 Conciliation

Conciliation is similar to mediation with the exception the conciliator conducting the procedure is more coercive and will give opinions for the consideration of the parties. The personality and mannerisms of a conciliator does not play a part in the process with the process itself creating the distinction. Whereas a mediator may prompt disputing parties to suggest ideas towards settling a dispute, a conciliation officer will make suggestions and/or give contrary opinions to the ideas put forward. This is carried out during the facilitation and evaluation stages in an attempt to lead the parties to a joint resolution. As a result, a conciliation officer's recommendation may not be a surprise, whereas a mediator may present options with innovation leaving the parties to ponder over items pending a further round of mediation. Conciliation is understood to have a more direct approach when dealing with a dispute, with officers usually appointed because of their expertise in specific matters for a certain type of project or works.

3.7.6 Dispute Boards

A Dispute Board (DB) typically comprises three persons engaged prior to the commencement of construction works that are appointed to perform an overview of the execution of a project and the parties' performance under the contract. The objective of a DB is to assist contracting parties to avoid disputes and, where unavoidable, assist them with a cost-effective dispute resolution process. To be effective, members of a DB must be relevantly experienced in the type of project under construction, have a thorough understanding of contractual issues and be acceptable to the contracting parties. The DB provides regular and ongoing forums for discussion at technical and contractual level akin to a project, which is necessary to help avoid disputes between the parties. Moreover, the establishment of a DB from inception means DB members become an integral part of the project team and are trusted to be fair and impartial in their opinions with the role not involved in decision making that could impact the performance of the project. The selection process involves the employer nominating one member and the contractor nominating a second, with each party having the right of reasonable objection over the other party's selection. A chairman is also usually nominated as DB executive to the agreement of the parties.

The inclusion of a DB on a construction project must be stated in the contract documents, with the DB provided regular progress reports during the course of the works and duration of the contract. This permits the DB members to remain familiar with the contract, project and participants from the beginning and to have up-to-date knowledge of relevant issues while the project and contract is active. Members of the DB make regular site visits and inspections with representatives of the employer and contractor present, and are briefed on site progress and any potential problems. The aim of the DB's site visits/meetings is to facilitate communication between the contracting parties and encourage the resolution of any issues of conflict at site level to prevent them manifesting into actual disputes.

Either contracting party has the right to refer a dispute to the DB and, following a referral, the DB will hold a hearing. This will involve questioning witnesses and evaluating any requested design and/or documentation in order to provide a reasoned determination in the shortest possible time. The contractual effect of the DB's determination can differ, and depends on whether the DB is appointed as a Dispute Resolution Board (DRB), formed from a USA model, or a Dispute Adjudication Board (DAB) formed from the DRB model. A determination by a DRB is a recommendation to the parties for the resolution of a dispute, but is not binding unless the parties agree to implement the decision. However, a DRB's recommendation would not usually be challenged because of the existence of a DB on a project because challenging the decision may defeat the purpose of a DB on a project in the first place. The Dispute Resolution Board Foundation (DRBF) endorses the use of the DRB recommendation model, as it is based on a track record of success and considered an effective, speedy and cost-effective method of ADR. By contrast, determination by a DAB is contractually binding on the parties and may be formally disputed, usually through arbitration or litigation. If a party wishes to dispute a DAB's determination the contract will usually specify a time limit for the issue of a notice, which the parties must recognise. In the absence of such notice, the DAB's decision is contractually binding even if one party issues a formal notice for the final resolution to be determined by arbitration or litigation, which may have to be deferred until the end of the project. Until such time of a resolution, the DAB's decision is binding on the parties.

3.7.7 Early Neutral Evaluation

When disputing parties elect to commence a win/lose judicial process such as litigation, the outcome is uncertain. However, the tentative result can be advised to the parties through early neutral evaluation, which to be instrumental must be carried out prior to commencement of the proceedings and not while they are in progress. The person appointed to carry out the evaluation must be a legal expert and experienced in matters relating to judicial decisions. Furthermore, the person must be competent and qualified to provide an impartial view, focusing on the legal aspects of the dispute towards a resolution and, to a lesser degree, with the industrial context. The process involves a hearing of limited scope once the dispute has crystallised with the facts known, and is held in a tribunal court with an evaluation of probable outcome issued for advice that is neither binding nor a guarantee. The evaluation can be a good investment for the disputing parties, as it may prompt each to settle their differences prior to commencing the dispute resolution pathway and 'cut their losses', which is an advantage of the process. However, if a pending case involves a large amount of documentation, it will require time and effort to compile the evidence for the evaluator's consideration, and is a obvious disadvantage of the process. The early neutral evaluation concept may be selected as part of a negotiation procedure in the early stages of ADR which may strengthen the parties' understanding of a potential outcome, yet is unlikely to be included in a contract as a type of ADR as the procedure is for advice only.

3.7.8 Expert Determination

Expert determination involves the appointment of an impartial expert to assess a dispute, who applies a determination based upon facts using legal and industrial means. A characteristic of this type of ADR is that it is binding; if the parties elect for this method, they can only do so if it is part of the contract. Once commencing the pathway, the expert and/or disputing parties may request a hearing to outline the facts surrounding the dispute. A hearing will involve the expert considering each party's version of the dispute that is carried out in a non-intrusive manner without cross-examination, as the intention is for the expert to fully understand the nature of the dispute towards making a decision. The expert is appointed by a nominating body and is a contractual adjudicator with legal training, credentials and experience appropriate to the technical matters surrounding the dispute; and as with other forms of ADR, the appointment must be acceptable to the parties. Expert determination can be a speedy form of ADR, and is one reason for selecting and including the process in a contract as it can save disputing parties time and money, especially if involving a live project where works are in progress when it can be accessed at any time. This route is advantageous to disputing parties when:

- answers to questions raised upon points of law are straightforward and require little or no research;
- the contract between disputing parties is comprehensive; and
- the parties have confidence in the expert's ability to understand the surrounding facts of the dispute.

As the outcome is final, any attempt by a losing party to overturn a determination will take time, effort and money, and is a disadvantage of the process. Moreover, any appeal can only usually be initiated if the losing party considers aspects of contract law have not been considered, for example, the determination did not enforce certain expressed terms because of their implications elsewhere in the contract.

Expert determination is not to be confused with the services of an expert witness, when disputing parties appoint an expert to provide an independent report addressed to an arbitration or litigation court to aid a decision on the outcome of a dispute. Neither should the process be confused with an expert appraisal, which is an impartial independent assessment prepared to assist a dispute resolution process that focuses on technical aspects of a specific nature which is issued for advice. However, the appraisal may be supplied as evidence for any type of ADR including expert determination, for example if a contractor is in dispute with a project client regarding defects to a completed building. The appraisal can be used as evidence in this case, which is usually restricted to technical matters with any legal comments or experience considered for reference only.

3.7.9   Adjudication

Also referred to as statutory adjudication, this type of ADR empowers an independent adjudicator to issue a provisionally binding determination based upon facts and law as stated in the Housing Grants, Construction and Regeneration Act 1996 and the Local Democracy, Economic Development and Construction Act 2009. The former is omnibus legislation addressing issues associated with private sector housing grants, amendments to law in relation to construction contracts and regeneration schemes involving development grants. The latter is a piece of legislation placing a duty on local authorities to promote understanding of the functions and democratic arrangements of the authority among local people.

An adjudicator is appointed when legislated payment procedures are considered breached. Lodgement of an action is usually initiated by the party owed money, who becomes a claimant, and the party defending the right not to pay the full amount becoming the respondent. Grounds for a claimant activating adjudication occurs when the respondent does not pay a debt in time without communicating a reason(s) or the respondent rejects a payment request (in full or in part) that the claimant considers is in breach of legislation. Of all the types of ADR, adjudication is most likely to be the method a quantity surveyor could encounter on a construction project, as it is overarching legislation aimed at dealing with the flow of payments to and from the contractor.

Standard forms of main contract (and subcontract) may include a provision for the parties to a contract to nominate an adjudicator from a selective list of Adjudicator Nominating Bodies (ANBs). However, the parties can access an ANB to appoint an adjudicator if not written into a contract because adjudication is a statutory right. The appointment process commences when one party notifies the other of an intention to commence adjudication. Once the party giving notice makes contact with an adjudicator or an ANB, the adjudicator or ANB issues a notice to the parties within one calendar week of receipt of the request advising of an intention to either accept or decline the request. If accepting, the adjudicator then has four calendar weeks to make a summary judgment. During this time, the adjudicator seeks information from the claimant and respondent and gives each party the option to reply; if electing to do so, each party must copy the other with the information provided. The most widely used communication procedure is documentation only. However, and at the adjudicator's request, interviews can take place with the parties present and, where applicable, the assessment period can be extended by up to two calendar weeks subject to the party initiating the process agreeing. The adjudicator's decision is issued to each party, which is usually at the same time and provides reasons for arriving at the decision.

This type of ADR is advantageous to the disputing parties, as it is quick and relatively inexpensive at approximately 10% of the cost of litigation. A disadvantage is a possible lack of faith with justice if an adjudication applies rulings that appear to be shallow and weighted in favour of the adjudicator's instinct. Rulings may have flaws in their relation to law, which might be unknown to the disputing parties and adjudicator because of the vigorous timing imposed by legislation. When a claimant wins an adjudication, the loser may consider they have been ambushed by the system and could refer the matter to the appropriate ANB if an ANB appointed the adjudicator. If a losing party seeks to open a determination for scrutiny, the ANB may investigate the evidence or uphold the decision; this will take time, and may prove futile. However, as the determination is not finally binding, the loser can resort to arbitration or litigation to avoid making payment. The drawback here is that decision makers with arbitration and litigation may resist involvement with an appeal and may only become involved if it appears legislation has been overlooked. However, as the financial amounts included in adjudication determinations are usually minor in comparison with litigation and arbitration, it is unlikely such extremes of the appeal process would be sought.

3.7.10   Arbitration

When disputing parties seek to settle a conflict revolving around one or both parties' performance under a construction contract where a lot is at stake in terms of time, money or both, arbitration can be the most suitable ADR process to litigation. Arbitration is perhaps the oldest form of ADR which is regulated by legislation and used worldwide, and occurs when consenting parties agree to appoint a neutral third party to settle a dispute that is binding. It has an advantage over litigation in that solicitors, barristers and expert witnesses are chosen for their combined knowledge of law and technical expertise, meaning that dealings are balanced, or possibly weighted towards, industrial technicalities instead of legal issues.

The law applicable for arbitration on a global scale is the United Nations Commission on International Trade Law (UNCITRAL) Model Law on International Commercial Arbitration, drafted by the core legal body of the UN system in the field of international trade law. The UK is one of the few developed nations that does not adopt the UNCITRAL model law, with the governing act for arbitration in England, Wales and Northern Ireland being the Arbitration Act 1996 and in Scotland the Arbitration (Scotland) Act 2010. The creation of these acts steers disputing parties to choose either arbitration or litigation as a dispute resolution process and not both which, if contradicted by a clause in a contract, would result in the clause being struck down because of the inference of the acts. If a contract makes no reference to arbitration, it is still possible for disputing parties to refer their dispute to arbitration should they elect to do so. However, due to the costs and time involved, parties should seek legal advice and consider their options prior to commencing the route.

An arbitration ruling is called an award and involves a losing and winning party. Appeals are limited and, in the event of a party wishing to lodge an appeal, proceedings can only commence based on law. Thereafter, a case can only be litigated through the courts making it public in the process which may be contrary to the parties' desires. When requested to enforce an award because of the reluctance of a losing party, a court may only assess the arbitration procedures conducted during the hearing to ensure the act(s) have been adhered to. In general, courts have a high regard for arbitration because the process is regulated and continues to stand the test of time with fair determinations, which also eases the burden on the courts due to the number of cases lodged within the legal system at any time.

A distinction between arbitration and litigation is that arbitration is consensual and cannot be imposed by one party, whereas with litigation one party can serve notice on the other without warning. The parties to arbitration do not require legal representation and can represent themselves if they so wish. However, and in reality, disputing parties tend to see this as a brave step and usually resort to legal assistance instead. Arbitration can be referred to at any time during the construction phase of a project, which the parties can agree to leave until after the date of sectional or practical completion; that may also be a condition of contract. However, if a contractor is in dispute with a subcontractor without project client involvement, and in the interest of time (subject to the conditions of the agreement), arbitration proceedings can commence after completion of the trade works if the date of sectional or practical completion is considered too distant.

If a project is of a long duration and there is a dispute between the contractor and project client during the construction phase then, and subject to the terms and conditions of contract, arbitration proceedings can commence. During this time the contract is not vitiated and the parties are obliged to continue their performance under the contract, meaning works and payments continue as normal. If works are suspended because of a dispute with arbitration, the stated method of ADR in the contract that results in an award to the contractor, then the contractor should be able to recover costs together with an extension of time to the end date. On the flip side, if the client wins the award the contractor must recommence works and accelerate the programme at their own expense so the project can complete on time.

4

Project Commencement

4.1   The Project Team

A contractor will usually initiate a project when senior management of the business are satisfied an agreement to carry out the works exists with a client. This stage is reached after the formal execution of a contract, or upon receipt of a Letter of Intent or Letter of Acceptance from the client or client's agent (see Chapter 3, Section 3.6.1 for a discussion of these Letters). The recognition of an agreement triggers the contractor's pre-commencement period, which is the period of time before the contractor takes possession of the site to commence operations. During this time, the contractor will need to organise a project team to manage and deliver the scheme, which ideally comprises names and job titles stated in the contractor's tender, including the quantity surveyor. However, after submitting a tender, a contractor may substitute some or all of the names because of role changes, permanent departures from the company or the anticipated start date, which is usually discussed with the client before the start date on site.

Under a usual arrangement, the contractor's project manager will organise a pre-start meeting with the contractor's project team to act as an icebreaker, where personnel reacquaint themselves from a previous project or meet for the first time. A pre-start meeting is an excellent method for introducing team members as it promotes collaborative working and addresses key components of the project, including:

- the scope of works and duration of the project;
- client and consultant contact details;
- schedule of site accommodation and where the accommodation is to be located;
- specific issues regarding the project (e.g. site boundary lines, existing services, site access, tree protection orders from the local authority, condition of existing buildings, site hazards, etc.); and
- the availability of working designs and information.

Ideally, the meeting is held on site and includes a walk-over survey, which is beneficial as it permits personnel to acquire a feel for the physical presence of the project. Alternatively, the meeting may be held at the contractor's head office chaired by the project manager or member of senior management. At the meeting, the manager may issue a construction programme for advice to team members stating the project activities as well as the early work requirements, (e.g. site accommodation and any early works required by the programme). This may involve activities within the pre-commencement period which is usually for the contractor's benefit, such as the

Construction Quantity Surveying: A Practical Guide for the Contractor's QS, Second Edition. Donald Towey.
© 2018 John Wiley & Sons Ltd. Published 2018 by John Wiley & Sons Ltd.

demobilising of plant from one project for mobilising to the new project and works such as demolition or other site clearance operations. During the briefing, the project manager will also advise team members of the human resources required for managing the project. This is to enable an understanding of the management structure of the contractor's organisation, which also aids those that are new starters with the business, and is best demonstrated in a flow chart. Where applying, the flow chart should be the same as that included with the contractor's tender or any post-tender negotiations, as it demonstrates the hierarchy arrangement for the delivery of a scheme and how back-up resources from head office assist the project team. Figure 4.1 depicts a flow chart where the project team is semi-autonomous with the contractor's head office. In this scenario, head office resources comprise personnel that visit site to assist the project team and provide supplementary and supportive roles, while not having a direct role in the day-to-day running of the project.

A characteristic of this meeting is that it should be exactly as the name suggests, that is, a meeting before works start. This is because the date at which the contractor takes possession of the site is the date that triggers the start of the construction programme which is usually part of the agreement with the client and, once commenced, causes the pre-commencement period to cease.

After the pre-start meeting, it is customary for members of the client's team, including client's agent and project stakeholders, to become acquainted with the project team. This may be achieved by inviting parties to meet the project team at a mutual location such as the contractor's head office, or at a location preferred by the client/client's agent. This is usually organised by the project manager, and may be held at the client's/client's agent's offices or first formal site meeting once the site accommodation is established.

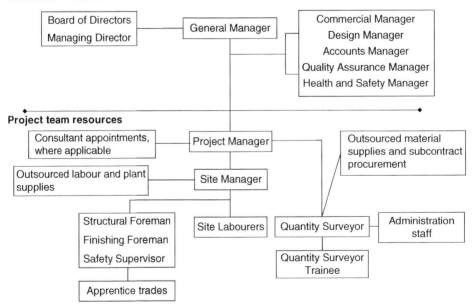

Figure 4.1 Project resources chart.

4.2   Pre-Construction Handover

At the pre-start meeting or at a date soon after, as part of pre-commencement activities, the contractor's estimator involved with the tender process will normally produce a set of lever-arch files comprising hard copy prints of the contract documents (where a formal contract has been executed) as well as design and documentation used to prepare the estimate. As part of the estimator's handover, files are issued to the project manager with the quantity surveyor usually provided copies for his/her own use. The contract documents could differ from the tender documents received as part of an invitation to tender, as they may include amendments after a tender is submitted (e.g. where value management is carried out), resulting in the design and documentation being altered that changes the tendered amount and becomes the contract sum, or schedule of rates if the price is not fixed. The estimator is the most suitable person in the contractor's organisation to compile these files, as he/she will be aware of any negotiations and changes since submitting the tender and will ensure the information supplied is that which forms the basis of the agreement.

It is important for the project team to refer to these as contract documents and not 'tender' or 'for construction' information, as they represent the basis of an agreement that is frozen in time and become the focal point from which any changes are assessed. For example, if architectural drawing number A100 revision 'A' is a tender drawing that underwent change to become revision 'B' with the contract price agreed upon that revision, this would make revision 'B' a contract drawing and is usually referred to in the contract. If a further and final issue is released as revision 'C', it becomes a construction release drawing; the basis of any change will then be the comparison between revision 'C' and 'B', and possibly a contract variation. The issue of contract design and documentation is for administrative purposes only; such information must not be considered suitable for construction purposes, as it is customary for design consultants to issue a set of working designs and documentation when they become available.

The number of design drawings and amount of documentation that form part of a contract can be considerable, with the contents depending on the project in terms of scope, finance and complexity. The information included in the hard copy handover files is also usually stored electronically in a site file created on the contractor's computer server, with a typical issue comprising the following:

- bill of quantities (if applicable);
- document/drawing register with a list of revision numbers that form a part of the contract;
- designs and documentation (may only be available electronically due to the size);
- construction programme;
- contractor's proposals for the works;
- copy of the executed contract (if available), usually with each page initialled;
- negotiation letters, forms and correspondence that form a part of the contract;
- tender clarifications and qualifications if not appended to the contract;
- any value engineering that is included in the contract;
- list of subcontractors named in the documents by the client; and
- contract sum breakdown.

In addition, the estimator may provide supplementary information used to prepare the estimate to aid the project team's understanding of financial allowances included in the contract sum. These are of no interest to the client and do not form part of the contract; they are primarily for the benefit of the project manager and quantity surveyor, and include:

- builder's quantities (instead of a client's bill of quantities);
- copies of subcontractors, material suppliers and plant hire companies quotations;
- copies of consultant's fee proposals (where applicable);
- analysis of received quotations and fee proposals;
- risk identification;
- cost plan allowances;
- cost coding structure;
- tender submission;
- information received from the client after tender submission;
- temporary works designs;
- minutes of meetings; and
- site visit reports.

As the contract issue is of a commercial nature it is for the project manager and quantity surveyors' use, meaning it is unnecessary to overwhelm the site manager and foremen with the details. The site manager and foremen's time is best spent driving the project instead of being side-tracked with commercial matters such as percentage allowances for material waste and details of supplier's quotations. Naturally, the site manager and foremen should show interest in commercial matters and budgets, and act in the interest of the project. However, it is the quantity surveyor and project manager that drive the commercial and contractual aspects of a project, and who jointly carry responsibility for their management.

4.3 Office and Site-Based Roles

Once a project is initiated by a construction company, senior management will usually decide if the duties of the contractor's quantity surveyor are to be carried out on site or from a head office. The choice is influenced by either the contractor's policy, project value and/or complexity of the works. However, the sector of the construction industry can also influence the choice. For example, national house builders operate from a number of regional offices around the country with each constructing possibly hundreds of dwellings per year across a number of projects within their region. Typically, each office employs a team of quantity surveyors to concurrently manage a number of projects of varying values with the cost of their employment part of each project's preliminaries. These businesses have established procedures in place and generally rely on income from speculative selling to the public, which makes the business free from contracting with corporate clients. However, on occasion a national house builder may enter into a formal contract with a housing association that would purchase properties from standard house type designs offered by the house builder. As a result there is seldom a main contract to administer, releasing some of the quantity surveyor's time to deal with cost management and the procurement of subcontract works. Quantity surveyors engaged by national house builders are aided by bonus surveyors that measure and value works completed by labour-only subcontractors (e.g. brick/block layers) and process their payments, usually on a weekly basis. The nature of contracting however is somewhat different as there is a main contract to administer. The small- to mid-tier-sized contracting businesses may adopt a policy of not basing a quantity surveyor on site and exclude the cost to employ from the preliminaries with the expense considered an off-site overhead and included in the contractor's margin. This will mean the contractor is competitive with tenders when pricing projects because the preliminaries charges are reduced. However, some clients may perceive the margin applied to cover offsite overheads and profit as high when part of a tender breakdown, and possibly query the amount with the contractor after a tender is submitted. The largest-sized contracting businesses have established offices to assist a project team for the running of a project, and typically include:

- a material buying department to purchase materials and hire plant;
- a plant department that stores, maintains, delivers and collects contractor-owned equipment;
- accounts department to process payments;
- IT department;
- human resources including quality control;
- occupational health and safety;
- legal department; and
- design managers who drive the consultant's criteria for design-and-build projects.

This is an integrated style of business where a head office is fundamental to the running of a project. Where this exists, the quantity surveyor may be a visitor to site and not permanently based on the project. However, this criterion is not set in stone and will vary between contractors and each project undertaken. When employed by this type of contractor, a qualified quantity surveyor may be capable of administering one or more projects concurrently of low value based from a head office. Alternatively, a quantity surveyor's role may be absorbed by a single project if it is complicated or the contractor is new in business, or becoming established in new sectors of the industry and going through the motions of introducing management systems. High-valued projects usually require one or more quantity surveyors to have a permanent site presence during the construction period and may receive assistance from support staff and junior surveyors on an *ad hoc* basis with the cost to employ each team member part of the project preliminaries.

4.4 Construction Programme

As a general rule, the project client gives possession of the site to the contractor on or before the start date stated in the contract. Typically, from this date the contractor commences operations and agrees to complete the works by the date of practical completion (sometimes referred to as the 'taking-over date', with the preferred title stated in the form of contract). The duration between the date the contractor takes possession of the site and the date of practical completion is the construction period from which working activities are planned in a construction programme. The programme is usually prepared by the contractor's project or planning manager, and describes each working activity by duration within the construction period. When assessing this period, and as part of a tender, the contractor can do either of the following:

- set a commencement date and sequence of activities that creates a date of practical completion and, in the process, defines the construction period; or
- take into account the client's preferred construction period and plot working activities to determine if they can be accommodated.

Once a construction period is agreed and included in a programme, it is common to include the programme in the contract as it binds the parties for the delivery of the scheme. The document is of interest to the quantity surveyor because any changes that alter the construction period can influence the cost which, if occurring, must be recognised as being a responsibility of the contractor or client.

The format for presenting a programme can be produced in a number of forms with the choice possibly stated in the invitation to tender. The most commonly used formats in the construction industry are programme evaluation and review technique (PERT) charts and Gantt charts. PERT charts demonstrate task sequences that display the duration of each task which must be finished before the next can start, known as the critical path (e.g. site topsoil strip and stockpiling the material which must be finished before the reduced level and trench excavations can commence). These are shown as numbers in a circle or shaped indicators with links between describing the works along the critical path and their duration. As an alternative, Gantt charts display each critical path by duration using a series of lines or bars measured along a horizontal axis displaying each working activity and duration. To demonstrate, Figure 4.2 shows a PERT chart for piling operations with explanations of a network arrangement, and Figure 4.3 demonstrates a Gantt chart for the same works created from Microsoft Project software (similarly found with Oracle Primavera P6 software) showing 'finish to start' processes that create a number of critical paths.

There are pros and cons for each chart, summarised as follows.

- Gantt charts are abundantly used in the construction industry, more so than PERT charts. The understanding of Gantt charts therefore aids the flow of communication when addressing time, as there is a common understanding of the format.
- Gantt charts are more suitable for small projects as numerous details are possible to reproduce. However, this can be overwhelming on a large project, and can be simplified with a PERT chart that summarises works as a network.

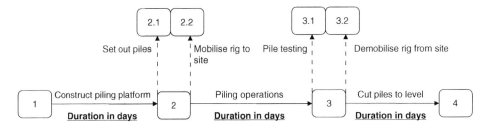

The numbers in the rectangles are networks representing each milestone and commencement of the critical path. The description along the directional arrows are tasks that require completing in sequence and are secondary to each network milestone. Any concurrent works separate to the piling operations will splinter off the appropriate milestone as shown.

Figure 4.2 PERT programme.

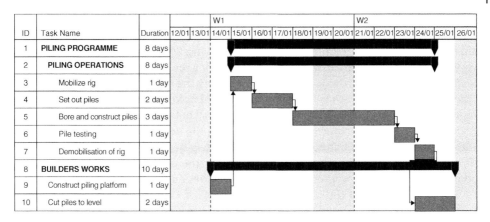

Figure 4.3 Gantt chart.

- In order to produce a Gantt chart, the activities must be known in detail. This is suitable for a construction programme, but inappropriate for early programmes used with cost planning (the latter is best demonstrated as a network as it does not need to be as detailed).
- Updating Gantt charts means accessing existing details, possibly meaning it is better to start again, where PERT charts can be converted with ease. However, with Microsoft Project, Gantt charts can be updated in a dynamic manner which is controllable. The downside here is that it requires intimate knowledge of the software workings in order to exploit the benefits that may be best managed by a consultant.
- PERT charts explicitly define visible activities and dependencies with shorter lists than found with Gantt charts.
- For those unfamiliar with PERT charts, details of tasks and their duration may not be so obvious because the information is less detailed than found with Gantt charts.
- Gantt charts demonstrate time to scale, whereas PERT charts are networks that are not to scale.

A well-prepared construction programme will outline both client and contractor responsibilities, and experienced parties to a construction contract would normally agree to include a programme in a contract that reflects each party's responsibilities. However, there may be a perception by either party that a programme should display the arrangement of construction activities only, which in effect means the management of all stated activities would become the contractor's responsibility. This is an inaccurate perception because a programme should be considered a statement of the order of construction activities, including the timely release of design and documentation that may be a clients' responsibility. While construction activities are at the core of the programme, the omission of dates for the release of design and documentation can create the basis for a dispute. For example, consider a scenario where a contractor is constructing a residential apartment building, and the contract includes a provisional sum for the supply and installation of bathroom sanitary ware to apartments as the client wishes to have flexibility with the selection. Here, the programme should state the latest date the selection must be received by the contractor as well as a date for commencing installation. If the programme shows the installation date to commence on week 60, yet is silent on the date for selection with the actual choice made one week before the stated date for commencing installation, it is unlikely materials will be scheduled, purchased and delivered to site to meet the programmed requirements. In a scenario like this, a contractor may fail with a claim for an extension of time to prolong the construction period because the programme has no provision for a date of issuing the sanitary ware selection. This is because an extension of time may only be granted based upon the information in the programme, assuming of course the programme is a contractual document. An elaborated programme that includes a date for selecting the sanitary ware would therefore address a client's responsibility and help avoid any dispute.

If a programme is a contractual document, it is the contractor's responsibility to update and monitor the progression of stated activities during the construction phase, as it becomes a guide to the performance of the works. However, a contractor only needs to comply with a request to update a programme if it is an obligation of the contract. If updates are not part of the contract, a vigilant contractor will measure the project performance anyway by updating the document regularly, say every three months, to take into account completed works and works yet to be completed. This is of interest to the quantity surveyor as it is important to understand if a project is scheduled to meet the completion date, and applies at any time during the construction phase as time is linked with cost which impacts the main contract preliminaries. The information provided in an updated programme prompts the contractor to act, as the project team may be unaware if a project is either ahead of schedule, on track or lagging behind, with the information necessary to avoid any delay to the agreed end date.

A construction programme as a contractual document is vital for the parties to a construction contract as it can be used as evidence to assess any delays and/or financial claims. It is also of benefit to the quantity surveyor as it aids an understanding of the timely issue of trade contractor's awards, permitting sufficient time to plan ahead and ensure awards are issued before works are due to commence. Similarly, it also assists trade contractors to plan their works that may be issued with the master programme to help prepare a specific programme to suit the main contract works schedules. Trade contractors providing this information usually limit the supply of a programme to large-sized projects of durations in excess of a year, as these businesses need to employ staff with programming knowledge to meet the contractor's expectations.

4.4.1 Programme Float

Float is a term used to describe a free period (or contingency time) in a programme released by the early completion of an activity on the critical path. In reality, and without contractual provisions stating otherwise, a contractor owns the float and the client does not usually gain the benefit of saved time as it may only be temporary. It is for this reason the term is usually excluded from a contract to dispel assumptions that it means an earlier finish than the agreed date of practical completion, as it is the completion date the client and contractor are most interested in.

Programme float may be intentionally included in a programme as a risk contingency that may become enhanced when risk expectations have less impact than envisaged. However, float may also occur naturally due to ongoing project activities that start and complete in less time than expected, leaving the contractor with surplus intervals of time. The 'floating' time created may be managed by absorbing the released time and prolonging the next activity by the same duration, therefore not jeopardising the end date. For example, a contractor may envisage it could have one month float available at the end of month 20 of a 30-month project, and possibly finish before the agreed date of practical completion. However, if this float is lost between months 20 and 30 by the prolongation of some works which is a responsibility of the contractor, it could result in no harm to the end date as the contractor is still able to complete on time. The golden rule here for the quantity surveyor and project manager is to not ignore any justification for issuing an extension of time request to the end date when an occurrence giving rise to such a request is not of the contractor's making, even if float is available to mitigate the delay. Programme float is for information only, and any extension of time to the end date caused by an occurrence beyond the contractor's control should be treated in the same manner as if the float did not exist, and applies even if a client is aware of the float. For example, consider that during the construction phase of a project a natural flood occurs. Repairs are necessary, delaying the end date by 3 months, and there is a 2 weeks float envisaged by the programme at the time of the flood. If the contract has a provision for the contractor to request an extension of time to the end date due to a natural flood, the contractor must still make the request for a 3-month extension. This is because if the float is absorbed by other means after the repairs are complete with the extension of time request not made or, if requested, is reduced by sacrificing some or all of the float or the request has not been approved, it may mean the project will complete late, which would be to the detriment of the contractor. Whether or not the contractor could claim for a retrospective extension of time after the flood repairs are complete is a matter for the client and the conditions of contract.

4.4.2 Programme Acceleration

If works in progress are lagging behind schedule, the future works may require accelerating to put the programme back on track. In effect, acceleration is the opposite of float because no surplus time is available with the programme in time deficit. A lagging programme is of concern to a project client and contractor because the end date may be in jeopardy; this requires remedial action, which may be possible to remedy with acceleration. Acceleration involves the contractor addressing various methods of recovering lost time to achieve the agreed end date. If the contractor considers that works are behind schedule with the float absorbed and there is no extension of time entitlement, the delay becomes a contractor's responsibility. In this scenario, the site and project managers will normally schedule the requirements needed to recover lost time to enable the contractor's quantity surveyor to calculate the costs. This is often referred to as 'crash costing' and includes methods of assessing additional expense for increasing the required amount of labour and plant to within a shorter period than programmed, as well as out-of-hours working and extra supervision by the contractor to put the programme back to a position it needs to be in so the project can complete on time. Where the delay is a contractor's responsibility this may be confirmed in an updated programme which, subject to the terms and conditions of contract, may be for the contractor's use only. However, a diligent contractor would issue the programme to the client/client's agent together with an additional programme demonstrating the mitigating measures using acceleration that the contractor proposes to implement. This would be a responsible approach by the contractor, as it is acting in a humble way and demonstrates how the problem is to be rectified. At worse, the programme may be rejected if it cannot meet the end date, which may mean a client's lack of confidence with the contractor and possibly grounds for termination of the contract.

A contract may include an acceleration clause which, if applying, is usually for the client's benefit. Such a clause would give the client's agent the right to request an increase in productivity if seeking to bring the date of practical completion forward. However, this is usually subject to limitations that may be stated in the contract (e.g. the timing of a notice of a request to the contractor and the earliest date the acceleration is to start and the client's preferred date of practical completion that the contractor can agree to). Where programme acceleration is a condition of contract, and where requested by the client's agent, the contractor must demonstrate methods it will use to comply with the request and issue a revised programme for approval; this is usually accepted if it demonstrates the end date can be brought forward. By contrast, if works are delayed because of an acknowledged responsibility of the client, or the client seeks an earlier finish and the works are not in delay, the contractor is only obliged to acceler-ate the programme if there is an acceleration clause in the contract and must advise of the additional costs if acceleration is requested. Here, the quantity surveyor will advise the project manager of the 'crash costing' amounts as part of the advice. Where an acceleration clause does not exist in a contract, a contractor can only oblige the client by accelerating the programme if requested to do so and issue advice on the cost. In such a scenario, and if the client cannot agree to the cost, the contractor has no obliga-tion to accelerate the programme and, if the delay is created by the client, the contractor should seek an extension of time to the end date with costs if the contractor will incur additional expense because of the delay.

4.5　Project Administration

The contractor's quantity surveyor would be wise to take advantage of available time during the pre-commencement period by creating project administration tools because if not dealt with before establishing a site presence on site, it may result in inefficient administration once the site works commence. A site presence means the physical pres-ence of the contractor's accommodation during the construction phase, which is usu-ally removed prior to the client occupying the building. It can also mean a site presence after section(s) of a project are handed over to a client while other sections are still in progress until completion of the last section.

The pre-commencement period is usually an active time for the quantity surveyor who may be preparing a final account on a complete (or near-complete) project and has the dual task of being involved in a new project. During this period, the project and site managers will manage numerous activities involving administration for the establish-ment of the new project with the quantity surveyor assisting in the process. The admin-istration involves:

- obtaining appropriate project insurance policies or cover notes (usually obtained from the contractor's accountant, financial controller or business development manager);
- obtaining bank guarantees and security bonds for the client (usually obtained from the same source as the project insurances);

- applying for permits/licenses from authorities (e.g. heritage, local authority, building control, etc.);
- obtaining proof of planning permission from the client, including any conditions;
- attending to the client's site accommodation requirements where applicable;
- arranging for mobilisation of the contractor's site accommodation and equipment;
- disconnecting building services supplies if demolition of existing buildings is required;
- applying for temporary service connections for the site accommodation from statutory authorities;
- arranging the collection of keys to gain access to land or existing buildings;
- ensuring the construction programme is available;
- compiling the health and safety file;
- registering the project with the Health and Safety Executive if under the rules of CDM 2015;
- ordering site administration materials and consumables (e.g. document files, stationary, etc.);
- confirming the site welfare/amenities plan with client approval if required;
- reconfiguring the site office plan to suit the project requirements;
- completing a scaffolding plan showing building perimeter requirements and shade cloth areas;
- creating a material handling plan and method statement of working operations, including crane and hoist plan where applicable;
- issuing courtesy notices of intention to construct where applicable (e.g. informing neighbours);
- arranging for a new geotechnical report if necessary;
- obtaining dilapidation reports into the conditions of any surrounding properties, preferably signed off by the local authority and property owners, to identify the condition of the land/buildings prior to a start on site (this will safeguard the contractor if there is a nuisance claim from owners that site construction activities have caused damage to their properties);
- where applicable (and in England and Wales only), ensuring notice of intent to carry out the works is in accordance with the Party Wall Act 1996 (this is legislation coming into force when a contractor carries out works where a party wall exists and applies when works involve modifying a wall, constructing a party or boundary wall between properties and any excavations that are to be carried out within certain distance(s) of a neighbour's structure to a lower depth than the existing foundations; any disputes with adjoining owners must be resolved prior to commencement of the works, which is usually a matter for the client);
- obtaining hazardous materials reports;
- obtaining an environmental report/management plan;
- compiling a quality assurance manual;
- creating a traffic management plan where applicable, regarding how road traffic will be controlled;
- obtaining temporary work designs and permits if required for early works (e.g. façade supports);
- applying for permits for a hoarding/construction zone (e.g. scaffolding, gantries, parking, etc.);

- lodging a voluntary Considerate Construction Scheme (CCS) when applying initiatives for the CCS Site and Company Considerate Codes of Practice aimed at improving industrial image, or a local authority alternative (UK model);
- organising IT for the site accommodation;
- liaising with the Health and Safety Officer for safety measures, first aid, signage, clothing etc.;
- when a project is a design-and-build contract, confirming the appointment of a Building Control Body (BCB) that has the authority to approve the design as complying with the Building Regulations (e.g. a local authority, approved inspector or competent person who can provide self-certification);
- creating a list of design consultants including contact details; and
- collating 'for construction' working design and documentation.

The quantity surveyor must be aware of administrative requirements for a project during the pre-commencement period, even if not having a direct role in all of them, as well as an understanding of the importance of the timely supply of 'for construction' information. Needless to say, a contractor cannot be expected to start works if it does not have the correct information. If a lack of supply delays the start date, it could prolong the construction period; this must be recognised by the quantity surveyor to determine where the burden of responsibility rests. For example, if there is a delay with the supply of information on a design-and-build contract, it is generally a contractor's responsibility. By contrast, if a project involves adjoining owners and the triggering of the Party Wall Act, it is generally a client's responsibility to conclude the negotiations. A team approach will help to identify any of these matters to ensure the project gets off to a good start.

4.5.1 Cash Flow

As with any business, a contractor is required to produce a regular expenditure report that demonstrates the movement of cash (or bank deposits) showing income, expense and profit over a defined period for business efficacy. For a construction project, this information is produced by the quantity surveyor in a cash flow report produced at intervals of the construction period. The report shows the status of payments received from the client, payments made by the contractor and the margin retained as a result. Cash flow reporting has the following benefits to a contractor:

- it evaluates amounts and the timing of income required to pay creditors;
- it aids a process for understanding a rate of return and if a project is running to programme (e.g. the timing of income received compared to that envisaged in a programme);
- it determines any problems with the liquidity of the business;
- it is a measure of received business profits that can be used for purchasing assets and paying loans;
- it is a measure of evaluating risk (e.g. if income generated is sufficient); and
- it aids advanced planning for financial drawdowns from reserves or loans to fund a project.

A favourable cash flow is one that is positive where the income received exceeds outgoings, with the difference being a contribution to the contractor's offsite overheads and profit. Conversely, cash flow becomes negative when income received is less than the expenditure. This unfavourable scenario can occur when payments are received late and creditors are paid on time, putting a strain on the business meaning the contractor may need to rely on a loan to remain solvent. However, negative cash flow can be planned, which can be due to the type of business if accustomed to spending capital over a defined period before receiving any income. This is a common business practice with developers that purchase land and pay contractors to carry out construction works, and only yield income after the date of practical completion when purchasers occupy or lease new buildings. Where applicable, developers usually express negative cash flow as capital expense as it can be defined as a loss of financial interest on reserves it would otherwise generate if invested. However, contractors often work to lean margins and require regular income that must exceed outgoings if the business is to remain solvent, and tend to consider positive cash flow as king over profit. This is because the value of projects undertaken may require the contractor to financially underpin the business with a loan that may be hard to repay due to late payments, with the contractor content on sustaining a low margin to secure work subject to payments being received on time.

At the beginning of a project, it is necessary for a contractor to forecast monthly income and expense in line with the construction programme. This can be achieved by firstly breaking down the contract sum (or approximate value if a schedule of rates is used) into parts, as shown in Table 4.1. In this example, the project is valued at £3 million with the works to be carried out over 6 months and includes a £273,000 margin for the contractor's offsite overheads and profit. Assuming each interim payment will be paid monthly, the amount for cash flow purposes is ascertained by calculating the percentage of each component that will be complete each month the project will be running, which can be obtained from the construction programme. To this sum, a pro rata amount is added for the margin that produces a gross amount due each month. With this knowledge, it permits the contractor to forecast the timely receipt of payments and to plan for the payment of creditors, while also retaining a margin.

The information in Table 4.1 is an estimated value of works by time and useful for demonstrating cash movements in a chart that can be produced from either construction industry software or the Microsoft suite. The basis for creating a chart is to plot a suitably scaled horizontal 'X' and vertical 'Y' axis and insert the details in order to derive the information on each axis. The project cash flow in Table 4.1 is provided as envisaged income/cumulative cost in Table 4.2 and depicted in Figure 4.4, which creates an 's'-curve. The contractor's margin is the difference between the dashed line (representing the contractor's income) and the solid line (representing the contractor's expense), which is clearly shown. The income/cash receipt is usually one month in arrears to the value of works because of credit terms, with one month being usual in a construction contract. The information on this type of chart is normally issued to the contractor's financial controller with the information used to assist with financial arrangements (e.g. agreements with lending institutions where funds are drawn down to pay creditors). For this reason, the quantity surveyor should prepare these management tools as early as possible and ideally before the works commence.

In the absence of a construction programme, it may be possible to produce a cash flow chart using a rule of thumb assessment when the project value, contract period and

Table 4.1 Projected progress of works

| Description | Sum (£) | Allocation (%) | Month number and envisaged status of completed works (%) | | | | | |
			1	2	3	4	5	6
Substructure	138,000	100	50	50	–	–	–	–
Columns	27,000	100	–	100	–	–	–	–
Upper floors	207,000	100	–	25	40	35	–	–
Stairs	21,000	100	–	100	–	–	–	–
Roof structure	69,000	100	–	–	50	50	–	–
Roof finishes	28,000	100	–	–	–	100	–	–
External walls	167,000	100	–	25	25	25	25	–
Windows	108,000	100	–	80	10	10	–	–
External doors	15,000	100	–	100	–	–	–	–
Internal walls	15,000	100	–	–	80	20	–	–
Internal screens	48,000	100	–	–	25	75	–	–
Internal doors	41,000	100	–	–	–	50	50	–
Wall finishes	83,000	100	–	–	25	25	25	25
Floor finishes	108,000	100	–	–	25	25	25	25
Ceiling finishes	87,000	100	–	–	–	40	30	30
Fitments	231,000	100	–	–	25	50	12	13
Sanitary fixtures/plumbing	174,000	100	–	–	25	50	12	13
Water and gas supply	39,000	100	–	–	25	50	12	13
Gas and electric heating	153,000	100	–	–	25	50	12	13
Fire protection services	18,000	100	–	–	25	50	25	–
Mechanical and ventilation	135,000	100	–	–	25	50	–	25
Electrical power and lighting	111,000	100	–	50	–	50	–	–
Builders work with services	21,000	100	–	–	–	–	50	50
Site preparation and clearance	35,000	100	100	–	–	–	–	–
Paved and hard surfaces	68,000	100	–	–	–	–	80	20
Boundary fences and walls	28,000	100	10	–	–	–	90	–
Soft and hard landscaping	28,000	100	–	–	–	–	–	100
External services and connections	52,000	100	–	–	10	90	–	–
Roads and external drainage	45,000	100	90	–	–	–	–	10
Design fees	190,000	100	80	–	–	–	10	10
Preliminaries	237,000	100	20	20	15	15	15	15
Contractor's margin	273,000	100	12	12	23	23	15	15
TOTAL	3,000,000							

Table 4.2 Projected cash flow

	Envisaged income		
	Value (£)	Cash receipt (£)	Cumulative (£)
Month 1	380,000	0	0
Month 2	360,000	380,000	380,000
Month 3	600,000	360,000	740,000
Month 4	910,000	600,000	1,340,000
Month 5	410,000	910,000	2,250,000
Month 6	340,000	410,000	2,660,000
Month 7	0	340,000	3,000,000
Contract sum	**3,000,000**	**3,000,000**	

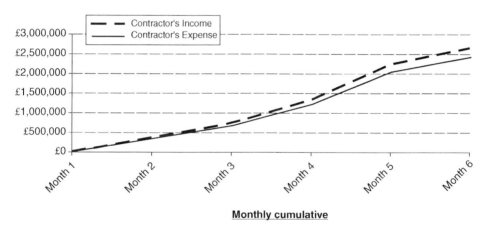

Figure 4.4 Cash flow graph.

contractor's margin are known. Construction projects usually produce an 's'-curve on a cash flow chart, and a theoretical rule of thumb to produce this is based on the following:

- one-quarter of the cost of a project will be incurred during the first third of the contract period;
- one-half of the cost of a project will be incurred during the second third of the contract period; and
- one-quarter of the cost of a project will be incurred in the last third of the contract period.

If a cash flow chart does not resemble an 's'-curve there is usually something wrong with the input and the data should be checked for any errors.

Once a project commences, it will be necessary to monitor income and expense for advice to the contractor's financial controller as cash flow is influenced by the rate of progress on site which, in reality, may differ from the initial programme. A method of determining if a project is running to programme is with earned value management (EVM). This is a project management technique where a project's planned performance is measured against actual progress, which can be demonstrated in a graph. For reporting purposes this can be similar to the cash flow chart as shown in Figure 4.4 with the exception that planned value (PV) and actual cost (AC) are plotted to provide a snapshot of the variance. When updating cash flow information, income received and expenditure incurred are set as a baseline from which works yet to be completed are assessed to create a cash flow prediction for the remainder of the project. For example, in Table 4.1, the roof finishes are shown as being 100% complete by the end of month four, derived from the construction programme. If the works are actually completed by month three, it will mean the contractor can claim and be paid for 100% of the roof finishes one month early. In effect this would increase the cash flow amount for month three, and conversely reduce the amount from month four. It can therefore be seen that changes to construction activities influence the rate of cash flow to and from the contractor.

4.5.2 Cost Targets

Cost targets are a series of predetermined monetary amounts a contractor can afford to pay for trade works (or parts thereof) set as budgetary sums that, when paid, should not exceed the amount payable to the contractor. Typically, each cost target is flexible and set to a range of 70–90% of the amount payable to the contractor. This range is aimed at ensuring a contractor has adequate contingencies to cover risks associated with material waste, trades procured as labour-only packages and unforeseen items of expense that cannot be claimed from the client (e.g. incentive payments for the timely completion of works). Cost targets usually apply to trades where the labour component is not tendered with the targets created from current market rates (e.g. brick and block laying), and where a contractor can provide adequate supervision and source the materials and plant efficiently. The site and project managers play an active role with labour-only procurement, as they usually possess knowledge of labour resources available in the locality of the project under construction. They may also possess knowledge of existing projects run by the contractor and the availability of labour on other projects, which can permit the negotiation of fair prices and attract a workforce at short notice. However, cost targets can also be predetermined for labour and material domestic subcontract works; this is a relatively straightforward exercise to carry out and involves creating a register of individually priced trade bills/schedules for comparison with received quotations to determine whether the prices are within budget.

Cost targets for split-trade packages are created by the quantity surveyor or estimator who may provide details with the pre-construction handover documents. The targets are derived by obtaining the quantities in a trade bill of quantities/schedule and charging them at affordable rates. To asses an affordable rate, the unit rate payable to the contractor in a bill of quantities/schedule is divided into fixed components, that is, labour, plant, materials and subcontractor charges if applicable. From this, each component has any risk contingency identified, which is expressed as a percentage of the unit rate. The maximum amount a contractor can afford to pay as a cost target is the unit rate less the contingency allowance charged against the quantity, as anything more could mean the contractor operates at a loss. To demonstrate, Table 4.3 shows an assessment in tabular form when the quantity surveyor wishes to cost target a labour price for the block-laying trade.

From the information provided in Table 4.3, it can be seen that a rate of £80.00/m^2 will be paid to the contractor for the block wall construction plus an allowance for overheads and profit. Of this amount, £45.00/m^2 is the cost of labour for which the quantity surveyor can target at 85% and set a cost target rate of £38.25/m^2, allowing for contractor's risk. The cost target price is therefore:

$$950 \, \text{m}^2 \, @ \, £38.25 / \, \text{m}^2 = £36,338.00$$

Once assessed, each cost target is relayed to the project and site manager for reference when sourcing labour-only subcontractors, and is advantageous as it sets the basis for discussing prices. Cost targets for material and plant components of a trade price are for information only, and the quantity surveyor will procure the goods and equipment at agreed prices with suppliers and hire companies using predetermined budgets derived from each cost target.

When cost targets for labour-only works are considerable amounts, it is prudent for the site or project manager to negotiate a rate only to a maximum quantity and not as a fixed price. This is because the initial labour-only subcontractors may not start and complete the works and, in the example above, may involve numerous block-laying gangs meaning it may be wise to pay for works progressively at an agreed rate to the maximum cost target. Advantages of cost targets to a contractor include:

- they help identify affordable expense to avoid over-expenditure;
- they can be a useful monitoring tool to aid the financial management of resources for a project;
- considered flexible that can be modified;
- site management can plan works to suit the availability of a labour force;
- they act as a guide for providing incentives, bonuses and penalties for late completion; and
- they permit labour-only contractors to be aware of a contractor's culture and working practices.

The quantity surveyor normally issues cost targets to a site manager before the site is established, where they can be used to assist the procurement of early trades. This arrangement promotes a positive working culture, as it entrusts the site manager with a degree of commercial awareness instead of being totally reliant on the quantity surveyor and project manager.

4.5.3 Works Package Scheduling

A contractor tasked with the responsibility of constructing and delivering a project must have an effective procurement strategy for managing the scheme which, in this context, means methods for the appointment of subcontractors, material suppliers and plant hire companies needed to carry out the works. It may also include the appointment of design consultants where the contractor has design responsibilities and other consultants not involving design (e.g. programme planning). The scheduling of works package awards is usually initiated by the quantity surveyor with the decision to approve

Table 4.3 Cost target assessment

Bill nr 6: brick and block walling

Description	Quantity	Labour	Plant	Materials	S/contract	Total*	Factor (%)	Target rate
a. Block walls, 100 mm thick	950 m²	45.00	5.00	30.00	0	£80.00 m²	85	£68.00 m²
b. Form cavity 50 mm wide, including ties	510 m²	3.00	0	2.00	0	£5.00 m²	85	£4.25 m²
c. Partial cavity fill rigid insulation	510 m²	10.00	0	10.00	0	£20.00 m²	85	£17.00 m²
d. Proprietary lintels 200 × 150 mm	56 m	35.00	0	65.00	0	£100.00 m	85	£85.00 m
e. Proprietary lintels 200 × 250 mm	16 m	40.00	0	100.00	0	£140.00 m	85	£119.00 m
f. DPC horizontal cavity tray	28 m²	20.00	0	55.00	0	£75.00 m²	85	£63.75 m²
g. Adjustable stop ends to DPC	72 nr	5.00	0	2.50	0	£7.50 each	85	£6.38 each

Block walls, 100 mm thick

Labour contingency allowed in the rate, 17.5% (17.5/117.5 × 45.00)				£6.70 m²		
Plant contingency allowed in the rate allowed in preliminaries				£0		
Materials wastage allowed in the rate, 15% (15/115 × 30.00)				£3.91 m²		
				£10.61 m²/£80.00 = 13.3%		

Form cavity, 50 mm wide including ties

Labour contingency allowed in the rate, 10% (10/110 × 3.00)				£0.27 m²		
Plant contingency allowed in the rate, not applicable				£0		
Materials wastage allowed in the rate, 15% (15/115 × 2.00)				£0.26 m²		
				£0.53 m²/£5.00 = 10.6% etc.		

* Excludes overheads and profit.

In this example, the maximum a contractor should target to pay for the brick and block wall trade is 86.7–89.4% of the unit rate depending on the component of the trade. To simplify this, the factor is set at 85% to err on the safe side. With this factor, the contractor has a risk allowance of 15% that places the price of the construction operations to be marginally better than the contract allowance.

resting with the project manager. However, a contract may name or nominate certain suppliers or subcontractors that must receive awards from the contractor; if they do not, the contractor will be in breach of contract. A project can warrant scores of awards, with the actual number depending on the financial value and complexity of a scheme. Due to the number of suppliers and service providers to be procured it is necessary for the contractor to keep track of procurement activities, which is possible with a letting schedule. This is a document that schedules trade and material suppliers for a project in a logical format including any consultant appointments that may be required. Alternatively, consultants may be outsourced by a design or project manager with inclusion on the letting schedule optional.

The construction programme will describe a sequence of works, and is a good starting point for the quantity surveyor to identify trade works to procure for a project for inclusion in a letting schedule. A well-prepared programme may also state the timing of awards to ensure subcontractors and suppliers are appointed before their works are due to commence on site. A construction programme is not a procurement tool however, and only assists in identifying scope and the timing of operations, which requires tracking on the letting schedule for reference. Ideally, descriptions of works on a letting schedule should resemble the project manager's preferred procurement. For example, a project requiring the construction of an in-situ reinforced concrete frame will involve work packages for the formwork, steel reinforcement and in-situ concrete trades that can be procured with the services of one or more domestic subcontractors, labour-only subcontractors, plant hire and material suppliers, or a mixture. Where the division of trade procurement is undecided, or a programme is unavailable, the quantity surveyor can use the bill of quantities/schedule as a starting point to create the letting schedule. If a bill of quantities/schedule is unavailable, the form of tender may provide a suitable trade breakdown. A letting schedule should be considered a flexible management tool, because procurement selection may alter from time to time during the construction phase. A primary reason for this is because of trade availability, which may fluctuate, or if domestic subcontractor's quotations exceed the cost target/budget, meaning the contractor may have to alter the procurement plan to preserve the budget. The format of a letting schedule for a new build project is shown in part in Table 4.4.

From viewing the table, it can be seen a part of the letting schedule is preserved to distinguish items procured by either the builder or client. The inclusion of this part is discretional and of use where the client is responsible for the placement of awards excluded from the contract with the contractor. Alongside each trade listing is a start date obtained from the construction programme. Once this is ascertained it is necessary to consider lead time, which is the time required for mobilising to site, organising a labour force, product manufacturing times and material delivery periods. To understand this concept, it is necessary to appreciate the sequence of construction activities and follow an order of precedence from the project programme that varies with each trade. For example, at the beginning of a new build project there is not usually a requirement to award the floor finishes package urgently, as this will be required later in the programme. By contrast, the requirement to place an award for the piling operations by the earliest date in a project programme should be prioritised. In order to determine lead times, answers to the following questions may aid decisions.

Table 4.4 Letting schedule

Project:
Date of last update:
Commencement date:
Date of practical completion:

Item/Code	Trade package works description	Procurement	Start date on site	Lead time (calendar days)	Target award date*	Actual award date
101	Site accommodation	B				
102	Project sign board	C				
103	Temporary plumbing & drainage	B				
104	Temporary power	B				
105	Temporary communications	B				
106	Hoarding & fences	B				
107	External scaffolding	B				
108	Temporary roads	B				
201	Surveyor	B				
202	Ground piling	B				
203	Cutting bored piles to level	B				
204	Spoil removal from bored piles	B				
205	Excavation topsoil strip	B				
206	Excavation, bulk	B				
207	Excavation, civil	B				
208	Excavation, other detailed	B				
209	Roads and sewers	B				
300	Formwork	B				
301	Steel reinforcement supply	B				
302	Steel reinforcement sundry supplies	B				
303	Steel reinforcement fixing	B				
304	Concrete supply	B				
305	Concrete place, pump, finish	B				
306	Concrete testing	B				
307	Concrete sundries	B				
308	Concrete curing	B				

B: builder; C: client; *start date on site less lead time

- What early trades are required and when are they to start on site?
- Is there a design and specification available for the early trades to work to?
- If there is no design/specification available for later trades, when will they become available?
- Is the final colour/product selection available for finishing materials?
- If it is not possible to place an award as a fixed price because of lack of design, will a schedule of rates be suitable and acceptable?
- Are there any product samples to be provided that need approving before placing an award?
- Are there any products to be sourced from overseas that are on long delivery periods?
- Are there any materials to be purpose made and on long delivery periods?
- Which trades have long lead times because they need to coordinate their design with the contractor?
- Are there any subcontractors works written into the contract that must have awards placed by a certain date as part of the main contract agreement?

Lead times are for the benefit of material suppliers and subcontractors with each being the duration (usually in days) leading up to the start date on site. In addition to this time, the quantity surveyor will need to set procurement motions in place for the contractor's benefit and make appropriate allowances for administrative requirements, for example:

- the number of suppliers/subcontractor to be contacted who will price the work;
- the negotiation time required to agree a price for the works;
- the contractor's vetting process of subcontractors and suppliers; and
- the length of time it will take to process and execute an agreement.

The quantity surveyor must be aware that any material supplier/subcontractor or plant hire company is not bound to order or process materials, gather a workforce or mobilise to site unless in receipt of an award from the contractor, which must be factored into the lead time assessment. For example, let us say a programme for a new build project on a green field states a start date for the topsoil strip of 1 June, and the quantity surveyor consults a plant hire company and the site/project managers regarding the required lead time. If the plant hire company requires 10 calendar days from receipt of a written order to mobilise the plant and it is necessary to allow a further 3 days for processing the order by the contractor, it concludes as a 13-day lead time. The target award date is therefore calculated as follows.

Start date on site: 1st June
Less lead time of 13 days = > Target award date: 18th May

For flexibility, lead times may be on the cautious side and include time for the contractors' benefit, which is to be commended as it safeguards any delay in the process.

Once each award is issued, it requires to be logged on the letting schedule to demonstrate the matter as concluded. A golden rule for the quantity surveyor and project team is to ensure no trades are overlooked, as it would cause embarrassment to a project team if a trade award is missed only to become apparent when there is a need on the site.

4.5.3.1 Early Works and Procurement Scheduling

Certain trade works may need to start on site before the contractor has a site presence (e.g. demolition or bulk earthworks) and, if a project falls under the rules of the CDM Regulations 2015, a principal contractor must be appointed to oversee the working operations. This may present the main contractor with a problem if it cannot create a site presence because an existing building requires demolishing or the ground requires levelling beforehand. To comply with the CDM Regulations 2015, an applicable project must be registered with the HSE naming the principal contractor which must be approved prior to any works starting. This is carried out by completing an F10 form which, for efficiency, is best completed online on the HSE UK Government website for issuing to a local HSE office. A main contractor cannot complete and submit an F10 form, accept the role of principal contractor and then assign the duties to a subcontractor if electing not to or is unable to create a site presence to oversee the works. Here, an early works appointed subcontractor must make an application to the HSE using the F10 form and, once approved, accept the role of principal contractor while the works are in progress. If more than one subcontractor is required to carry out works before the main contractor can make a site presence, only one subcontractor is required to register for the role of principal contractor and provide the duties. Where applicable, the quantity surveyor must ensure awards are issued in a timely manner and include a condition in the appropriate award that the selected subcontractor tasked with providing the duties of a principal contractor provides proof of the approval from the HSE before mobilising to site. Once the works are completed and the main contractor satisfied the subcontractor has completed their duties, the main contractor takes possession of the site and acquires the role of principal contractor with the subcontractor's duties as principal contractor discharged.

Once the main contractor acquires the role of principal contractor, the creation of a site presence can commence or can be extended from an earlier presence that may have been created by the prior principal contractor. This usually involves the services of a field surveyor to set out the site accommodation and subcontractors to carry out temporary works, including:

- electricians to provide power and lighting to the accommodation;
- plumbers to provide pipework for water supply and a drainage system for the accommodation; and
- groundwork contractors to provide foundation bases, service and drainage trenches, paved areas, haulage roads and hardstand areas for material storage.

The requirement to place awards for these works puts the quantity surveyor in a precarious position because the letting schedule may state the only awards required to be issued at this time may be for demolition works and/or other site clearance operations. There is light at the end of the tunnel here, as contractors usually have working relationships with domestic subcontractors that are aware of a contractor's position during the early stages of a project. Here, appropriate subcontractors may submit quotations for the project manager or quantity surveyor to authorise as a minor works contract. Alternatively, they may be engaged upon a schedule of rates where completed works are measured, quantified and valued by the quantity surveyor or reimbursed as dayworks. These subcontractors may also have priced the main contract trade works at tender stage and provided separate quotations/schedules for establishing the site, which would usually be included in the handover file provided by the estimator prior to commencing the works. When placing awards for site establishment works it does not automatically mean each subcontractor will be awarded the main contract trade works, which an experienced subcontractor will understand. If the subcontractor is a new client to the contractor, a well-performed site establishment will act as an indicator of their performance and any offer for the main contract trade works should be seriously considered.

4.5.4   Material Supply and Plant Hire Registers

Once the various procurement methods are identified on the letting schedule, it will probably include some works to be procured with materials purchased by the contractor for installation by labour-only operatives. This can apply to facets of building works and/or some items of the preliminaries where construction cost management and the surety of competitive prices for the purchase of materials is possible, for which the skilled quantity surveyor is deemed competent to manage.

When placing material supply orders, some products may be ordered in bulk quantities on a master or bulk purchase order. Bulk orders are for the supply and delivery of specified materials, which are issued to suppliers to a given quantity at agreed rates for delivery to site over a defined period to suit the sequence of works, for example: 500 tonnes of steel bar reinforcement of varying diameters plus fixing accessories in a number of consignments up to a maximum of 500 tonnes over a 12-month period, with the chargeable rates fixed for the duration. Where this type of arrangement applies a register is required for each material, which ideally is created at the pre-commencement stage by the quantity surveyor. In general, material supply registers are created as electronic spreadsheets and stored in a project file on the contractor's server. The information they provide includes: the supplier's name(s); brief description of the goods; quantities requested; dates of each call-off; actual delivery date(s); docket or delivery note number(s); quantity delivered; and location of the building which the materials are for (e.g. doors D1–D6, Building A, Level 1 Corridor). A similar register needs to be created for items of plant hired by the contractor (not by subcontractors) that records the name of the hire company and item(s) hired as well as notes of the commencement and off-hire dates plus notes of any date(s) the hired mechanical plant is idle. It is important to note any dates of idle mechanical plant because a hire company is usually responsible for their maintenance and may need to supply replacement parts; they may therefore make an erroneous charge for ongoing hire while the plant remains idle. The register should also have capacity to record the quantity and dates of hiring bins/skips that hold waste and the dates they are removed from site, because charges are usually invoiced at a fixed rate per bin/skip that includes hire for a limited period as well as delivery and collection charges and waste disposal fees.

The quantity surveyor is not usually responsible for updating these registers, with the task usually carried out by site managers/foremen that schedule material deliveries and make requests for plant items while a project is running. However, it is important to ensure these registers are created as early as possible for managing, as their creation is part of project administration and may need to be referred to by the quantity surveyor while the project is under construction.

4.5.5 Design and Documentation Registers

A document register is a document created by the contractor showing the title and unique reference of each design drawing, specification and other documents prepared by consultants for a project. Generally, the register will state: each design and document title; reference number; revision; date of the revision; and a distribution list. As an option, it may include the date the contractor formally receives a consignment of new and/or revised information. The register is usually created by the quantity surveyor, and is a replica of the document transmittal notice issued by the document controller for the project. Under a traditionally procured project, the document controller is usually the client's agent or contract administrator. With a design-and-build contract the responsibility of document control rests with the contractor, with the document transmittal notice usually prepared by a design manager.

When design and documentation is new or revised and received by the contractor via a transmittal, the document register is updated to clearly demonstrate the current working information. The contractor then has a responsibility of ensuring relevant information is distributed to appropriate parties, that is, subcontractors and, to a lesser extent, material suppliers to ensure they are in receipt of the current information. The issue of such information should be accompanied with a copy of the document register to clearly show the contents of the consignment. Characteristics of a document register include:

- it must resemble the issue of the document transmittal notice prepared by the document controller;
- it must list the latest information;
- it is evidence that can trigger a change request notice, variation or claim from the contractor; and
- it is considered a formal issue when accompanied with new and/or updated information and is proof of the supply of the consignment.

A project can possibly involve hundreds of drawings and scores of specifications and reports; for effective project administration, the contractor must include a system of document control in the event of a dispute or there being any confusion with the working information. A measure of effective document control is with the use of a document register which must show each discipline (e.g. architectural, structural engineering, etc.) and be logical in format. For contract administration purposes, the register should also include reference to the design and documentation forming part of the contract, including dates and revisions; even though not working information, it is normal practice to include the information for reference. For clarity, the register should be partitioned to distinguish contract design and documentation and their revisions from 'for construction' issues. It should also show current revisions in bold or colour highlighted to ensure the contractor's personnel, material suppliers and subcontractors are aware of the current working information. Bolding or colour highlighting the contract issue is not a good idea as it may give an impression such designs or documents take precedence with the construction process, which of course they do not. Table 4.5 shows part of a sample document register created from computer software that would serve the purpose of communicating each consignment.

A duty of care is required with the creation of this document when the contractor undertakes a design-and-build project. This is because a number of drawings and

Table 4.5 Document register

Project title:				Project Nr:				
				Day	01	05	10	**15**
Project address:				Month	05	05	05	**05**
				Year				
Distribution:				Discipline: **ARCHITECTURAL**				
1.								
2.								
3.								
Design drawing title	**Drawing Nr**	**Current**	**Contract**	**For construction**				
Site plan	A100	**A**	A	A				
Ground floor plan	A101	**B**	A	A	**B**			
First floor plan	A102	**C**	A	A	B	**C**		
Document title	**Document Nr**							
Architectural specification	AS100	**2**	1	1		**2**		

specifications may be undergoing design development that are not normally released to the project team until they are considered suitable 'for construction' purposes. To aid this, design managers may create their own register as a tracking tool of the design development which is independent of the project team. When the design and documentation is suitable for construction purposes it is issued to the project team with a document transmittal; usually the quantity surveyor will update the document register on behalf of the project team so the two correspond.

4.5.6 Progress Claim Scheduling

Rarely will a contractor be paid a single payment upon completion of the works unless the works start and finish within one month, when payment can be made under a monthly credit term arrangement. Where construction works exceed one month, the form of contract should include provisions for interim payments to the contractor until the agreed sum is paid, which is endorsed by legislation under the Housing Grants, Construction and Regeneration Act 1996, widely referred to as the Construction Act. Section 110 (1) of Part II of the Act entitled 'Construction contracts' relates to construction contract reform and states that every construction contract shall:

a) provide an adequate mechanism for determining what payments become due under the contract, and when; and
b) provide for a final date for payment in relation to any sum that becomes due.

British construction contracts are worded to comply with the Construction Act and subsequent changes in legislation changing parts of the Act as well as Fair Payment Principles (a term applicable to public sector works following the guidance of the Fair Payment Charter, making fair payment in construction a mandatory requirement for central government projects). By relying on the Construction Act, a contractor will normally seek interim payments for works in progress. This will usually be on a monthly basis as 'on account' payments, a term referring to the issue of payments in a timely manner until the agreement of a final account.

When a contract is for a lump-sum price it requires breaking down in accordance with the agreement for progress claim purposes (the breakdown is usually appended as a form to the contract). If a contract is silent on the breakdown by only stating a contract sum and the client's commitment to the issue of timely payments, the contractor is at a disadvantage because there is no agreement for assessing works in progress for payment purposes. Where this applies, it is important during the pre-commencement stage for the quantity surveyor to provide the client's agent with a proposed detailed breakdown, known as a progress claim schedule, which is a document showing how progress claims for interim payment purposes will be submitted. It will also be necessary to ask the client's agent which entity progress claims are to be submitted to if not stated in the contract. This is usually the client's quantity surveyor/cost manager or other tasked with the responsibility for assessing a valuation of the works for inclusion in a payment certificate. When a bill of quantities is part of a contract, the progress claim schedule should be formatted to suit the trade bills arrangement. This is because the client's quantity surveyor/cost manager will assess each progress claim from the bills, which is a contractual document the parties are bound to. The contractor's quantity surveyor must not divert from this requirement by creating an alternative format using building elements such as floors, walls etc., as it may be rejected. Table 4.6 provides an example of a progress claim schedule that could be submitted for approval where a bill of quantities is a contractual document, citing the contract clauses entitling the contractor to payment.

When a bill of quantities is not part of a contract, the breakdown should be logical with the contractor normally driving the selection. Ideally, the format should be similar to that shown in Table 4.6 as if a bill of quantities is part of a contract. This can be created by referring to a builder's bill/schedule or tender submission if available, and breaking the price down into parts. Key factors to consider when creating a progress claim schedule include:

- the total of the trade bills, preliminaries and margin must not exceed the contract sum;
- provisional sums and prime-cost (PC) sums must be recognised as part of the contract sum where applicable;
- clauses must be specific to the contract and project; and
- the date of interim payment certificates must be noted, usually obtainable from the contract.

A characteristic of traditional procurement is that payment certificates are usually issued monthly. With a design-and-build project, however, the contract may include a priced schedule of works as milestones of physical completion for payment purposes known as activity schedules. Where applicable, activity schedules are defined in a schedule in the contract with a stated price for each (e.g. design fees, substructure, external walls, roof etc.), the total of which must equal the contract sum. A disadvantage of this system for a contractor is that a clause of the contract referring to the schedule may be silent on the part payment of a milestone when a part is complete, or may expressly state that entitlement to payment is only possible when each activity is fully complete, which is when it becomes a stage payment. If the quantity surveyor is involved

Table 4.6 Progress claim schedule

Project nr:
Project title:
Form of contract: JCT SBC/Q 2016
Commencement date:
Date of practical completion:

Client:
Date of issue of interim application:
Date of issue of interim certificate:
Date of issue of interim payment:
This progress claim nr: Date:

BOQ collection page	Trade	Contract sum	%	(A) Cumulative progress claim	%	(B) Less previous progress claim(s)	%	(A–B) This progress claim
	Excavating and filling							
	In-situ concrete: concrete							
	In-situ concrete: formwork							
	In-situ concrete: reinforcement							
	Brick and block walling							
	Carpentry first fix and roofing							
	Roof tiling							
	Plasterboard linings							
	Windows and external doors							
	Internal doors							
	Fitments							
	Door ironmongery							
	Rendered surfaces							
	Ceramic tiling							
	Carpets							
	Painting							
	Furniture and equipment							

(Continued)

Table 4.6 (Continued)

Project nr:
Project title:
Form of contract: JCT SBC/Q 2016
Commencement date:
Date of practical completion:

Client:
Date of issue of interim application:
Date of issue of interim certificate:
Date of issue of interim payment:
This progress claim nr: Date:

BOQ collection page	Trade	Contract sum	%	(A) Cumulative progress claim	%	(B) Less previous progress claim(s)	%	(A–B) This progress claim
	Sanitary fixtures							
	Fire protection services							
	Paving, planting and fencing							
	Rainwater disposal systems							
	Sewerage disposal							
	Pipe supply systems							
	Air conditioning							
	Electrical power and lighting							
	Preliminaries							
	Margin							
Clause 4.1	**Total (excluding VAT)**	£		£		£		£
	Other payment clauses							
4.3, 4.4	Adjustment of the contract sum							

Ref	Description			
4.5	Value added tax			
4.6	Construction Industry Scheme			
4.7	Advanced payment (10%)			
4.13	Contractor's right of suspension			
4.14–4.16	Site materials and listed items			
4.17–4.19	Retention (Bond in lieu)	Not applicable	0	0
4.20–4.24	Loss and Expense	Not applicable	0	0
Schedule 7	Contribution, levy and tax fluctuations, deleted			
	Total (including VAT)	£	£	£

with a design-and-build contract where stage payments apply, preparation of the progress claim schedule is a straightforward exercise and involves preparing the schedule from the activity schedules in the contract. The stage payment procedure may appear draconian from a contractor's perspective, as it could mean months could pass without receiving payment, placing stress on cash flow. For this reason, a contract may include provisions stating a contractor will be paid a proportion of the works on an activity schedule if the same proportions of works are complete. Alternatively, the contract may provide for monthly payments instead. When payments are by time, the format of the progress claim is usually at the contractor's discretion. This may be by trade as shown in Table 4.6, or created by breaking down the contract sum into elements as activity schedules and submitting the information to the client's agent for approval.

With design-and-build projects there is no contract bill of quantities with the client's quantity surveyor/cost manager usually absent, meaning the contractor will submit progress claims for payment to the client/client's agent. However, as these projects vary in size and complexity, the client/client's agent may elect to appoint a quantity surveyor/cost manager to assess the value of works in progress for payment purposes, and instruct the contractor to submit progress claims in the same manner as a traditionally procured project.

4.5.7   Cost Management Systems (CMS)

Once a contractor initiates a project, the financial controller of the business usually allocates the project with a job or project number on a computer database for identification. This sets the basis for financial control and the methods for recording project expenditure, which is possible with a cost management system (CMS). The type and style of CMS used by contractors varies, and is influenced by the company's financial turnover, the number of staff that will operate the system, the value of projects undertaken, frequency of use and the level of detailed information the system is expected to provide. Characteristics of an effective CMS include:

- the use of logical coding with clear definitions of each description of work;
- it is easily serviceable;
- it is reliable to self run and can produce a variety of reports;
- processing times are rapid, without problems involving computer memory capability;
- operating manuals are available from software writers;
- training for use of the system is available with workshops;
- the system identifies financial project risk;
- methods are available for accessing data in relation to subcontractors, suppliers and project activity;
- data can be produced in an environmentally friendly manner (i.e. paperless); and
- there is back-up support for dealing with technical problems.

The quantity surveyor is not usually responsible for selecting a project number or the creation of project information, and usually relies on the financial controller to advise when the project is live and accessible on the CMS. Once the CMS is activated and includes a new project, the quantity surveyor is usually responsible for setting the parameters of the project budget with the first task involving the entry of trade budgets against cost codes. The intention of having cost codes in a CMS is to create a system where codes abbreviate word descriptions that may be otherwise too long to produce on reports. Coding is usually a mixture of numbers and letters written into a database by a software provider. For example, a primary code may be prefixed with 'A', this being an abbreviation for the substructure. Under this primary code, a series of sub-codes may be used, for example 100, meaning A100 will be the code for excavating and filling. This may be followed by a series of secondary codes to the sub-code, for example 'A100-1, A100-2, etc.' representing trade work descriptions falling under the category of excavating and filling within the substructure. The sum of budgets entered as secondary codes must equal the sub-code and, following the example of 'A100', a total budget will be set for the excavating and filling works when all secondary codes have sums inserted. Further sub- and secondary codes can also be created for other trades, for example 'A200-concrete', 'A300'-steel reinforcement' etc., with the summary of secondary and sub-codes prefixed with the letter 'A' making up the substructure component. This process then continues with other elements, for example 'B' for brick and block walling, 'C' for carpentry first fixing and roofing, 'D' for roof tiling etc., making up the project budget.

When setting budgets, the coding structure may have been pre-set by the estimator. However, this may have been carried out using software specifically for estimating purposes that does not form part of the CMS which, if applicable, may mean the quantity surveyor will need to manually input budgets. It is worth noting here that coding is a database for use on any project a contractor may undertake with the listings suitable for all types of projects. Codes used for each project should therefore be selected to suit a method of procuring trade package once a project is running. To demonstrate, Table 4.7 shows options for entering budgets for the excavating and filling works.

From reading the table, the estimator's allowances have been made against codes inserted as Option 1. By selecting Option 2, the arrangement is streamlined in favour of using fewer codes that could resemble the procurement selection. The use of Option 2 is advantageous for the quantity surveyor as it aids the payment process to suppliers and subcontractors, involving fewer codes than used for Option 1. An important factor to consider when entering budgets is that the sum of all cost codes must equal the contract sum. To avoid any oversight in this respect, a well-designed CMS may require the contract sum entering into the system prior to setting the budgets; this aids the checking process as it ensures the total of the budgets does not exceed the allowance. If the system does not have this facility, a manual check is required to ensure the contract sum is not exceeded.

4.6   Site Establishment

The pre-commencement period involves the project team organising the site establishment, and it is important for the quantity surveyor to become familiar with the administrative requirements needed to make this happen. To ensure this, a progressive check list confirming the status of administrative requirements could be used as shown in Table 4.8, which is not exhaustive, and covers the topics discussed in this chapter.

The completion of the site establishment is a rewarding experience for the project team as it means site operations can develop under the direct observation of the team, giving control to the contractor for the day-to-day running of the project. The

Table 4.7 CMS budget entry

Project Nr: 1888
CMS Code series: 'A' - Substructure

A100 - Excavating and filling

Cost centre	Standard description	OPTION 1 Budget input (£)	OPTION 2 Budget input (£)
A100-1	Remove overgrowth	1,000	0
A100-2	Topsoil strip	6,000	7,000
A100-3	Reduced level excavations to 1 m deep	3,000	3,000
A100-4	Reduced level excavations over 1 m deep	0	0
A100-5	Filling to level to 1 m deep	65,000	65,000
A100-6	Filling to level over 1 m deep	0	0
A100-7	Basement excavations	0	0
A100-8	Detailed trench excavations	15,000	20,000
A100-9	Detailed pad excavations	5,000	0
A100-10	Detailed lift pit excavations	0	0
A100-11	Earthwork support	2,000	2,000
A100-12	Deposit spoil on site	3,000	0
A100-13	Dispose spoil off site	15,000	18,000
TOTAL	**A100 - Excavating and filling**	**115,000**	**115,000**

completed physical establishment should be comparable with the site accommodation plan agreed between the contractor and client, as deviations could incur additional expense which the quantity surveyor must recognise as part of cost management. Hopefully, the layout should be comparable with the agreement with the costs incurred within the realms of expectations. If the layout has changed and increases the anticipated cost, the quantity surveyor will need to understand the reasons and determine whether it is to be at the contractor's expense. It would be wise to deal with this matter at the earliest stage instead of leaving it until the project is up and running, when it becomes less of a priority.

4.6.1 Site Induction

A site induction is the test of a persons' understanding of their risks and responsibilities in relation to health and safety matters and how it affects the individual, fellow workers, the public and working environment on a project, and is applicable to site operatives, managers and administrators whose normal place of work is based on a project or are regular visitors to site. There are different levels of induction, the style of which depends on the scope of works or building type, project duration, nature of the working environment and any specific project client requirements. An induction is usually project

Table 4.8 Site establishment administration checklist

| Project title: | | Project nr: |
| Date of possession of the site: | | Client: |

Project team:

Project manager:	Secretary:
Site manager:	Leading hand 1:
Structural foreman:	Leading hand 2:
Finishing foreman:	Site labourer 1:
Quantity surveyor:	Site labourer 2:
Quantity surveyor trainee:	Apprentice carpenter:

	Description	Completed (Y/N)	Action
1.0	**Estimating handover: issued to the team:**		
1.1	Contract drawings		
1.2	BOQ/Builder's quantities		
1.3	Specifications and Reports		
1.4	Main contract agreement		
1.5	Subcontract/supplier quotations		
1.6	Cost plan allowances		
1.7	Contractor's proposals to client		
1.8	Document transmittal receipts		
1.9	Contract negotiations		
1.10	Other: specify		
2.0	**Site planning**		
2.1	Material registers created:		
	(i) Bar and mesh reinforcement		
	(ii) Concrete		
	(iii) Skips		
	(iv) Sundry mechanical plant hire		
2.2	Scaffolding plan		
2.3	Materials handling plan		
2.4	Crane and hoist plan		
2.5	Construction programme		
2.6	Waste management plan		
2.7	Health and Safety file		
2.8	Site induction forms prepared		
2.9	Other- specify		

(Continued)

Table 4.8 (Continued)

3.0	**Authorities and notices**
3.1	Insurances obtained
3.2	i) Contractor's 'All Risk'
3.3	ii) Professional indemnity
3.4	iii) Other - specify
3.5	iv) Licenses and permits - specify
3.6	Bank guarantee obtained
4.0	**Administration: QS**
4.1	Cash flow chart created
4.2	Cost targets set
4.3	Contract sum entered in CMS
4.4	Letting schedule created
4.5	Document distribution register compiled
4.6	Site drawing racks in place
4.7	Progress claims scheduled
4.8	Main contract conditions reviewed
4.9	Electronic and hard files created
4.10	'For Construction' information on site
5.0	**Administration: other**
5.1	Correspondence to/from files (QS/PM/SM)
	(i) Client
	(ii) Consultants
	(iii) Subcontractors and suppliers
5.2	Books to site
	(i) Site diary and Site Instruction pads
	(ii) Requisition and Dayworks sheet pad
	(iii) Health and safety induction pads
	(iv) Other: specify
6.0	**Corporate site notice board erected**

specific and involves the inductee being briefed on the project's requirements by an inductor, such as the contractor's project manager or health and safety representative, prior to the inductee undergoing a competency test. Commencement of the test usually requires the inductee to acknowledge the contractor/project client's health and safety and environmental and quality assurance policies by way of their purpose and value. Thereafter, questions and answers revolve around an understanding of health, safety and the working environment as well as any site-specific criteria, and may be conducted in private, as part of a group or online. For example, questions may arise regarding the prohibited use of alcohol and drugs, the correct use of personal and protective equipment, and the recognition of out-of-bounds areas (e.g. areas where flammable gases and liquids are stored). Competency is tested by the inductor acknowledging the inductee's response to a range of optional answers, for example, 'The use of alcohol and drugs is tolerated on site…. (A) always; (B) sometimes; (C) never; or (D) when the project manager says so', with the answer obviously being (C).

Upon completion of the questionnaire, the inductee returns the form(s) to the inductor who will mark the responses and advise the inductee of any errors. Failure to recognise responsibilities considered 'hard subjects', such as access to areas that require authorisation because they are a hazard, may be a point of concern, leading to failure of the test. However, being incorrect on items that are 'soft subjects', such as an understanding of where the offsite car park area is located, may not cause concern for the inductor who will possibly grade the completed questionnaire as a pass and advise the inductee of any shortcomings. Upon satisfactory completion of the assessment, the inductee is issued a booklet, card or certificate confirming competency. Failure of the competency test may involve the inductor referring the inductee to the commencement of a course of study in order to gain an understanding of the requirements, whereupon after a suitable time the testing process is repeated. Normally, anyone not in receipt of an induction pass whose capacity is to provide a service stemming from site is not allowed to work on site.

4.7   Review of the Main Contract

The contractor's representative executing the contract is usually a director of the business and, once executed, the original is usually kept in the contractor's head office with a copy issued to the project manager and/or quantity surveyor. In general, the contractor's quantity surveyor is responsible to the project manager for administering the commercial aspects of the project, with the project manager responsible for contract administration and delivery of the project on the contractor's behalf. The delegation by the project manager can be rewarding for the quantity surveyor, as a project manager's background may not originate from contract administration and may welcome advice on commercial and contractual matters during the construction phase. As part of the review of the main contract, the quantity surveyor must become familiar with the form of contract as well as the series of drawings and documents driving the contract sum (or schedule of rates where the price is not fixed) prior to the works starting. This means gaining an understanding of the structure of the agreement for navigation purposes; this does not mean memorising details of clauses, schedules, drawings, etc., as this is unrealistic. Where a contract is not concluded prior to the start on site, the Letter of Acceptance or Letter of Intent and draft contract should be reviewed, and a check made upon receipt of the executed contract for any changes. Before the project commences, the quantity surveyor would be wise to spend time understanding what the contract means for the contractor, as a role of the quantity surveyor during the construction phase involves acknowledging the importance of the agreement as well as being aware of the contractor's risks, obligations and the rights it is entitled to. The contract should always be accessible and not merely placed at the back of a drawer, as it sets the pathway for administration procedures and provides answers to many questions that may arise during the construction phase of a project.

The RICS publishes a survey of contracts in use in the UK from time to time, which is a series of surveys determining trends in the use of standard forms of contract and procurement methods. The most recent report was prepared by Davis Langdon (rebranded as AECOM in 2013) on behalf of the RICS Built Environment, and issued as a 12th edition based on surveys carried out on the results of completed questionnaires sent to private practices and public authorities. The survey found that over 97% of construction projects use a standard form of contract, albeit sometimes with client or consultant amendments, with 88% of projects using one of the JCT family of contracts. This pattern is consistent with earlier surveys, meaning it is likely that a quantity surveyor involved with the administration of building contracts for contractors in the UK may encounter a JCT form of contract at some time. The discussion on contract review for the remainder of this chapter is therefore focused on enabling the reader to understand the structure of an agreement where a JCT standard form of contract is to be administered.

By default, JCT forms of contract address the professional title of members involved on a construction project:

- 'Employer', that is, the client;
- 'Contractor', that is, the builder/main contractor (or Principal Contractor under CDM 2015);
- 'Architect', also referred to as 'Contract Administrator', that administers the contract on behalf of the 'Employer'; and
- 'Quantity Surveyor', that is, the 'Employer's cost manager dealing with financial aspects of the works with the responsibility for authorisation resting with the 'Architect'.

The title 'Architect' may be substituted with 'Contract Administrator' because the 'Architect' may be a designer only, leaving the pathway open for other professionals to administer the contract. The representative administering the contract on behalf of the 'Contractor' is not mentioned by title, as the responsibility is generally viewed as one that applies to the 'Contractor' in name only.

4.7.1 Expressed and Implied Terms

When reviewing the main contract the quantity surveyor and project manager must become aware of the impact of expressed terms, which are specific terms written in a contract that leave no doubt regarding the intention and scope. This has its obvious advantages; however, on occasion it can work against the contractor. For example, a situation could occur if the quality of installed goods fails to meet the architect's expectations with the architect issuing an instruction to replace such goods and referring to a clause in the contract with an expressed term, for example: 'Clause x.xx ...works are to be carried out to the satisfaction of the Architect'.

This will be of concern to the contractor, especially if the installed products are considered by the contractor to comply with the contract with the architect having a different opinion. To avoid this unpleasant scenario, a suitable contract clause will include obligations for the contractor to provide product samples or produce a full-scale mock-up room (as found with the likes of new hotel construction). These prototypes are representations of the final product that may require technical testing for sound proofing, etc., and inspecting for aesthetics, which the architect must formally approve to give the 'green light' for production so there is no disappointment with the final product. However, if there was failure to seek approval because approval is not an expressed condition of the contract, it means the contract has no remedy for such a situation. In this scenario, the contractor has no obligation to supply samples for approval and, even if electing to supply samples, the architect has no obligation to approve them. In an attempt to resolve such a matter, the situation may be referred to statute, as normal commercial arrangements expect the quality of products (which can mean products, methods of installing products and finally installed products) to be satisfactory, fit for purpose and to the minimum standards in the absence of an expressed term stating otherwise. Here, a contractor can refer to the Supply of Goods and Services Act 1982 to resolve any misunderstanding regarding the quality of an installed product, which may involve seeking legal advice if the terms and conditions of the contract are not considered breached.

The parties to an agreement may be bound to a contract that is silent on certain matters that either party may concede makes the contract difficult to work with, for example no dispute resolution clause, with the parties reluctant to access statutory procedures such as litigation. This can occur if a standard form has been edited and erroneously excludes the methods for addressing such matters. Remedy for this drafting shortcut can be made with implied terms that only a court can imply into a contract to give the arrangement business efficacy. When implying a term, the new term must not contradict an existing expressed term and must have clarity, as a court will be reluctant to reinterpret was included unintentionally. Once activated, it is implied in accordance with statute and case law which is created over time, meaning a court will not usually consider anything else that would leave either party in a different bargaining position already created. In the scenario mentioned above, where there is a failure to approve product samples that are rejected after installation, it is highly unlikely a court would introduce an implied term to make product approval a condition of contract, as it could be construed as an attempt to change the bargaining powers of the parties already in place.

Where the parties to a contract wish to include new terms in a contract yet wish to streamline the arrangement by not involving a court, they could create a supplementary agreement. This has an advantage to the parties as the arrangement can be completed quickly without third party involvement. However, a degree of caution should apply here, and each party should seek legal advice as the agreement may include terms that are unintentionally illegal or may contradict existing clause(s). Moreover, the parties may not recognise any assignment clause in the contract that could change the names of the parties at a later date that may be flawed in the supplementary agreement, with the flaw only coming to light during a dispute resolution process where a third party exposes the inconsistency.

4.7.2 Articles of the Agreement

The Articles of the Agreement is an introductory part of the form and precursor to the body of the contract, setting the basis to the agreement. The parts comprising the Articles of the Agreement are described in the following sections.

4.7.2.1 Agreement

This is the date of the agreement between the employer and the contractor, stating the name of each entity creating the contract, company registration numbers and registered addresses.

4.7.2.2 Recitals

Recitals are an explanation of matters or facts that exist in an agreement. When adopting a standard form, clauses in the body of the contract may not specifically define each party's expectations of the other's performance during the term of the contract, which may be clarified in the recitals. For example, a recital may state the employer has provided the contractor with a bill of quantities for pricing and returning. Moreover, the recitals may include an explanation of the works under contract, so that it leaves no doubt of the intention of the completed works, for example:

> … the construction of 10 nr two storey 3 bedroom properties of timber frame construction with brick veneer and tiled roof complete with partitions, utilities, fitments and external works including the provision of a fully serviced estate road to service the properties.

Recitals may also confirm any negotiations and clarifications regarding the party's conduct prior to executing the contract which the party's collectively acknowledge, for example:

> The Employer has supplied the design and documentation and the Contractor has priced the work in accordance with the consignment as well as supplementary information supplied after the tender period.

4.7.2.3 Articles

Articles are a confirmation of the nature of the agreement and include the names of project stakeholders that will interact with the contractor and employer for the delivery of the project. As the contract is between the contractor and employer, stakeholders are not bound by this agreement and are included for fiduciary reasons, for example 'Quantity Surveyor', 'Architect', etc., with each usually having a separate agreement with the employer. The contractor's quantity surveyor must be aware of this difference and recognise the fact that the contractor has no contract with any project stakeholder except the employer. In effect, this means a contractor's role with anyone other than the employer is collaborative only and, if wishing to seek contractual remedy for a stakeholders default other than the employer's, it must make be through the employer.

The Articles set out obligations in a simplified manner and are usually in single sentences, for example:

> The Contractor will carry out and complete the works in accordance with the design and documentation and the Employer shall pay the Contractor the sum of £xxxx for execution of the Works in accordance with the design and documentation.

Articles may also include reference to the proper law of contract and legal proceedings permitted by the law of the country in which the project is based.

4.7.2.4   Attestation

This is the execution and witnessing section of a contract completed by the signatories that should include the date the form is executed. A contract may be noted as either a simple contract or deed, yet legally does not need the inference (the difference is the limitation period for legal claims against the contractor for defective works, which is 6 and 12 years for a simple contract and deed, respectively). Contracts of high risk and value are normally executed under a deed as a matter of course.

4.7.2.5   Contract Particulars

These particulars are specific to a project, for example date of issue of interim payment certificates, date(s) for sectional completion for part(s) of the works and/or a date for completion of the whole works, etc. Where applicable, they are cross-referenced to clauses of the body of the contract for the details (e.g. Clause 2.4 of JCT SBC/Q 2016: Date of Possession - progress, regarding the date the site will be handed over to the contractor to commence operations).

4.7.2.6   Definitions

Clauses in a contract may include certain words or wording shown in italics so they can be easily identified for cross-referencing to the definitions, which is a series of key words and a glossary of wording in an agreement that when included in a contract is binding on the parties. A glossary may include words with capital letters appearing out of context with natural meaning, and the quantity surveyor must become aware of this when communicating anything in writing if referring to the contract definitions, which may apply to a person, title, group of persons (and what they represent), place or thing. For example, the contracting parties are the Contractor and Employer, and the Definitions are in the Conditions, all with a capital first letter. These definitions prevail over dictionary definition terms that form part of Legal English and lack normal punctuation, possibly including the use of some Latin words in the process.

Definitions may not be limited to wording and may clarify an ambiguity relating to a meaning, for example, if assessing the number of working days when the contractor will receive payment after receiving an interim payment certificate advising the employer to pay the contractor. Here, the contract may state that 14 working days may pass before a payment is due to the contractor, with the definitions clarifying that when calculating working days, they are Monday–Friday inclusive, and if any working day falls on a public holiday, that day is excluded.

4.7.3   Insurance

A construction project must have adequate insurance to protect the interests of the physically built works, risks to the parties of the contract, persons engaged on a project carrying out their duties and the existing built environment within the vicinity of the project. Project insurances need to provide adequate financial protection to cover the cost of rectifying an incident. This includes any incident created naturally or the result of an accident, or which may have developed from a situation involving risk or the creation of an occurrence associated with the project that would cause financial harm and loss to a person or business if liable for the incident. Because of this, the obtaining of suitable insurance can be a complex process, and a well-researched project will recognise insurance needs within the form of contract. The contract will not usually spell out the wording required in a policy (as that is a matter for the insurance provider); however, the contract will specify appropriate insurances needed to meet the objectives as well as the party's responsibilities for obtaining policies to ensure there are no gaps with the appropriate cover. A range of insurances as a minimum requirement for a construction project are discussed in the following sections.

4.7.3.1 Insurance for the Works, Including New Works and Existing Structures

Insurance for the works includes what JCT contracts refer to as 'Specified Perils' and 'All Risks Insurance'. A 'Specified Perils' policy is insurance cover for one or more events mentioned in the contract, for example, fire, lightning, floods, explosions. This excludes 'Excepted Risks' for incidents considered unlikely to happen that the parties to a construction contract accept as risks they are prepared to accommodate (e.g. pressure waves caused by aircraft and their affects). An 'All Risks Insurance' policy is suitable for cover against financial loss and damage incurred for any event not included as 'Specified Perils' or considered 'Excepted Risks'. There is a distinction between the requirements of these two policies, with the perils clause clear on the type of events and extent of cover required for each event. Conversely, an 'All Risks Insurance' policy must cover expense incurred for an event other than any perils event or excepted risk that causes physical loss of the constructed works. It must also cover the cost of replacing destroyed unfixed site materials, removing and disposing of debris and the propping or shoring of works as a result of an incident giving rise to an insurance claim. It does not apply to incidents resulting in the cost of repairing, replacing or rectifying works due to:

- wear and tear or the obsolescence of an item or part of an item;
- deterioration including rust or mildew;
- works or materials that are lost because of a defective design and/or specification criteria;
- loss or damage resulting from war, invasion, foreign hostilities and similar listed situations;
- the disappearance or shortage of works/materials unless an inventory is made and substantiated; or
- excepted risks.

A contract may include a clause or an appended schedule requesting the need for a policy to cover the effects of war or acts of terrorism which may not be included in a standard policy. Here, a policy provider may need to issue a separate policy that aims to fulfil the requirements of the contract where possible. A further request could be for Decennial insurance, which is a ten year warranty against partial or full collapse from the date of handing over.

Policies for the works under contract are usually required to be in joint names (contractor and employer) and provided by reputable insurance companies or brokers. A general policy obtained by a contractor for all of the projects it undertakes may be considered insufficient by the architect because the cover is usually an aggregate and to a maximum amount the contractor can claim in one year (or for the term of the policy), irrespective of the number of projects in progress. For this reason, the contract wording may state the requirement for a specific policy citing the project as covered as well as the joint names seeking the protection. Where required, if a general aggregate policy is issued by the contractor, it will probably be rejected by the architect and would need substituting. Moreover, any incorrect spelling of names or titles on a specific policy may also be rejected and can usually be corrected by the policy provider with ease. The JCT forms of contract schedule responsibilities for obtaining insurance under Options A and B and allow either the contractor or employer, respectively, to obtain insurance for new works. Option C is used where work is on or within an existing structure, including building extensions, with purchase of the policy usually an employer's responsibility.

The quantity surveyor must become acquainted with project works insurances and be in receipt of copy policies or certificates of currency/cover notes effective from the date the works commence. At the beginning of a project, it is a normal contractual require-ment for the contracting party responsible for obtaining insurance to provide the other with evidence of the existence of policies (in joint names where required) to confirm the project has adequate works insurance. Time is therefore well spent by the quantity sur-veyor in highlighting insurance requirements written into the contract with coloured markers to emphasise the extent of cover required. As a matter of course (although not usually contractual or mandatory), copies of the certificates of currency/cover notes should be submitted with each progress application for payment, as it demonstrates the existence of relevant cover at the time of making the request.

Policies usually run for one year; if a project is to run for a longer period, each policy must be renewed and be effective immediately upon the expiry date of the prior policy. Due to the number of policies that may be required for a project, it would be wise to create a register of insurances including dates of expiry to ensure there is suitable cover at the appropriate times.

4.7.3.2 Injury to Persons

Insurance is required for persons going about their business on a project in the unfor-tunate event of an incident giving rise to personal injury or death. In general, if a sup-plier of services only or goods and services is injured while carrying out their duties with the incident giving rise to an insurance claim, it is the contractor's responsibility to provide the insurance from which the claim is made. By providing this insurance, the contractor must indemnify the employer from liabilities caused by such events for all persons. However, contract clauses may distinguish between employer- and contrac-tor-appointed operatives that may influence the type of insurance and extent of cover. Employer's operatives are individuals engaged or authorised by the employer or employer's contractors to visit a project and perform specific duties, while contractor's operatives carry out a service directly for the contractor. Depending on the form of contract and the agreement of the parties, a contractor may need to provide a policy to indemnify the employer in respect of employer's operatives carrying out duties if there is an event caused by any such person(s) triggering an insurance claim. In such a scenario, the employer will only be liable if such operatives are negligent in their duties with the responsibility for insurance to persons carrying out their services with a duty of care resting with the contractor. Alternatively, the employer may provide insurance for any persons engaged that perform duties on behalf of the employer and, in the process, may indemnify the contractor in the event of their negligence where the contractor's insurance would be ineffective. It is highly unlikely a contractor would obtain insurance for an event involving negligence by the employer, as such an arrangement would be convoluted.

4.7.3.3 Damage to Surrounding Property

During construction operations, there may be risk resulting from events such as ground heave and subsidence caused by working plant that could cause damage to existing buildings due to vibration or impact. Here, contract insurance provisions usually require the employer to provide a policy to cover damage caused to neighbouring properties in the event of damage affecting the stability or integrity of surrounding buildings. A claim against this type of policy is effective when the contractor follows the design and procedures in accordance with the employer's requirements and acts with a duty of care. If carrying out excavations close to an existing building results in damage to a building, the contractor would therefore not be liable if complying with the conditions of contract and is found not to be negligent in their duties. Conversely, and subject to the terms and conditions of contract, the contractor may be required to provide a policy to indemnify the employer for damage to any surrounding property or properties insured by the employer in the event of damage caused by its negligence or by the default of one of its subcontractors.

4.7.3.4 Joint Fire Code

The Joint Fire Code considers the affects an outbreak of fire may have in relation to the construction methods used on a project, for example timber frame construction or the widespread use of acetylene within zones or locations where such works are carried out to facilitate a project. The Code is not associated with the completed structure itself, and is aimed at addressing risks associated with the construction processes from design through to completion of the physically built works. To obtain contract works insurance policies, insurance companies may request compliance with the Joint Fire Code due to the nature and risk of the works under contract. For this reason, mandatory compliance with the Code is included in a number of standard forms of contract. When called for, both the employer and contractor must provide policies indemnifying each other in the event of unintentional breach of the Code. If the Joint Fire Code does not apply, the standard contract clauses are usually struck through.

4.7.3.5 Professional Indemnity Insurance (PII)

When a contractor provides design for the works, a separate policy or policies is required to indemnify the employer in the event of errors, breaches or omissions that create a defective design(s) which is incorporated into the works and, over time, gives rise to an insurance claim. The intention of professional indemnity insurance is to protect the employer from incurring liability once such an event occurs which is attributed to a defect in the design(s). This may only become apparent years into the life of a building, for example the failure of constructed foundations or premature defective roof covering caused by the use of incompatible specified materials. The contractor must provide this type of insurance if providing a design service to the employer, and the wording in a policy should define the extent of liability and length of time the policy will run after a date of practical completion. Individual policies provided by consultants engaged by a contractor that provide a design service may not be acceptable if the employer is seeking a single policy provider to provide the indemnity. However, an insurance company may provide a single policy to a contractor once it receives satisfactory details of the design consultants to be covered in a policy it will provide.

4.7.4 Employer's Financial Security

With normal industrial practice, it is usual for a contractor to provide financial security for the employer in the event of its default under the contract. Default includes insolvency and an overall lack of performance to deliver a scheme, and there are various methods open to the parties of a contract to provide this assurance.

4.7.4.1 Cash Retention

When payments are made to the contractor for works in progress, the certified value and payable amount is temporarily reduced by 5% (or to a percentage written in the contract) for cash retention held by the employer. Subject to the terms and conditions of the contract, half of the retained amount is refunded at practical completion with the balance released at the end of the defects liability (or rectification) period. The retained sum from each interim payment is deposited into a trust account, which is usually opened by the employer and is independent of the employer's business, as in effect the money belongs to the contractor; the contractor has no access to the account while the construction and rectification periods run and the employer remains solvent. However, in the event of the employer becoming insolvent, receivers dealing with the employer's estate may not prohibit a contractor from accessing the funds. The trust account is not an investment for the employer, with any accrued interest kept in the account for the benefit of the contractor, which is released at the time of refunding the retention. The contractor does not usually have a say in the type of investment, as the objective is merely to retain funds in accordance with the terms and conditions of contract with any discussions on the type of investment being superfluous and unnecessary. The status of retention is shown on each interim payment certificate (excluding interest) and, because it is taken as a percentage of each payment, the retained amount increases progressively and may be capped to a maximum amount depending on the wording of the contract. As the quantity surveyor is normally provided with a copy of each interim payment certificate, it is vital the correct retention is withheld from each payment and does not exceed the capped amount if stated in the contract.

4.7.4.2 Bank Guarantee

A downside of withholding cash retention to a contractor is stress on cash flow due to the starving of income during the construction and defects liability/rectification phases. As an alternative the contractor can provide a bank guarantee which the employer may either accept or reject, with the option possibly included in the invitation to tender and subsequent form of contract. If accepting, the wording in the contract must state the bank guarantee is an acceptable alternative to cash retention (unless there is provision for both) that, when applicable, is usually issued as two certificates provided by the contractor's nominated bank with the institution acceptable to the employer. The amount on one certificate is usually equivalent to a cash retention sum that would be due for release on the date of practical completion with the guarantee effective to that date. The other certificate is to the value of cash retention that would be held at the end of the rectification period, and effective to that date. The process required for obtaining the guarantee and certificates requires the contractor to pay a fee to a bank for the purchase which, when paid, permits the release of the certificates with the bank acting as guarantor. The supply of the certificates assures the beneficiary (the employer) on the performance of the works provided by the services provider (the contractor) to predetermined amounts. Each certificate must be project specific and name both the contractor and employer. Usually, the bank issues the certificates to the beneficiary with copies issued to the services provider that filters through to the architect for information.

When the use of a bank guarantee is permitted, a review of the main contract by the quantity surveyor will ensure the effect of the certificates is understood and that a check is made on their suitability. This task must be carried out before the first payment is received; otherwise, the architect will withhold cash retention that may take effort to reverse after any interim payment certificate is issued. If the architect notices errors in the names, titles or description of the project on the certificates, however minor, they may be rejected with cash retention taken instead and stated on each interim payment certificate until such time the architect is satisfied the certificates are effective. Certificates must not be generic, and a parent or cross company guarantee from a contractor is not usually sufficient financial security as it does not involve a bank that is a suitable financial guarantor.

Certificates for bank guarantees are to fixed amounts, which means additional security may be required for variations that increases the contract sum. In a construction project where cash retention is used as security, an increase in the value of works would automatically create a *pro rata* increase in the amount of cash retention. Where a bank guarantee is provided instead this can cause a problem, as the sums on the certificates are to fixed amounts with a possibility of there being no remedy in the contract to deal with the matter. Unless worded specifically in the contract, a wise contractor would provide certificates that guarantee sums greater than the cash retention equivalent with expiry dates beyond that stated in the contract. This is to allow for increases in the contract sum and any extensions of time that may be granted, which the contractor can negotiate as and when any changes arise if the guarantees are low in value. Upon completion of the contractor's performance without call on the guarantor, and usually at the employer's request, the bank discharges the guarantor's services and notifies the contractor.

4.7.4.3 Performance Bond

This type of financial security is separate to cash retention and a bank guarantee, yet operates in a similar fashion to a bank guarantee except it is a single bond usually at 10% of the contract sum and normally issued prior to a start on site. It would be used on a project that will incur high expense during the early stages when the cash retention or equivalent in a bank guarantee is considered low security. The purpose of the bond is for the guarantor to reimburse the employer if the contractor fails to perform under the agreement.

4.7.4.4 Advanced Payment Bond

A contract may include a clause making it an obligation for the employer to make an advance payment to the contractor. This is similar to a loan or deposit that is interest-free and usually paid before the works commence, with the objective of positively cash-flowing the project to pay for site mobilisation costs and deposits for material purchases from suppliers. Where applicable, it is usually 10% of the contract sum with the contractor issuing surety in the form of a bank guarantee/bond before receiving payment. The advanced payment is usually recovered by progressively deducting 10% from the gross amount included in each interim payment certificate until the amount is repaid. The guarantee is effective until the date the advanced payment is fully recovered by the employer, which can be reduced periodically while the works are in progress to reflect amounts progressively recovered. During the construction phase, should the value of variations increase the original contract sum, the contractor may request a further advanced payment to the same percentage as the original. The architect may contest such a request if of the opinion that the initial advanced payment is all the contract permits. However, and in order to qualify for payment, if the contract sum is adjusted the contractor could argue the advanced payment clause is deemed to recognise the contract sum including adjustments. On the contractor's part, this will require a new or replacement bond to the value of advanced payments that can be reduced by the amount recovered from progressive payments at the time of making the request; this may in effect mean the existing bond could be sufficient if it has not already been reduced. Here, and to be effective, the beneficiary (project client) should seek clarification from the bond issuer.

4.7.4.5 Offsite Materials Insurance

Where a clause permits the payment of materials stored off site, the contractor must provide insurance for the goods if such a request is a condition of contract and will receive payment if a policy is in place and the conditions for payment of the goods are met (e.g. adequately protected, suitably stored and complying with the specification under the contract).

4.7.5 Carrying out the Works

The body of a construction contract comprises sections with terms, conditions and titled numerical clauses that may also include sub-clauses, for example JCT SBC/Q 2016 clause 2.3 Materials, goods and workmanship and sub-clauses 2.3.1 and 2 dealing with the contractor's responsibilities regarding materials, goods and workmanship. The titled sections of the body of a contract are discussed in the following sections.

4.7.5.1 Contractor's Obligations

The obligations refer to the execution of the works, which a contractor agrees shall be:

- carried out in a proper manner, including diligently developing the design where the contractor has design responsibilities;
- completed at an agreed cost;
- completed and commissioned on time to ensure the date of practical completion is met; and
- constructed to the quality specified in the contract documents.

The obligations have a distinct link to time, cost and quality. Matters relating to time and cost are straightforward, and is the contractor's agreement to deliver a project on time at an agreed cost the employer will pay. However, quality may be defined differently, and usually means to the standards prescribed in the specifications, design drawings and anything stated in the contract that refers to the design and documentation. If a term in a contract or note in any contractual information states that a product is to be manufactured of 'good quality', it may mean a material free from defects that achieves accreditation to British Standards and may quote the appropriate number and part of the Standard. This is the minimum acceptable standard a contractor is expected to deliver, and applies regularly in a design-and-build contract when a contractor's proposals are accepted by the employer who relies on the contractor's skill and judgment for product selection and installation methods. If a 'high' or 'superior' quality is specified, the employer may be expecting something beyond 'good' quality; the product selection and brands may be specifically mentioned in the contract in this case, together with any methods of workmanship used in their manufacture and installation on site.

4.7.5.2 Possession of the Site

The contract will usually state a date or timeframe by which time the employer must give possession of the site to the contractor to enable the contractor to establish the site accommodation and commence the contract works. However, there may be reasons for deferring the date or timeframe which is usually for the employer's benefit, for example: confirming planning permission; issue of an advanced payment to the contractor and receipt of the respective bond by the employer; or a delay in completing works carried out by the employer which is not part of the contract. Here, the contract may include flexible provisions allowing the employer to defer the date or timeframe to a maximum number of stated days or weeks beyond that stated in the contract. Where applicable, no breach of the agreement will exist if the date for possession of the site arrives and the contractor is hindered because of the employer's responsibilities when a suitable clause allows a period for deferment. However, once the deferment period elapses, and if the contractor is unable to take possession of the site through no fault of their own, at the extreme, and subject to the terms and conditions of the contract, the contractor may terminate the agreement. Where termination on these grounds is not a condition of contract, the contractor should formally request the architect to adjust the completion date which may extend the construction period. In such a scenario the contractor's project manager is usually the person advising of the situation, and writes to the architect exercising the appropriate clauses of the contract with the quantity surveyor copied in on the communications.

4.7.5.3 Supply of Design and Documentation Including Setting Out Information

The onus for the timely release of 'for construction' information will differ between a construct-only and design-and-build contract because the employer and contractor, respectively, drive the supply of information from consultants. Where applicable, the architect in a JCT contract has a fiduciary duty to provide working information in a timely manner, which is communicated via an Information Release Schedule, a document advising when working details will become available and released 'for construction'. This schedule is not usually annexed to the contract as it may not be available at the time of executing the agreement; however, the process could be confirmed in the Recitals by stating the date(s) for the timely release of information driven by the construction master programme. If the master programme or Recitals do not include a date(s), the contractor is at a disadvantage and must coordinate with the architect to agree the date(s). So what should the quantity surveyor do if is there is a delay in the supply of information on a construct-only contract that could hinder progress of the site? Suggestions include:

- refer to the contract or Letter of Intent/Acceptance and the approved programme to confirm there is a delay and is not hearsay;
- notify the contractor's project manager if unaware of the situation;
- keep records of the length of delay;
- consider where the burden of responsibility rests for the delay (i.e. employer or contractor);
- ensure a letter is issued to the architect citing clauses of the contract and works that are in delay, together with reasons; and
- ensure the architect is aware of the extent of the delay in days (an approximation will suffice).

The last two bullet points above would not usually apply on a design-and-construct contract as the contractor will be driving the consultants for release of the information.

One of the first activities on a project involving new works or an extension to an existing building involves the contractor setting out the works. This means working to datums, gridlines, levels and points of reference provided by the architect on behalf of the employer. Normally, the employer will engage a surveyor to identify site boundaries, topography, existing levels and the proximity of existing buildings in relation to the works under contract, and the contractor will work to those benchmarks. If the contractor exposes errors within the information supplied while setting out the works, an appropriate clause will explain what the procedure comprises and the architect will make a decision due to the inconsistency. If during the works the contractor notices an error in the setting out which is of their own creation, the contractor must rectify the situation at their own expense and follow the requirements of the contract that states the procedure in such an event.

4.7.5.4 Errors in Quantities

Under a construct-only contract where a bill of quantities is a contractual document and a discrepancy is found in the bill that alters the measured installed quantity, the contractor must seek an instruction to correct the error. Likewise, the same would apply if scheduled quantities in a contract specification or drawing differ from the actual requirements, as they may form an integral part of the bill of quantities. However, before seeking an instruction, the quantity surveyor must be confident the discrepancy is not the result of a variation changing the works under contract. Moreover, it is necessary to understand the effect of any terms in the contract that may place obligations on the contractor to check for inconsistencies, for example contractor's responsibility to check the design and documentation (including bill of quantities) and the consequences of finding any inconsistencies.

Where there is no doubt on the contractor's part and it is necessary to seek an instruction, the quantity surveyor should inform the architect of the error(s) and seek an instruction before commencing the works with the contract form outlining the procedure. However, a contract could be silent on a situation where the works proceed and the procedure for seeking and obtaining an instruction beforehand fails. If the contractor proceeds and completes the works in accordance with 'for construction' information, the contractor will have carried out a normal obligation to carry out the works which does not change the root of the situation, that is, an installed quantity differing from the contract bills. If this occurs, the quantity surveyor must issue the architect with a priced variation correcting the quantities and at the same time seek an instruction. To demonstrate, let us say a contract bill of quantities has a requirement for an area of $200\,m^2$ of 100 mm thick block walls which the quantity surveyor at some time discovers should be $400\,m^2$ without variations changing the scope of works under the contract or onerous responsibilities on the contractor to warrant the bills. Here, the variation and request for an instruction must seek an amendment to correct the error and not change the works under contract. Once the quantity surveyor has issued the information, the architect may request the client's quantity surveyor/cost manager to make an assessment. If the contractor's request is in fact correct, the conclusion is issued by the architect who may provide an instruction exercising the clauses of the contract to deal with the situation, for example:

> Architect's Instruction (AI) Nr 20
>> Pursuant to clause 2.14 of the Main Contract, the Contractor is hereby issued with this Instruction to amend the works as follows:
>> BOQ 6/3 (d)
>> OMIT
>> Walls in solid concrete blockwork ($1990\,kg/m^3$) size 440×215, skins of hollow wall, 100 mm thick with keyed face, bed and joint with fair finish in gauged mortar (1:1:6), $200\,m^2$
>> ADD
>> Walls in solid concrete blockwork ($1990\,kg/m^3$) size 440×215, skins of hollow wall, 100 mm thick with keyed face, bed and joint with fair finish in gauged mortar (1:1:6), $400\,m^2$
>> The value of any difference is to be assessed based upon rates in the Bill of Quantities.

Naturally, with a design-and-build contract, no such scenario usually applies as the contractor is responsible for quantities.

4.7.5.5 Inconsistencies with the Design and Documentation Supplied

If there is an inconsistency with details on drawings or specifications that contradict each other, or a construction operation is deemed unachievable due to a defect in the design, the architect must be advised. This is to allow the issue of new and/or updated information or for the architect to provide a written instruction correcting the inconsistency. Furthermore, if the contractor considers that additional details for working purposes are required, or believes the architect is unaware of the timing for the issue of additional information, a clause in the contract will permit the contractor to make a request in advance. Care should be taken by the quantity surveyor or project manager before making such a request, because some JCT contracts include terms where the contractor agrees to design 'discrete part(s)' of the works known as 'contractor's design portion' (CDP). Here, the architect or other designers engaged by the employer may complete the design to a certain stage for the contractor to complete, and it may be a wrong assumption by the contractor to make a request to alter a design to allow it to be completed for practical purposes. Where all of the design responsibility rests with the architect, there is a fiduciary duty on the designer's part to acknowledge such a request and furnish the contractor with new and/or updated information. By contrast, with a design-and-build contract, a contractor must rely on their own resources to drive consultants for the supply of the information.

The supply of new and/or updated information does not usually pave the way for a variation as, in effect, the works may not have changed. This is because the additional information may be issued to clarify ambiguities and amplify critical details to reduced scales, or is the result of the final coordination between consultants. The additional information may also be for works included as a provisional sum that the quantity surveyor must assess for the process of adjusting the contract sum.

An incomplete design is usually finalised with the creation of shop drawings, which are usually produced by subcontractors for the contractor. The supply of shop drawings and the seeking of approval is not the correction of inconsistencies in the master design, and is a process of good practice on the contractor/subcontractor's part to demonstrate how specific works will form a part of the whole works and can apply to temporary or permanent works. Where a contractor is appointed on a construct-only contract, if there is an inherent error in an initial design that goes unnoticed in shop drawings which are approved by the employer's consultant the contractor is not usually responsible for the affects. Furthermore, should there be a situation on a construct-only contract where goods are designed, produced or installed that inherit or may inherit problems that are observed by the contractor, the contractor must seek an instruction on the action they are to take. However, with a design-and-build contract, in the event of an error going to production from an approved drawing with the error only noticed once the product is installed, the contractor cannot normally claim for financial reimbursement for rectification, as in effect the contractor has committed the error.

Where there are inconsistencies with the conditions of contract, the contractor can only notify the architect of the situation and not seek remedy, as the architect is not usually empowered to issue an instruction involving ambiguous or contractual errors such as clause writing. Here, the matter would need to be addressed by the employer and contractor who may resort to legal remedy with the use of implied terms (see Section 4.7.1 above for details of expressed and implied terms) or Rectification (discussed in Section 4.7.5.7 below).

With normal industrial practice, the prioritised sequence of design and documentation for carrying out the works follows an order of precedence. This means that the contract may provide remedy where a specification and drawings conflict, for example where finishes shown on the drawings conflict with the finishes schedule in a specification, an order of precedence may state the drawings take priority, or vice versa. An order of precedence clause may also clarify inconsistencies with design disciplines. For example, works of an electrical nature stated in the electrical specification and shown on the electrical designs will take precedence over an architectural specification or drawing if including items of an electrical nature. By contrast, if the electrical drawings show fixing details to partitions that show wall insulation and the insulation does not appear on the architectural details then, and subject to the contract, there is no insulation because the requirement for insulated partitions is an architectural precedent. Similarly, if there is reference on an architectural drawing of a large scale, say 1:100 that refers to details on an architectural drawing to a reduced scale of say 1:5, the information on the reduced scale is normally the working document. However, the wording in the contract may override normal industrial practice and, if applying, the order of precedence in the contract must be complied with. If the contractor notices such ambiguities an instruction must also be sought, and an astute contractor will not rely fully on any order of precedence.

4.7.5.6 Divergences

Changes required in the design and documentation to comply with statutory requirements including building regulations are called divergences. With a construct-only contract, methods of dealing with such changes are usually stated in the contract with the contractor not enticed to actively seek divergences, yet has a duty to advise the architect if becoming aware of their existence. For example, if on a construct-only contract roof trusses are designed to be spanned 600 mm apart which are installed and later found to require additional bracing due to high wind exposure, the contractor will rely on the information supplied and does not need to question the inclusion of bracing in the first instance. This is because the design consultant is deemed competent to provide the information and, as such, the contractor would be entitled to extra costs for installing the bracing and should receive an instruction.

A contractor must not alter the design or change any completed works to comply with statutory requirements without receiving an instruction, as the contractor may be in error with their assumption (e.g. where planning permission is granted based upon an earlier version of the building regulations, with the changes unnecessary). An instruction may be issued as emergency works, for the rectification of works about to start or for works already completed to comply with the statutory requirements, which the contractor must oblige. However, if a drawing or specification contravenes the building regulations or other statutes with the works installed and the contravention going unnoticed, the contractor is not normally in breach of the contract because they failed to notify. The quantity surveyor needs to be aware of any divergences clause in a contract and the implications, as any changes that come into effect may involve additional expense. If the architect issues an instruction to change something without mentioning reimbursement, with the change involving expense, the quantity surveyor needs to assess the cost and seek a variation to the contract sum if the need for divergence is not due to the contractor's negligence. However, where divergences occur under a design-and-build contract, it is normally the contractor that accepts responsibility for compliance with the building regulations.

4.7.5.7 Vague Contractual Terms and Errors

Where there is a discrepancy with the wording in a contract, any genuine errors are deemed acceptable to change by common sense unless there is an expressed term stating to the contrary and what the procedure should be. This would apply to the likes of clerical errors and the use of wording that contradicts numbers (e.g. too instead of two), with the need to change not detrimental. Once a contract is executed, it is only possible to change any terms through a court of law, and a vague or ambiguous term inserted into a standard form may be serious to the contractor if overlooked. For example, a clause on payments may state:

approximately (inserted in ink)

The final date for payment pursuant to an Interim Certificate shall be ^14 working days from the date of issue of that Interim Certificate.

Here, a contractor could expect to receive payment no later than 14 working days after the issue of the architect's interim certificate. However, because the word 'approximately' is inserted, a situation may eventuate where no payment is received after 14 days, or the contractor does not receive notification of why the payment is withheld because the employer is relying on the word 'approximately'. If the contractor and employer have an oral agreement regarding a term such as the above, and there are notes of discussions that failed to be included in the contract, a court can amend the terms of the contract through Rectification. This is remedy where a form of contract has an appendix or alteration created in hindsight which is endorsed by the parties to say what was intended to be said in the first place that, by mutual error, was omitted from the form. Changes are granted by equitable remedy instead of legal alterations at the discretion of a court, as long as the Rectification is legal and the parties to the contract give consent. This makes it different from an implied term inserted by a court because it does not involve case law or statute and is merely a correction the parties mutually wish to address. In the event of a vague term for the payment of interim certificates being included in a contract as discussed above, an endorsed alteration may say '…approximately 14 days and no more than 21 days'. However, where an Entire agreement clause exists in a contract, this may nullify the chances of obtaining Rectification. The effect of an Entire agreement clause means that any notes, schedules, appendices, etc. relevant to a pending agreement that were available to the parties at the time of executing the contract will be ineffective unless they are included in the contract.

4.7.5.8 Unfixed Materials

A contractor may or may not be entitled to receive payment for unfixed materials with the outcome dependent on the conditions of contract. If the contract is silent on the matter, it will mean the contractor is not entitled to payment. However, and where permitted, entitlement usually applies to materials stored on or off the site pending installation as permanent works, and not temporary works such as scaffolding and formwork.

Contractors generally seek contract forms that include provisions for the payment of unfixed goods. The reason behind this is that payment aids cash flow, as the contractor will need to pay suppliers for materials once they are delivered to site, or stored in designated locations. The same applies to subcontractors that may only agree to accept an award from a contractor if they will be paid for unfixed materials, so they can pay their suppliers.

The amount to include in an interim certificate payment to the contractor for unfixed goods is subject to assessment and is usually carried out by the client's quantity surveyor/cost manager with contractor input. Usually, the contractor will qualify for payment if goods are purchased specifically for the project, adequately secured, not defective, comply with the specification as verified with test results and are not purchased prematurely. Not purchased prematurely means goods that can be installed within a reasonable timeframe after their purchase and delivery. However, what is a reasonable timeframe is rarely defined in a contract, and can be many months if goods are only made available to suit a specific manufacturing cycle. Materials considered as purchased prematurely may not qualify for payment if their installation date is distant without justification for their early delivery (e.g. internal doors, when the foundations are under construction with the programme stating doors are due for installation 12 months after constructing the foundations with the programme not ahead of schedule, or the doors on long delivery periods or only available because of a specific manufacturing cycle).

Materials stored off site may be kept in store for long periods, with the actual duration depending on the construction programme for which the contract may require the seller to provide insurance as surety. As the transition of goods from being unfixed to installed is an ongoing process, it can be difficult to determine their precise value at a given time for inclusion in a valuation, and the contract may state the contractor will receive a percentage payment of the assessed amount for interim payment purposes, which is typically 70–80% of the assessment. This errs on the side of caution in favour of the employer in the event of overstating/estimating the value and, where applicable, must be a condition of contract and not an afterthought. Once the contractor receives payment, the goods become the property of the employer and the contractor has a responsibility for their safeguarding.

4.7.5.9 Rectification Period

The contractor is responsible for defects in the completed works for a defined period. This is usually 6 or 12 months after the date of the certificate of practical completion which triggers a defects liability period or Rectification Period as referred to in JCT contracts. If the employer sells the completed building or rents space to tenants, the contractor is bound to the employer only for the defects and not the new owners or tenants as, in effect, the contractor has no agreement with these entities.

A defect generally refers to shrinkages, faults and inherited problems associated with design (where applicable), materials and workmanship, that are a responsibility of the contractor. During the Rectification Period, the employer is responsible for general maintenance and the cleaning of installed parts, and the contractor must understand the difference between general maintenance and defects. For example, grouting between ceramic tiles requires general maintenance by periodic cleaning due to natural wear and tear, and will not become a defect if the grout loses colour during the Rectification Period because of a lack of periodic cleaning. By contrast, if during the same period the glazed faces of the tiles show crazing, it may be due to a co-efficiency of expansion of the glaze over the tiles and will become a material defect. In this scenario, if tile supply is a contractor's responsibility under the contract, the defect will be a responsibility of the contractor. The fact the contractor may have subcontracted the works to a trade tiling contractor is irrelevant to the employer and the conditions of contract, and the contractor must attend to such defects with or without subcontractor involvement. The quantity surveyor must therefore ensure any trade package agreement between the contractor and subcontractor which includes the supply and installation of materials and commissioning of works by the subcontractor includes a condition that the subcontractor is directly responsible to the contractor for defects during the Rectification Period.

4.7.6 Delays in Carrying Out the Works

Time is crucial for the successful delivery of a construction project, and a contractor's best endeavours are to maintain the programme so that the project starts and completes on time. A contract will usually include a clause outlining the procedure to follow for an adjustment to the completion date when a delay is deemed the employer's responsibility. However, if the contractor has caused the delay without any employer involvement, the clause for an adjustment of time will not apply and the contractor must accelerate the programme to put the works back on track to achieve the end date. As the focus on this part of the chapter is a review of the main contract, the discussions below address the effect of prolonging the works that require the architect's involvement when the responsibility for delay rests with the employer.

4.7.6.1 Adjustment of the Completion Date

The term adjustment is probably a better word to describe the change to an end date, as it is possible to bring a date forward as well as extend it. For example, if a contractor takes possession of a site by a date included in the contract and the employer halves the scope of works soon after, it could mean the contractor may finish early. However, the contractor usually has no contractual obligation to issue an adjustment of time reducing the contract period even if such a request is made by the architect, as it is normally beyond the contract. However, and if requested, the contractor may formally reduce the construction phase in a collaborative fashion and, when doing so, must ensure the new date is realistic (e.g. halving the scope of works may not halve the project duration). Unfortunately, due to the nature of construction projects, requests to adjust the end date usually involve delays, which can only be requested by the contractor citing specific reasons in writing to the architect. The JCT Contracts include a procedure where the contractor must issue a notice of delay for each and every occurrence that would lead to an extension of time, which must be linked to one or more Relevant Events. Referring to clause 2.29 of JCT SBC/Q 2016, these events are:

- variations (i.e. any works that are a variation to the original works that delay the programme);
- the issue of one or more specific architect's instructions;
- the lapse of a deferment period hindering the contractor's possession of the site when an employer's responsibility;
- where a bill of approximate quantities is not a reasonable forecast;
- where the contractor suspends the works, is later vindicated and recommences operations;
- where the employer's design team or contractors impede, delay or default the works;
- delays caused by statutory undertakers;
- exceptional adverse weather conditions;
- loss or damage caused by 'specified perils' as listed under the insurance provisions;
- civil commotion (e.g. the use or threat of terrorism and imposed method(s) of dealing with them);
- UK central government, local and public authorities or statutory authority involvement;
- industrial actions (e.g. strikes or lock-outs effecting the supply of materials for the project);
- *force majeure*, French for 'superior force', generally referring to an 'Act of God' if not included as 'Specified Perils' or considered exceptional adverse weather conditions;
- insolvency of a named specialist; and
- delays associated with the management of fossils, antiquities and the like.

Where this form is adopted, only the aforementioned circumstances can give rise to an extension of time being granted with the architect/contract administrator only empowered to acknowledge such circumstances. If the contractor seeks an adjustment to prolong the end date, the Relevant Event(s) must be referred to in the request including documentary evidence (e.g. drawings, reports, photographs, citations, meeting notes, etc.).

Clauses of contracts are often reluctant to include provisions for prolonging the end date following the insolvency of one or more domestic subcontractor(s), as it is generally considered the (main) contractor's risk. The contractor must therefore safeguard any programme float for their own use to deal with the unexpected, including situations where a subcontractor(s) succumbs to business failure as the appointment for a replacement may take some time. The exception is if a subcontractor is mentioned in the JCT form where the architect has influence with the selection. Here, a contractor must issue an award to the said subcontractor even if there is concern with the subcontractor's business vulnerability and, in the event of insolvency, the contractor must follow the procedure stated in the contract. Here, any time delay in appointing a substitute subcontractor because of the insolvency of a first could fall under the category of variations, thus becoming a Relevant Event with the contractor possibly entitled to an extension of time, providing of course the process is not hindered by the contractor.

The JCT 2016 forms of agreement decline involvement with the use of nominated subcontractors, with the nomination process itself somewhat outdated. However, it is possible for a contractor's quantity surveyor to be engaged on a project where the nomination process exists under a suitable form and would need to address relevant clause(s) in the event of a subcontractor becoming insolvent and seek an extension of time as necessary.

4.7.6.2 Liquidated and Ascertained Damages (LADs)

Liquidated and ascertained damages (LADs) is a sum of money stated in a contract, usually at a rate per day or week or fraction of a percentage of the contract sum, that a contractor agrees to pay the employer if the contractor defaults during the course of operations resulting in a delay to the agreed date of sectional and/or practical completion. To be effective, damages are ascertained by calculations that reflect the anticipated loss the employer would endure through a delay over which it has no control (e.g. lost income from the rent of new properties because the project is delayed) which, subject to the contract, may be capped to a maximum amount. However, a contract may omit reference to the word ascertained, indicating the liquidated damages may be a general assessment without fine logic. Nevertheless, and where applying, these damages aim to represent an amount the employer considers will be lost because of a delay that can be recovered through the contract.

LADs do not need to be proven and are usually included in the invitation to tender documents for advice to competing contractors, so there is no surprise when executing the contract. In the event of a delay attributable to the contractor, the computations for applying damages are straightforward to calculate and are assessed by the architect. For example, if LADs written into a contract are £500 per calendar week and the contractor completes six calendar weeks later than agreed, the assessment will be:

$$6 \text{ weeks} @ £500.00 \text{ per week} = £3,000.00$$

Contractors try to avoid LADs by completing on time and, as the quantity surveyor is involved with cost aspects of the project, it is necessary to monitor site progress by reviewing the construction programme in comparison with the works in progress (which can be by periodic visual inspection). Where the contractor has not caused a delay the quantity surveyor must exercise contractual obligations and, in conjunction with the project manager, seek an extension of time to the end date to mitigate the possibility of LADs being applied, even if the contractor is considered proportionally responsible for the delay.

Where an architect grants an extension of time, it may not be to that requested by the contractor. Time is said to be 'at large' where there is disagreement between the contractor and the architect/employer regarding the revised date of completion, and applies when the employer is responsible in full or in part for a delay and cannot agree with the contractor what the end date should be. In this scenario, if time is 'at large' at the date of practical completion liquidated damages cannot normally apply, meaning a contractor may possibly avoid them. Where occurring, a contractor must not see this as an advantage and exploit the situation if the 'drop dead' end date is in dispute, as any delay will most probably incur a cost to the contractor's preliminaries that is not recoverable through the contract. Furthermore, a contractor may incur increased charges from material suppliers and subcontractors as they are completing late, and would need to recoup inflationary amounts as they are generally not directly responsible for the delay.

4.7.6.3 Delays in Commencement and Effect on LADs

Care is required with understanding the date of commencing a project written in a contract and the date of practical completion because a delay to the date of commencing does not have the same consequences as delaying the date of practical completion. Contract forms may omit to state the construction period in weeks and allow them to self-calculate from a stated date of commencement through to the date of practical completion. For example, a works programme may show the overall operations as commencing and finishing in 52 weeks irrespective of dates and, if included in a contract, would bind the parties. To be instrumental, the programme must include a start date which is driven from the date the contractor takes possession of the site. If this date is delayed, the contractor must make a request for an extension of time to the end date, making allowance for any stated deferment period. In resolving this, the architect may extend the start and end dates by the maximum permitted deferment, thus not altering the construction period. For example, if the permitted deferment in an agreement is up to 4 weeks and there is no change to the construction period which is to run for 52 weeks, the consensus will be:

> ORIGINAL: Start 1 January and finish 31 December = 52-week construction period
> REVISED: Start 1 February and finish 31 January = 52-week construction period.

This is an important distinction because the contractor may perceive the delay as giving rise to a 56-week programme that creates a one-month delay and may make a claim for staff and preliminary expenses because of a longer programme. This is clearly not the case because only the dates have altered and not the construction period; the contractor would be liable for LADs after 31 January if failing to complete on time because of their own performance.

4.7.6.4 Partial Possession by the Employer, Project Completion and Lateness

Clauses in a contract may include for partial or sectional completion of a project for taking over by the employer prior to a date of practical completion of the whole of the works. Such provisions allow the employer to conduct business operations, albeit restricted to the appropriate section(s), while the contractor still has a site presence during the construction phase. This can occur with a newly built building that has commercial businesses on the ground floor operating as retail outlets and residential apartments on upper levels, with the ground floor handed over before the residential units. Here, a precondition would be for the architect to issue a sectional completion certificate to permit the commercial businesses to operate.

A sectional completion date(s) must be written into a contract stating the part or parts and cannot be introduced as an afterthought. The quantity surveyor will need to recognise this date(s) because proportionate LADs may apply for non-completion. When a sectional completion date as stated in the contract or an approved revised completion date arrives, it is not ignored by the architect if the works are incomplete. Here, the architect issues a certificate of non-completion, or a notice of non-completion by the employer's agent if a design-and-build project. The irony here is that if an extension of time to that date exists, it automatically cancels the certificate of non-completion; once the revised date of sectional completion arrives and the works are complete, the certificate of completion is issued.

Practical completion occurs when the contractor completes a project, which also includes any previously certified sections when, in the architect's opinion, the completed works complies with particular contract clauses. In a similar fashion with sectional completion, once the agreed date of practical completion arrives, the formality for issuing a certificate of non-completion applies if the works are incomplete, which is automatically cancelled if an extension of time exists with the occupational/practical completion certificate only issued once works are complete. So, what happens if either a sectional or practical completion date arrives and an extension of time has been requested but not granted, and the architect does not issue a certificate of non-completion? In short, the answer can be found in the wording of the contract. Under clause 2.28 of JCT SBC/Q 2016 for example, up to 12 weeks after the agreed date of sectional or practical completion, the architect is granted discretionary powers to issue an extension of time. During this 12-week window, the employer cannot deduct liquidated damages as in effect there is no certificate stating the works as incomplete. The architect however has a contractual duty to issue a certificate of non-completion after an inspection and, at the same time, may issue an extension of time.

There is a fine line between the 'drop dead' date for sectional or practical completion stated on a completion certificate and the actual date the employer takes over. Although the procedure is systematic, the physical occupation of a building is a somewhat different process. The employer may not be in a position to take over on the date of the issue of a completion certificate, which may be because of timing on the employer's part or due to final testing and commissioning in progress, with the contractor possibly in the process of demobilising the site accommodation and making good the area disturbed. It is therefore not unusual for one month to pass after the certified date of completion before the employer occupies a building as the project in essence is complete, albeit for final testing and clearance of the site accommodation.

4.7.7 Control of the Works

A construction project requires managerial control of the works as part of good practice which, in its various forms, is driven by the selected procurement route that influences the degree of involvement from the contractor, employer, architect and other employer-appointed parties.

4.7.7.1 Representatives and Right of Access to the Works

Clause 3.1 of JCT SBC/Q 2016 requires at all reasonable times for the contractor to permit the architect, or persons authorised by the architect that are representatives of the employer, to access the works. This is subject to any reasonable objections the contractor may make to the architect/employer to protect proprietary rights, that where applicable must be respected by visiting parties. In addition, the employer and employer's representatives must also respect the contractor's responsibilities regarding health and safety by complying with the site requirements (e.g. use of protective clothing and reporting to the contractor upon arriving on site by signing the visitors book). Moreover, and at all reasonable times, the contractor must appoint a person in charge deemed competent to accept the delivery of instructions on behalf of the employer. Subject to the contract not stating otherwise, a competent person can be any member of the project team and can include the quantity surveyor if based on site.

During the term of the contract, representatives of the employer may be replaced at the employer's request. However, under JCT SBCs 2016 clause 3.5, this is limited to the architect/contract administrator and client's quantity surveyor/cost manager. When activating this clause, the replacement professional is not empowered to reverse any previous instructions, certificates or variations. However, the contractor loses this right if the employer is a local authority and the nominated replacement is an official of the local authority.

4.7.7.2 Assignment

Assignment is the legal transfer of the rights and duties of the contract to another, and can be activated by the contractor, employer or both at any time the contract is active. As a rule, construction contracts discourage assignment, which may be forbidden by an expressed term (silence on the subject also means assignment cannot take place). Where permitted however, it may be included as a provision or for future planning, and to be effective must be expressly included in the contract without the need to mention the name/names of the new party/parties, as such criteria may be concluded under an alternative contract. When reviewing the form of main contract, the quantity surveyor and project team will need to be aware if assignment is included in the agreement; if permitted, it may mean one or both parties to the agreement could be replaced during the construction phase or period thereafter, when the contractor is responsible for defects.

4.7.7.3 Subcontracting

Subcontracting is the delegation of a duty or duties by the contractor to another under the contract, which by performance does not absolve the contractor from its obligations and duties to the employer. This means that when subcontracting the works, responsibilities are not legally sublet with subletting possibly expressly forbidden by the contract. Unless nominated under the contract to enter into a tripartite agreement with the employer and contractor, subcontractors are domestic subcontractors to the contractor and may be:

- selected by the contractor only;
- selected by the contractor only with consent from the architect;
- selected from a preferred list in the form of contract;
- named in the form of contract for mandatory appointment (JCT IC 2016 only);
- listed in the contract bills for selection by the contractor (e.g. clause 3.8 JCT SBC/Q 2016); and
- listed as named specialists (e.g. item 9 of Schedule 8, JCT SBC/Q 2016).

Once appointed, domestic subcontractors will rarely have any contact with the employer or architect, and agree to carry out their services to the contractor with the contractor supervising and coordinating the works. However, it may be a condition of contract that a contractor also supervises and coordinates works to be carried out by contractors engaged by the employer, that JCT refers to as 'Employer's persons'. These are not subcontractors in the real sense, as they are not bound to the contractor because of their private agreement with the employer.

4.7.7.4 Statutory Control

Works under contract must comply with statutory regulations and be in sync with the appropriate standards and building regulations for a project to be legally compliant, which is usually confirmed in the contract documents as a matter of course. Where a part or parts of a project (which can mean design, works or a mixture) is potentially in breach of legislation because of an error(s) in the design, or the works are completed with the error(s) included, such error(s) must be rectified using the Divergences clause of the contract (discussed above in Section 4.7.5.6). Statutory control also applies to the CDM Regulations, which means any project falling under the umbrella of the CDM Regulations must be acknowledged by the employer and the contractor.

4.7.7.5 Instructions

Due to the level of employer control on a construct-only contract, clauses of the agreement contain provisions permitting the architect to issue instructions to the contractor. This does not mean to say a contractor must comply with every instruction as there may be grounds for disagreement, yet the contractor must acknowledge the instruction by either complying or objecting. Rarely will a contractor wilfully ignore an instruction; however, if the contractor is not diligent, a clause of the contract may be triggered for the ignoring and what the action involves. For example, clause 3.11 of JCT SBC/Q 2016 empowers the architect to give the affect for non-compliance, meaning the employer can appoint others to carry out the works and recover monies incurred from the contract sum. Situations where instructions can be issued include:

- variations, including the expenditure of provisional sums;
- postponement of the works;
- notification of the contractor's compliance and non-compliance with existing instructions;
- intentions on works not considered in accordance with the contract;
- the removal of work and/or exclusion of person(s) from the works; and
- the right of the architect or appointed clerk of works to carry out inspections including opening up and testing of the works.

While most standard forms of contract include clauses for the issue of instructions, they often omit to state the mode of issue. This can be detrimental to the contractor, meaning it may lose the right to a variation if an instruction is oral and not confirmed in writing or the confirmation received is different from the understanding of the oral instruction. In practice, an oral instruction from a titled person authorised in the contract to issue an instruction triggers the appropriate clause. The authorised person is usually (but not limited to) the architect. If a contractor complies with an instruction issued by someone other than the architect (or other named) and carries out works to that instruction, they do so at their own risk. This may be a bitter pill for the contractor to swallow, especially if dealing with a number of entities appointed by the employer. For this reason the quantity surveyor and project manager must ensure the project team are aware of the entity or entities authorised to issue instructions.

Instructions from consultants, representatives of the employer including clerk of works and even the employer may have no effect unless they are named in the contract as an authorised person to issue an instruction. So, what happens if a contractor purchases and hangs doors in accordance with an approved design and the employer visits site, instructs the doors to be replaced, and verbally advises the contractor it will be paid for the replacements, with the employer not empowered to issue an instruction under the contract? Here, the contractor should do nothing until instructed by an authorised person named in the contract and politely advise the employer of its position. However, if an instruction is issued by anyone representing the employer that merely enforces the contract because the issuer is of an opinion that part of the works fails to meet the contract (e.g. a wall that is the wrong thickness), the contractor must comply if the facts are correct. This is because the instruction is a reminder to the contractor of their obligations under the contract. If the issuer of an instruction enforces the contract and cites a clause for non-compliance, the issuer can only do so if permitted by the contract; otherwise, they must contact the authorised person to issue an instruction.

In practice, anyone issuing an instruction may do so in the interest of the project and should inform the architect of their actions; if not, the contractor must make contact with the architect as soon as possible advising of the contents of the instruction if not from an authorised person. For example, a building inspector may require foundation trenches excavating to a deeper level than shown on the drawings and may have issued an oral instruction to that affect. Here, contractually, the oral instruction may have no affect and will need confirming by the architect; a contractor may actively seek such confirmation by mail, telephone or email if agreeing with the instruction. With this scenario, the building inspector is a person with delegated authority with the architect relying on his/her impartial judgement and will usually issue the instruction. Here, the architect could issue an instruction for the additional works and at the same time cite the implications, as there may be a provisional sum for additional excavation and concrete filling included in the contract sum which may be unknown (and generally of no interest) to the building inspector.

It is important the contractor understands that receipt of an effective written instruction must be authorised by an approved signatory, which may not be defined in the contract. For example, JCT standard forms of contract cite the architect/contract administrator as having the authority, and that is all that is said. In a scenario where the architect/contract administrator is a professional practice, it may lead the contractor to conclude that only the proprietor of the practice can sign an instruction. This may be a wrong assumption and, at the offset of a project, clarity should be sought and must come from the practice that would divulge the name(s) and title(s) of practice members with the delegated authority to sign an instruction; in construction projects this is usually in the form of an Architect's Instruction (AI). An AI is usually numbered and dated on the architectural practice's letterhead and, to be effective, must be signed by an approved signatory on behalf of the practice. Where a project is of a salient engineering nature, works are normally supervised by an engineering consultancy that issues Engineer's Instructions.

The contractor must be aware of the conduct of representatives of the employer empowered to issue an instruction that is neither oral nor written. For example, if on a construct-only contract a set of revised or new drawings is issued to the contractor that adds another level to a building without a signed instruction, is it still an instruction? On the face of it, it would appear the drawings are in fact an instruction to treat the works as a variation. However, the quantity surveyor and project team would be unwise to proceed with the works unless the drawings are issued with an instruction signed by an authorised person. It is likely in this situation that the conduct is actually a request for a quotation only, with the works not to be constructed until the quotation is approved which must be issued with an instruction. For this reason, any type of written instruction involving expense must be retained by the quantity surveyor for assessing as cost variations and their possible effect on the project in terms of time.

4.7.8   Cost Variations

Variations are alterations or modifications to the works under contract initiated by the project client/employer that add, delete or make substitutions to the works, and can be instructed before or after the original works are complete. Subject to the contract, a variation can also involve a change of working sequences at the employer's request, as well as a change of obligations or the imposition of restrictions (e.g. night shifts or weekend working). More often than not these result in cost variations borne by the employer, and also applies to the evaluation of dayworks, contractor's design portion (CDP) and assessment of the expenditure of any provisional or prime-cost (PC) sums included in the contract. By contrast, a variation may not involve cost. For example, a specified wall tile that is not ordered, delivered to site or fixed can have the colour changed providing it is of an identical size, pattern and manufacture and can be procured to meet an unaltered programme. However, in this scenario the requested change requires recording as a nil cost variation to ensure the materialistic change occurs, that is, the variation must be communicated as accepted by the architect, usually via an instruction for the change to happen. Moreover, cost variations do not usually apply if the architect issues an instruction to change works already constructed or to replace stored materials that do not comply with the works under contract or approved variations. In this scenario, the contractor must usually correct the errors at its own cost.

With a construct-only contract, potential cost variations are valued by the employer's quantity surveyor/cost manager for cost advice to the employer who may be exploring opportunities to change the works to provide benefit to the completed building or make changes to suit statutory requirements. Depending on the timing of a change request, the contractor may be asked to provide cost advice, especially if of a specialist nature such as the internal reconfiguration of elevators where a vertical transportation contractor has been appointed by the contractor that could provide a quotation. The scope and number of change requests made by the employer can be significant and will only involve the contractor when implemented, which the architect confirms to the contractor with an instruction. In general, cost advice steering a decision towards issuing an instruction (which can increase or decrease the contract sum) is of no interest to the contractor. However, the employer's quantity surveyor/cost manager will wish to avoid the embarrassment of cost shocks if the contractor's variation quotation is somewhat different. The contractor's quantity surveyor must therefore not be influenced by knowledge of any budgetary amount for a change and must value the variation quotation appropriately, and has a duty to provide evidence to substantiate the price submitted. Furthermore, where the contractor has provided prior cost advice at the request of the client's quantity surveyor/cost manager, the variation quotation should bear resemblance if the actual instruction does not deviate. It is equally important to ensure the client's quantity surveyor/cost manager complies with clauses of the contract that define rules for assessing quantities and rates that drive the value of variations, an example being the correction of errors in a bill of quantities as discussed under Section 4.7.5.4 above.

Once the employer's quantity surveyor/cost manager makes an assessment of the contractor's variation quotation it is issued to the architect, and usually approved or amended for including in interim and final payment certificates. Where a contract is a lump-sum price, the architect adjusts the contract sum to a revised value that includes the total of approved cost variations. To communicate this information, it is normal to provide the contractor with a notice of the revised contract sum which may be carried out after each variation assessment. Each contract sum adjustment is numbered for reference with the amounts stated as cumulative. For example:

> Contract Sum Adjustment Nr 8
> Original Contract Sum: £5,000,000.00
> Value of Approved Variations: £ 500,000.00
> Revised Contract Sum: £5,500,000.00

In this example any amount included in earlier contract sum adjustments will be superseded, and the revised contract sum will be the employer's financial liability in a final account once the said works are complete.

The architect will only issue a contract sum adjustment based upon the appropriate clause(s) of the contract, which in general with JCT contracts involves:

- the value of work cost variations and dayworks as described above;
- the permanent or temporary withholding of money in relation to: (a) fees and charges in respect of the works; (b) the cost of carrying out inspections and tests following an architect's instruction; (c) cost and expense associated with recommencing works following a suspension of the works with the contractor exercising the right to make a claim; (d) fluctuations in payable amounts to named subcontractors; (e) fluctuations in payable amounts due to levies, contributions and revised taxation; (f) loss and expense claims; (g) insurance purchase for the works under the Insurance Options Schedule 3 (either by the contractor or employer); (h) restoration costs borne by the contractor following an incident where the employer purchases insurance for the

works; (i) incorrect supplied details regarding levels/setting out information that alters the scope of works; (j) the cost for works carried out by others because the contractor ignores instructions; or (k) insurance for terrorism cover (policy extension and premium adjustments);

- the deduction of provisional sums or the value of any works assessed with approximate quantities in the contract that are reassessed as work cost variations;
- the cost to repair defects after a date of practical completion which are not carried out by the contractor; or
- changes in premiums payable where the Joint Fire Code applies.

Ideally, the value of approved variations should match that submitted by or agreed with the contractor. If there is disagreement the contractor may either accept or reject the amount and, if rejecting, may refer the matter to dispute resolution. Normally, in the event of a dispute, the works must be carried out as intended with the dispute procedure not vitiating the contract.

The arrangement for approving works variations in a design-and-build project is usually different because there is no employer's quantity surveyor/cost manager during the construction phase. Here, the employer may deal directly with the contractor or issue the employer's agent (if engaged), with an approval notice committing to an agreed amount that adjusts the contract sum for issue to the contractor.

4.7.9 Payments

The first payment a contractor normally receives is the advanced payment, providing the contract has such a provision and the contractor provides the appropriate bond(s) to qualify for payment. The mechanism driving this is receipt of an interim payment certificate with the amount not subject to retention. Thereafter, clauses in the contract set the framework for timely payments to the contractor, which include dates for the issue of further interim payment certificates and a final payment certificate prepared by the architect stating an amount due. A characteristic of construct-only contracts using the JCT model forms is the level of involvement from the employer's quantity surveyor/cost manager that assesses a value of the works on behalf of the architect required for the preparation of each payment certificate, and can be with or without contractor input. The contractor's quantity surveyor must not become complacent with such situations, and must assist the process by providing interim applications for payment as well as a final account in a timely manner.

In order for the contractor to qualify for payment, an architect's payment certificate is issued to the employer with a copy sent to the contractor; to be effective, a certificate does not need signing by the architect in charge and can be authorised by a person acting as payment certifier on behalf of the architect, which is usually in the business name of the architectural practice. However, the architect in charge should advise the contractor of the authorised names/entities empowered to authorise such certificates. Should the employer not contest the amount due on a certificate, there is a duty to pay the sum within the terms and conditions of the contract. A contractor may assume receipt of the architect's payment certificate is a guarantee for payment and may proceed to make plans based upon this assumption. However, this is not the real situation because contracts have clauses permitting the employer to advise the contractor if the amount stated as due on a certificate will be reduced, and if the reduction is to be temporary or permanent. The requirement to provide a 'pay less notice' is legislated under the Construction Act and the Local Democracy, Economic Development and Construction Act 2009. This latter act amends the Construction Act with the requirements of both acts reflected in standard forms of contract. A pitfall for the employer here is that a 'pay less notice' must be issued and received by the payee within a defined period, with failure to comply resulting in the full amount becoming due. If the employer complies with the contract and legislation by issuing a 'pay less notice' within the specified timeframe, the contractor's remedies upon receipt include:

- accepting the change by acknowledging any oversight by the contractor and architect;
- rejecting the change and referring the matter to dispute resolution as a contractual right; or
- suspending the works in full or in part using the contractor's right of suspension clause.

If a contractor elects to suspend the works and is later vindicated and paid in full with the suspension lifted to allow the works to recommence then, subject to the terms of the contract, the contractor can issue a claim for loss of interest on the outstanding payment and request an extension of time to the end date.

The quantity surveyor should be committed to ensuring payments to the contractor are received on time, as it is a reflection on the work in progress and a reward for the effort that goes into preparing each interim application. Of course, the income maintains the solvency of the contractor and is also the financial lifeblood for the project.

4.7.10   Termination

A successful way to discharge a contract is with performance when both parties complete their obligations with the project satisfactorily concluded. However, it is possible to terminate a contract any time it is active and, in the event of this unfortunate occurrence, contracts usually include clauses to address the situation. Where applying, contract clauses list grounds for termination (and the consequences) when sought by the employer, contractor or either party. Following clauses 8.4–8.6 of the JCT SBC/Q 2016, the employer may terminate the agreement before practical completion because of any of the following:

- contractor's insolvency;
- default by the contractor, including situations where the contractor is responsible for the whole or substantial suspension of the works;
- failure by the contractor to perform regularly or diligently with the progression of the works;
- contractor's negligence in repairing defective works and/or not complying with certain instructions;
- the contractor is not complying with the CDM Regulations;
- the contractor has assigned the contract without consent;
- the contractor fails to procure subcontract works in accordance with the contract; or
- the contractor is guilty of corruption and, where a public employer, impact of the Public Contracts Regulations 2015, 73 (1) (b) where the contractor should have been excluded from the procurement procedure.

Under the same form and clauses 8.9 and 8.10, the contractor may terminate the agreement because of any of the following:

- employer's insolvency;
- default by the employer if the contractor receives no proper payment;
- the employer interferes or obstructs with a payment certificate;
- the employer does not comply with the CDM Regulations;
- the employer fails to issue certain instructions covered by specific clauses;
- the employer assigns the contract without consent; or
- the works are suspended for a defined period resulting from specific instructions.

Termination by either party with this form under clause 8.11 can be the result of:

- a *force majeure*, releasing a contractor from its obligations caused by an event;
- statutory undertakers default or negligence;
- loss or damage to the works caused by 'Specified Perils';
- delays invoked by authorities dealing with civil commotion and the use or threat of terrorism;
- central government or local, public or statutory authority intervention; or
- in the case of a public employer, where the contract has been subject to substantial modifications which would have required a new procurement procedure (an impact of the Public Contracts Regulations 2015, 73 (1) (a) and (c)), or the contract should not have been awarded to the contractor in view of a serious infringement of the obligations under the Treaties and the Public Contracts Directive.

Where termination is sought, it is necessary to observe clauses outlining the notice procedures that the party seeking termination must give to the other and the consequences of termination. As part of the review of the main contract, it would be prudent for the quantity surveyor to be aware of these notice periods in order to understand their repercussions and ensure appropriate procedures would be followed that do not constitute a breach of contract. Moreover, the quantity surveyor would need to be open-minded if unique circumstances arise paving the way to termination not covered by clauses of the contract, as termination may then only take place using common law principles. This is beyond the quantity surveyor's skill, and legal advice would normally be sought by the contractor's directors in the event of such an occurrence. This may also apply if works are carried out in the absence of a contract when the contractor relies on the preliminary agreement that, for whatever reason, fails to transpire, leading to the permanent cessation of the works by the contractor. In such a situation, if a dispute arises, resolution could commence when the party seeking remedy issues a formal notice to the other of the intention to seek intervention through litigation or alternative dispute resolution (ADR). For a discussion on litigation and types of ADR, refer to Chapter 3, Section 3.7 on remedies for breach of contract.

4.7.11 Warranties

A warranty is a contractual term meaning the promise or pledge of security by the contractor to the employer to repair or replace any defective component or part of a component in the event of its failure to perform the intended function. If a building owner/occupier or a part building owner/occupier inherits a defect, the contractor has no obligation to acknowledge the warranty if the employer is insolvent and the rectification period has expired. This leaves the owner/occupier in the unfortunate position of possibly having to repair or replace the component at their own expense or sue the contractor in tort (a civil wrong) for negligence through a civil action and seek damages in an award. To avoid this unpleasant scenario, owner/occupier(s) can be named in a schedule of the contract as beneficiaries of collateral warranties that a contractor and/or appropriate subcontractors can provide, thus bypassing the employer and the contractor where applicable in the event of a defect manifesting. In other words, the party with the defect can deal direct with the installer or supplier to repair the defects. To be effective, collateral warranty clauses in a contract are linked to the contract particulars which identify the name(s) of owner(s), tenant(s) and any project stakeholder(s) known at the time of executing the agreement that are to become beneficiaries, and list(s) the part(s) of the works requiring the warranties.

An employer can only seek collateral warranties through the contractor. However, where legally sublet, the contractor can only obtain warranties from trade contractors prepared to legally assign their bound rights and obligations with the contractor over to the employer and owner/occupier and/or any other named party. This can be a minefield for the contractor to arrange, which will involve communicating with appropriate subcontractors before awarding trade packages to confirm they can comply with the requirements. By merely agreeing to the employer's request without seeking input from subcontractors, a recipe for disaster can be created if it transpires certain warranties cannot be provided (e.g. if seeking a warranty for a fully equipped heating and cooling system integrated with a building management system designed to observe the performance of the installed equipment, with neither systems suppliers guaranteeing compatibility because of technical issues).

When researched sufficiently and completed successfully, collateral warranties take effect at the date of practical (or sectional) completion or end of a period when the contractor is no longer responsible for defects. To be effective, the issuer of any warranty including collateral warranty may make it conditional that, when agreeing to provide a warranty, certain obligations must be met. For example, a vertical transportation contractor may agree to issue a warranty and collateral warranty for a stated period after installation of the elevators on condition the elevators are serviced by the installer under a comprehensive maintenance contract, which is to run for the term of the warranty. In such a scenario, the employer or future building owner must create a separate maintenance contract with the vertical transportation installer for the warranty to be effective, which can be direct or via the facilities management contractor.

When reviewing a main contract that includes provisions for collateral warranties, it is important for the quantity surveyor to consider which part(s) of the work(s) are to have collateral warranties supplied and the trades involved. This is because the contractor must select subcontractors capable of providing the appropriate warranties that are also licensed to supply, install and commission the products.

4.7.12 Contract Schedules and Special Provisions

The parties to a construction contract may elect to include contract schedules that are either part of the standard form or specific to a project which are appended to the form. A range of schedules includes:

- rules and methods for the engagement of named subcontractors (JCT IC 2016);
- rules and methods for the engagement of named specialists (other JCT forms);

- design submission procedures where contractor's design portion (CDP) applies;
- rules for Codes of Practice to assist in the management of identifying works as not in accordance with the contract;
- schedule of optional insurances for the works;
- schedule of bonds (e.g. bank guarantee, advanced payment, etc.) as employer's financial security;
- fluctuations option where prices are subject to change and the rules applicable to calculations;
- termination payments;
- third party rights, granted by the contractor and/or subcontractors that can limit liability with collateral warranties and apply capped amounts for liabilities and other losses suffered by a purchaser or tenant;
- bonus incentive payments for early sectional or practical completion;
- payment schedules by works stages (usually applicable to a design-and-build contract);
- health and safety bonuses (benchmarked for the lowest number of reported accidents);
- construction programme;
- variation and works acceleration quotation procedures; and
- contractor's proposals and design obligations under a design-and-build contract.

Special standard provisions have grown in use with standard forms of contract following the resurrection of industrial buzzwords and phrases used in the construction industry derived from legislative and cultural changes. For example, JCT SBC/Q 2016 Supplemental Provisions Schedule 8 promotes collaborative working as an optional expressed term that requires parties to work 'in a co-operative and collaborative manner, in good faith and in a spirit of trust and respect'. The form also promotes collaboration regarding health and safety going beyond the statutory duties, seeking to encourage good practice 'in which health and safety is of paramount concern to everybody involved with the project'.

Modern standard forms of contract also adopt collaborative methods aimed at improving project performance, including provisions for contractors to use IT systems effectively, provide cost savings and value management or suggest improvements to designs to benefit the employer. Furthermore, as contractors become more involved with sustainability affecting design portions, there is encouragement for contractors to participate with the design process and offer suggestions regarding economic and viable methods for the employer's consideration. Other special provisions include the use of performance indicators created for and monitored by the employer. Here, the contractor is obliged to provide information regarding their performance to the employer in order that the employer can assess the contractor's performance, with the employer reserving the right to advise the contractor if they consider the contractor's performance does not meet the requirements.

The above-mentioned provisions are driven by industrial change, which drafters of modern contract forms recognise. However, some forms may require special provisions unique to a project that need to be appended to a form to make the contract workable. This includes specific health and safety requirements when a contractor is working on a project of high risk with the criteria driven by the employer, contractor or both, which must also comply with the CDM Regulations. Special provisions may also apply because of an employer's type of business and any special project undertaken, for example nuclear power stations, working at height, and structures requiring specific work method statements such as 'super tall buildings' (buildings 300–600 m in height, as defined by the Council on Tall Buildings and Urban Habitat in their publication *Criteria for the Defining and Measuring of Tall Buildings*). Furthermore, such provisions may apply to projects requiring confidentiality, security and/or respectful reasons where a project's identity to the public and media is restricted, and can apply to embassies, consulates, places of worship, military bases, security buildings and data centres.

5

Supply Chain Procurement

5.1 The Supply Chain

Supply chain procurement is an important aspect of managing a construction project and could have been included in Chapter 6 on *Running the Project*. However, as the subject is an integral part of commercial and project management, a discussion on the topic is worthy of its own chapter.

Supply chain procurement refers to the systematic appointment of businesses, individuals, resources and technological requirements obtained from a range of suppliers that combine to deliver a product to the customer and applies to any industry. For the purpose of this book, the products to deliver are of course newly constructed and refurbished buildings. To enable a construction project to function properly, the timely placement of awards is necessary for any or all of the following:

- labour-only contractors;
- trade contractors that supply and fix materials and commission the works;
- material suppliers;
- manufacturers that produce goods for material suppliers to sell or fabricate as components;
- raw material resources for supply to manufacturers;
- plant equipment hire;
- design and specification consultants; and
- the main contractor's in-house resources that need mobilising to a project.

The placement of awards with these supply chain members is filtered through linear channels of responsibility which is driven by price, custom, preference, delegation, risk, skill sets, industrial frameworks and the degree of control each purchaser holds with their supplier. To be effective, each award must be formed upon mutual agreement, collaboration and understanding regarding intention and scope. For example, a project client may elect to place an award with companies to provide and install artwork that can include paintings, plaques, sculptures, coverings, etc. excluded from the main contractor's works, an arrangement that may benefit the parties for the following reasons.

- Artwork is expensive and the client may not wish to pay for risk, overheads and profit if procured through third parties.
- Aspects of artwork are beyond the contractor's expertise, with the involvement possibly futile.

Construction Quantity Surveying: A Practical Guide for the Contractor's QS, Second Edition. Donald Towey.
© 2018 John Wiley & Sons Ltd. Published 2018 by John Wiley & Sons Ltd.

- The main contractor may not be able to provide insurance for the artwork if installed prior to the date of practical completion.
- Artwork suppliers may not wish to be involved with customs akin to the construction industry (e.g. the withholding of cash retention).
- The client wants control of the selection of artwork and the timely delivery.

Where works outside the contract are procured by the client for installation by the date of practical completion but this arrangement fails because of the late issue of an award by the client, the contractor can seek an adjustment to the completion date subject to the terms and conditions of the construction contract providing that the contractor is not responsible (e.g. failing to advise the latest date the artwork is to be installed when part of a programme or programme update if a condition of contract). By default, this can also apply to the supply of design, specifications, materials, labour only and the supply of labour and materials (or a mixture) and statutory authorities to provide external services. However, as part of their responsibility the main contractor must endeavour to procure a supply chain in a timely manner to fulfil its obligations and scope of the main contract works. The quantity surveyor plays a vital role in this process, as it requires a commercially aware person to drive activities and ensure awards are placed on time and within the parameters of the main contractor's budget. On a large-sized project valued at say £10 million to run a year or more, the first three months of the construction phase are the busiest time for procurement. Where possible, the quantity surveyor should aim to award most trade packages and material supply orders within this timeframe. On the flip side, a project of lesser value and shorter duration requires awards issuing in a shorter timeframe and must be carried out within the time available using the construction programme as a guide and resources used for preparing the tender.

A contractor might be tempted to delay the placement of awards if made aware works variations are imminent and halt procurement activity, an approach not to be recommended. To demonstrate, let us say a contractor has secured a contract to construct a dozen or so houses based upon a set of drawings and specifications and, following receipt of an instruction, is advised the floor plans are to be enlarged with revised information to follow. Here, the contractor must endeavour to place awards based upon 'for construction' information so the works can theoretically commence on time, and should not wait to receive the revised information before issuing awards. If awards to the supply chain are based on the contract design, the contractor should instruct each supply chain member to suspend the works until the revised information becomes available, thus mirroring the instruction from the client/client's agent. In this scenario, any variation communicated to the contractor would reverberate through the supply chain.

It is possible for scores of supply chain members to be engaged on any project and, for this reason, it is necessary for the quantity surveyor to maintain the letting schedule and create a contact list citing the names and contact details of entities in receipt of awards which should be updated as and when awards are placed. The list should include subcontractors, major suppliers, plant hire companies and consultants as well as any client-engaged suppliers and contractors. This is to ensure the project team is fully aware of contact details to enable the coordination of construction activities.

5.2   Labour-Only Subcontractors

A main contractor may elect to split awards for certain trade packages and purchase materials for installation by labour-only subcontractors. This type of procurement provides the contractor with the following advantages:

- payment is made for the works at rates a contractor can afford;
- there is little or no employment costs, meaning payable labour rates are competitive when tendering for the main contract works;
- requirement for hired labour is on a needs basis;
- the hierarchy management of labour-only contracting businesses is omitted or reduced with less bureaucracy to deal with;
- labour appointments are usually rapid;
- benefit of full trade discounts for material purchases;
- more control and say with the working operations, material procurement and site storage areas; and
- there is no contractual duty to pay for labour if works are hindered when the contractor is not responsible.

Disadvantages include:

- closer scrutiny of material wastage as labour-only subcontractors are not usually responsible;
- possibility of additional payments for non-productive time to the workforce due to the late delivery of materials that is beyond the control of the labour-only subcontractor;
- it is possibly harder to source a labour-only workforce than labour and material subcontractors;
- more site supervision is required (e.g. coordinating materials/plant requirements);
- material scheduling is required because labour-only subcontractors are not responsible for the quantities delivered;
- additional administrative requirements due to the number of subcontract accounts to process and maintain (e.g. weekly or fortnightly payments);
- limited control over the reliability of labour to arrive on site (a written contract will provide protection); and
- drain on cash flow because labour payments are regular (e.g. weekly or fortnightly with the contractor paid monthly or by stage payments on an Activity Schedule).

As demonstrated above, there are pros and cons for adopting the use of labour-only subcontracting. When electing to procure works using this concept, the contractor must be in a strong position to manage the process in terms of resources, finances and experience.

5.2.1   Methods of Engagement and Reimbursement

A method of procuring labour-only contractors as subcontractors is with an invitation to tender, and usually applies when the works require plant and sundry materials that the main contractor does not normally hire or purchase (e.g. mechanical and non-mechanical plant for compacting and finishing placed concrete). The methods to adopt for obtaining tenders is no different from sourcing trade contractors that provide a labour and material service, which can be on a fixed price or schedule of rates (discussed further in Section 5.3 below). However, with the likes of carpentry, bar reinforcement fixing and brick/block-laying works, the contractor's site and project managers may procure labour-only trades by offering market rates, bypassing the need to seek tenders. Here, established contractors may have a database of labour-only subcontractors to choose from, and it is usually the site and project managers that initiate contact and engage them. Alternatively, the contractor may place adverts in local newspapers and trade journals, or seek the services of one or more recruitment agencies. Due to the competitive nature of labour-only subcontracting, advertisements might limit the advertised information by simply describing the type of project (e.g. industrial building(s)) and the locality, with the contractor advising any recruitment company acting on its behalf to do likewise. Moreover, advertisements may state that rates or prices are 'subject to negotiation' or that 'competitive rates apply' in order to keep the information confidential.

When negotiating payable rates, site and project managers should be in receipt of cost targets set to maximum amounts a contractor can afford to pay that are determined by the quantity surveyor. The subject of cost targets is discussed in detail in Chapter 4 (Section 4.5.2). The use of cost targets is widespread on construction projects where the main contractor can effectively manage the process which, to be instrumental, must rely on a strong supply chain for the procurement of materials. This is a usual business activity of national house builders and, due to the volume of housing stock generated annually, it attracts commercial interest from material suppliers. In determining a price for the labour-only works, national house builders create a series of cost targets for selected trades by house type (e.g. brick/block construction). For example, builders could create a cost target for cavity wall construction commencing from damp proof course (first course) to 21st course or top of joist height, another commencing from 22nd course to 43rd course or course directly below the lowest bedroom window sill, and thereafter to wall plate height (approximately 62nd course) and the gables. These cost targets are issued as price lists prepared by the quantity surveyor, and distributed to site and project managers that appoint tradesmen upon agreement with the prices. With this approach, labour-only subcontractors claim an amount from the cost target on a weekly basis to a maximum amount and 'book in' their works with the site manager or bonus surveyor for issuing to the accounts department for processing as weekly or fortnightly payments.

5.2.2 Contractor's Risk

Once a project commences, a main contractor might struggle to find a continuous labour-only workforce to satisfy the programme of works due to any or a combination of the following:

- scarcity of the number of trade persons existing in a locality;
- limited existence of specialist trades for which no rates exist; or
- small quantities of work or the work is so complex that pay at normal rates would be a pittance.

This can create a headache for the contractor, more so if demand for trades outstrips supply, making workforce retention a challenge with the possible need to apply incentives to be productive. For example, a timber frame cavity wall with a brick veneer reduces the wall construction value for brick/block layers as the frame substitutes the inner block wall. This means the payable amount to brick/block layers is low in comparison with traditional brick/block cavity wall construction and, when demand for this trade is high, operatives may elect to work for others that carry out brick/block construction as their earnings are higher. Here, a contractor must deal with the situation as best as possible because they have probably committed to delivering a project on time at an agreed cost, even though a scarce labour force may not have been a point of concern when the project was tendered. A method of mitigating these circumstances could be to increase cost targets and/or apply incentive mechanisms for timely completion. As a last resort, the workforce could be reimbursed on hourly rates. However, this last resort provides the contractor with the following disadvantages:

- the contractor does not know what the final expense will be;
- there is no incentive to expedite the works that could become prolonged;
- additional site supervision is required to authenticate the works and record time; and
- subcontractor complacency.

The advantage to a contractor of paying hourly rates is the appointment itself that triggers the start of the works and, if the contractor can be reimbursed on dayworks, the risk is transferred to the client. However, in order to warrant client involvement with dayworks, it must be a condition of contract as the client will not usually have an obligation to pay extra because of market conditions that are not in the contractor's favour.

When labour-only works are tendered, and upon receipt of tenders, the quantity surveyor must consider any tender qualifications that state inclusions, exclusions and clarifications. This also applies to trade contractors that provide labour and material tenders, discussed in Section 5.3 below. Furthermore, any labour-only/trade contractor may state that its tender is based upon assumptions and list the assumptions to avoid any doubt of what is included or excluded from a price. Moreover, a tender may also provide a schedule of attendances that a contractor is expected to provide to facilitate the works (e.g. materials supplied free of charge to the labour-only contractor of a sufficient quality and quantity to maintain a productive workforce, free use of scaffolding, free power supply for tools, etc.). As these are the contractor's responsibilities, it is usual to list such items in the terms of an invitation to tender stating they are by the contractor. However, it is not unusual for tendering companies to regurgitate the list in a tender for there to be no doubt of the expectations. When these items are qualified in a tender as excluded or clarified as being the responsibility of the (main) contractor, they are usually accepted as a matter of course because the cost for their supply is part of the contract sum payable by the client under the main contract.

The appointment of labour-only subcontractors is not suitable when the testing and commissioning (both progressive and final) of installed works is required, and includes the likes of civil works, electrical, plumbing and mechanical systems that the subcontractor must also warrant. This is because the contractor cannot take the risk with these trades for supervision, works coordination, material scheduling and ordering of goods, as it is beyond the contractor's expertise.

The appointment of labour-only subcontractors brings risk to the contractor due to additional supervision and general labour requirements compared to a project fully subcontracted through labour and material packages. Furthermore, the management of material wastage must be controlled, which can mean arranging 'just in time' (JIT) deliveries to coincide with the programmed installation dates, thus aiming at keeping a low inventory. When successful, this avoids the prolonged storing of goods that may otherwise become damaged, possibly giving rise to additional purchases and spiralling costs that could make the whole concept of labour-only procurement futile. On the other hand, when efficiently managed this type of procurement is financially viable and has stood the test of time by contractors that wish to remain competitive with the projects they deliver.

5.3 Labour and Material Subcontractors

Labour and material subcontractors are sourced from three origins:

- those selected and appointed solely by the contractor as domestic subcontractors (i.e. subcontractors that create a binding agreement with the main contractor);
- those named and inserted into schedules, specifications or a bill of quantities forming part of the JCT Intermediate Form of Contract (IC 2016) and named specialists in other JCT forms, subsequently appointed by the main contractor when they become domestic subcontractors; or
- those nominated in the main contract that are involved in a tripartite contractual arrangement with the client, subcontractor and main contractor.

Under the UK's Construction Industry Scheme (CIS), contractors deduct money from subcontractor's payments and pass it to HM Revenue and Customs (HMRC). The deductions count as advance payments towards the subcontractor's tax and National Insurance. Contractors must register for the scheme; however, subcontractors do not have to register, but deductions are taken from their payments at a higher rate if they are not registered. To avoid deductions made by the contractor, a subcontractor (applicable to labour-only and labour/plant and materials subcontractors) can apply for gross payment status by registering with the CIS, which permits registered subcontractors to pay all tax and National Insurance contributions by the end of the fiscal year. The CIS does not apply to employees, material suppliers, plant hire companies (with no labour), carpet fitting, architecture and surveying, and work on construction sites that is clearly not construction (e.g. running site facilities).

5.3.1 Domestic Subcontractors

The methods used for selecting these subcontractors includes:

- accessing the contractor's database of companies;
- recommendations from other contractors;
- contractor's staff referral;
- client referral;
- those that submitted prices to the contractor when preparing a bid for the project;
- those involved with creating the site accommodation;
- advertising the project in the press and requesting trade businesses to contact the contractor;

- trade contractors cold-calling on a project either in person or communicating in writing with an expression of interest once the site presence is established, which after a tender and vetting process results in a subcontract award; or
- Yellow Pages and online business listings, for example Yell.

Where a project is procured under JCT IC 2016, the contractor must abide by the terms and conditions of the contract and appoint those named in the agreement in a timely manner with the arrangement stemming from the time the contractor prepares a tender (see Chapter 3, Section 3.6.2.4 for more information), and applies even if neither party have had prior dealings. Notwithstanding the named subcontractor concept, it is important for the contractor to appoint subcontractors that can adequately service a project, and a method of testing this is with a vetting process. This is a quality control procedure where the quantity surveyor completes a questionnaire assessing each trade contractor's capability and appropriateness for a scheme prior to any subcontract arrangement coming into existence. Small-sized contractors tend to be informal regarding this procedure, and rely on word of mouth or only apply a vetting process to unknown trade contractors. However, larger contractors may endorse a policy of issuing a pre-tender questionnaire to any trade contractor, which must be completed to the satisfaction of the main contractor prior to the issue of an invitation to tender, and applicable to new subcontractors and possibly those engaged on past and current projects in order for the information to be current. The questionnaire should be project-specific, outline the project scope of works and list the commitments expected from would-be subcontractors, and is issued to potential subcontractors for their return. Alternatively, a less formal approach would be to conduct a series of questions and answers over the telephone with the form completed by the quantity surveyor. Information required about a potential subcontractor includes:

- registered business title and address;
- the current business financial turnover;
- minimum and maximum value of works undertaken;
- type of works undertaken; and
- name of two referees on current or completed projects and their contact details.

In the absence of a formal vetting procedure, it would be wise for the quantity surveyor to retain notes learnt of a trade contractor's capabilities after making initial contact, especially if new to the contractor, as the objective is to not waste time by approaching businesses that are unable to service a scheme. For example, if a contractor wishes to award a plumbing and drainage package worth over £1,000,000, there would hardly be any point approaching a company for a quotation that only carries out minor domestic works as the works they undertake would be incompatible. By contrast, a small building extension under construction would not be suitable for a company that only involves itself with large schemes, as their price would be uncompetitive. A simple vetting process would quickly identify the situation and 'make the shoe fit' for appropriate trade contractors to be selected as suitable for a scheme. Contractors have differing policies regarding the awarding of trade packages, with some leaving selection to the devices of the project team and others applying company policy and procedures. There are certain business ethics driving these objectives, described in the following.

- If the project is small in value or requires an immediate start after executing the main contract, and time is not on the contractor's side, those subcontractors that provided quotations during the initial tender period should be contacted first (the exception being any subcontracting business that has since become insolvent).
- The contractor may have reduced its margin to obtain the main contract award, meaning budgets are lean. This will result in risk to the contractor and, by inviting trade tenders during the construction phase, financial savings may be produced to mitigate the risk.
- Quotations received during the initial tender period may be open for a period of acceptance and, because the works may not be required for some time afterwards, the quotations will become invalid and require updating.
- Where a bill of quantities is a contract document, it forms the basis of the main contract lump-sum price and a contractor's policy may be to issue awards to those trade contractors that priced the bill of quantities.

If a bill of quantities is not a contractual document for a scheme and the project team's consensus is to tender the works anyway (time permitting), the quantity surveyor will initiate the process by issuing invitations to tender to obtain quotations. This process is similar to that carried out during the initial tender period, which may have been carried out by a quantity surveyor involved with the estimating team as discussed in Chapter 3 (Section 3.5.2). Naturally, any solvent trade contractor that submitted a price to the contractor during the initial tender period must be given the option to sustain or update their price.

A failure of a tendering procedure during the construction phase, which is often unnecessary, occurs when a contractor overwhelms trade contractors with information or provides irrelevant information, and omits relevant details in the process. For example, an electrical contractor will require the electrical specification and drawings as a minimum requirement to price the electrical works. If an invitation to tender includes architecturally designed partitions because the contractor considers they provide information on the location of power outlets which state their height from the floor, there is no need to include this information if the information is stated in the electrical specification. If there is conflict between the electrical and architectural information, the problem needs clarifying by the architect. Here, the quantity surveyor should make a request to the client's agent to clarify the point and, with time permitting, delay the invitation to tender pending an answer if the discrepancy will influence the price of the works. Ideally, an invitation to tender consignment should include:

- a set of contract or 'for construction' design and documentation;
- a written scope of works;
- the master construction programme; and
- a draft of the proposed form of subcontract agreement.

A scope of works is a written statement of requirements and is usually drafted by the quantity surveyor with input from the site manager(s) for approval by the project manager. The scope informs trade contractors of the works required plus any specific requirements the contractor wishes competing trade contractors to include in their prices that may not be obvious in the design and documentation. Ideally, the scope should be tabulated and include:

- an outline of the trade works, cross-referenced to the design and tender documents;
- staging and phasing of the works;
- items the main contractor will provide the subcontractor as builders work in connection;
- items the main contractor will provide as attendance;
- subcontractor involvement and coordination with other trades;
- subcontractor compliance with CDM 2015; and
- site-specific requirements.

A scope of works is not a statement telling a trade contractor how to carry out the works, as that is a matter for each tendering company to determine. It should however be considered a statement of requirements needed from a subcontractor describing the coverage of works that forms part of the price. When issuing a scope of works document, the main contractor's objective is to ensure there are no gaps in the requirements that leave minor parts of the works unaccounted for as this could lead to a later claim for additional time, works and payment that the contractor possibly cannot recover through the main contract.

A contractor may be managing a sequence of working activities to achieve the construction programme that involves coordinating different trades and, if applicable, notes to this affect will need including in each appropriate scope of works. For example, if a contractor is inviting electrical contractors to submit tenders for power and lighting and the client is to appoint a separate trade contractor to supply and install remote access controls using an 'intelligent system', the scope of works must state the successful electrical subcontractor is to coordinate their works with the client's contractor. This statement is required to allow coordination between trades to test and commission the system for the project.

When inviting tenders, it is necessary for the quantity surveyor to advise a tender period. This needs to be of a suitable duration to allow sufficient time for businesses to assess the design and documentation, seek prices from suppliers, ascertain risks and prepare and issue quotations for the works. A reasonable duration would be two working weeks if works are clearly documented, of an uncomplicated nature and a bill of quantities provided. This can be extended up to four weeks when works are complex and a bill of quantities is absent, meaning competing trade contractors are responsible for producing their own quantities. The design and documentation provided in the invitation to tender needs collating and scheduling accurately, and must include a cover letter and document transmittal that lists the title, number and revision of each document. The cover letter must state the method of submitting a tender which should be addressed to the quantity surveyor or project manager and can be by post, email or uploaded to a host website.

5.3.2 Nominated Subcontractors

The appointment of nominated subcontractors is the extreme of a client's control and involvement with the selection of subcontractors on a project, and involves the client facing two contractual fronts. The first is where the client enters into a binding agreement with a subcontractor of their own choosing using a predetermined form. The second is under an agreement with the main contractor that is contractually obliged to appoint and reimburse the subcontractor because of the tripartite agreement, thus contractually binding the subcontractor to the contractor.

The nomination process is not usually an afterthought and is normally included in the invitation to tender for the main contract works to allow competing contractors to either accept or reasonably object. At this stage, a contractor may decline to submit a tender or may submit an offer outlining its objections as a condition of tender for the client's consideration that could be subject to negotiation post-tender. Any objection(s) must be clearly stated together with reasons, for example unfavourable credit term requests from the nominated trade contractor or a nominated trade contractor's reluctance to indemnify the main contractor from claims against the client. With this arrangement, the client takes a lion's share of responsibilities and must have a strong desire to control a portion of the works because, in general, the client is agreeing to protect the subcontractor from default by the contractor.

A main contractor has a responsibility to deliver a project on time and may see the nomination process as risk that can cause tension and is a reason the nominated provision is seldom included in standard forms of contract. However, if nomination does apply on a project and is included in the contract agreement, the quantity surveyor needs to observe the rules of appointment and comply, as the subcontractor will have recall from the client because of the separate contract.

5.3.3 Tender Periods and Openings

During the time when new and/or updated prices are being sought via a series of multiple tender periods, the quantity surveyor could be approached with queries from trade contractors that can include and be actioned as follows:

- *Problems with the building process*: queries revolving around items such as areas for storing materials, site access, etc., which can be provided by the contractor.
- *Design errors*: any design error(s) must be referred to the client's agent or appropriate consultant to answer the query after the contractor confirms the observation is valid.
- *Inconsistency with design*: here, the order of precedence should apply. For example, if the thickness of a precast concrete floor is shown as 250 mm deep on the structural drawings and 200 mm on the architectural drawings, the structural drawings will take precedence because the floor thickness is critical to the structure of a building. However, the quantity surveyor should seek clarification from the designers, and in the meantime exercise the order of precedence concept by instructing trade contractors to provide quotations based upon the structural drawings and clarify the matter in their tenders. It would be a waste of time to request optional prices for different thicknesses and would be best answered with a clear instruction.

If one or more trade contractors of the same trade raises a query and the response is to amend part of the invitation to tender consignment with an addendum notice, each trade contractor issued with the invitation to tender must be provided the same response, including those that did not raise the query. This is to enable the parties to submit quotations on an equal basis and maintain healthy competition. If a response takes some time to obtain and stifles the tender period, the contractor should grant an extension of time when requested by more than one trade contractor. When doing so, the contractor must consider what impact this will have on the duration of negotiations after the tender period and timing for the placement of the award.

During any tender period, there may be situations when a trade contractor makes a request to the contractor to price works beyond that stated in the invitation to tender. For example, a formwork company may wish to include a price or schedule of rates for placing steel bar reinforcement in addition to the formwork because the company also carries out this service. Here, the quantity surveyor may elect to accept the request and instruct an extra optional price to be submitted for consideration. However, it would be wise to advise any trade contractor making such a request that, by accepting the request, the contractor reserves the right to accept any offer or part of an offer based upon the initial enquiry, and acceptance of any part of the trade contractor's offer is not contingent that the trade contractor must receive an award for the initial enquiry plus other works. If the number of trade contractors issued with an invitation to tender alters because some either decline to submit a price or new contractors are added, the project and site managers need to be informed so they are aware of the number of tendering companies pricing each trade package. This is best tracked on a tender schedule created on a spreadsheet that lists the trades and name(s) of companies tendering the works, together with the tender due dates.

Situations can occur during a tender period when a trade contractor cannot or does not wish to comply with a tender enquiry and seeks direction from the contractor. For example, contrary to the request in a tender enquiry, an earthworks company may wish to exclude the disposing of excavated material off site and offers to place the material in a designated area on site for disposal by others. Here, a degree of flexibility is required by the quantity surveyor who may wish to receive tenders and could accept the request if there is storage area on site to temporarily accommodate the material, and instruct the company to qualify the exclusion in their tender. However, any company seeking a change should be advised they are in competition and that competing businesses are expected to submit quotations in accordance with the conditions of tender; by making a decision to alter the scope, they do so at their own risk with their tender possibly disadvantaged.

Where competing trade contractors are to include materials in their tenders, they must price for the supply of such goods in accordance with the design and specification provided; on a construct-only contract it is not acceptable to substitute any product without the approval of the client's agent or title stated in the contract documents with the authority to instruct a change. However, a standard form of main contract may provide flexibility and actively promote input from the supply chain to improve the value of a project with suggested alternatives, which the contractor should encourage. Where applicable, such tenders must be considered as alternatives to the complying price. In the case of a design-and–build project, the contractor can elect for such collaborative methods to assist with the project deliverables.

Receipt of trade contractor's tenders should be to date(s) stated in the invitation to tender or any revised date. Tenders received after this time should not be considered, even if appearing favourable, as failure to comply with the request lacks respect for the tendering procedure. However, if letting of the award is solely the contractor's responsibility, such action is discretional.

5.3.4 Tender Comparisons

Trade price quotations submitted to the contractor during the initial project tender period may be amended by the offeror(s) after the contractor is awarded the project. Legally, the offeror(s) is entitled to do this on the proviso any offer has not been accepted as, in general, acceptance would bring a contract into existence. There are reasons why prices may fluctuate with time, such as those listed below.

- A trade contractor may have provided a price within a short timeframe and issued a 'guestimate' to assist the contractor during the initial tender period and, following the issue of 'for construction' information, there is time to examine the design and documentation in detail.
- The price of labour and materials may have increased since the original quotation was issued.
- As the contractor has secured the works, there is the realisation the project has become real and a trade contractor may discount its profit margin to make the offer attractive.
- The scope of works may have changed.
- The period of acceptance of the earlier offer has lapsed.

The quantity surveyor must ensure submitted tenders comply with the contractor's enquiry, including the effect of any subsequent addendums issued during each trade tender period. All offers must be in writing and any oral quotation should be disregarded if there is a failure to confirm the discussion. Written offers should be described as a quotation or an offer to carry out the works and not an estimate. A quotation/offer is a firm price and intention to commit, whereas an estimate is an idea of probable cost that is not binding unless written verbatim as an amount into an agreement. If a fixed price is not on offer and a schedule of rates applies, the offer must state the rates are firm and not an estimate and that reimbursement to the subcontractor will be based upon a measure and quantification of the completed works charged at the stated rates included in a contract.

A problem can occur when receipt of tenders for a trade package exceeds the budget allowance. This can happen when the contractor's budget is based on the procurement of separate labour-only and material packages, as this is generally a cheaper option than combined labour and materials packages. This situation can occur if a contractor secures a project on lowest price because it elected to include budgets for certain trades procured as split packages. However, once the contractor secures the project and becomes responsible for trade contractor selection, the procurement method can be altered to suit the needs of the project. Such changes can incur costs in excess of the project budget if the project manager overturns a decision made by the estimator at the initial bidding stage and wishes to place single labour and material packages instead of split packages. Reasons for this decision include:

- a scarcity of labour-only trade contractors;
- an abundance of labour and material trade contractors;
- a lack of suitable supervision available from the contractor;
- unavailable or unacceptable credit terms from material suppliers; or
- the project and site managers are of the opinion the risk is too high to manage.

If there is conflict with the project manager's preferred procurement method and that elected by the estimator for inclusion in the project tender and subsequent contract, the cost difference must be brought to the project manager's attention. Here, the skill of the quantity surveyor as cost manager is put to the test and involves providing advice regarding the financial consequences for changing the procurement method. This is best demonstrated in a cost analysis by comparing the budget allowance to a cost estimate for carrying out the works using either procurement method. To demonstrate, let us say the blockwork trade for the construction of an external frame to a new building has been tendered and the favoured procurement method is to award a labour and materials package. However, received quotations exceed the budget, and the favourable trade contractor has provided a schedule of rates advising it will accept an award for either labour-only or labour and materials in order to secure the works. Table 5.1 shows a cost analysis using the budget as a benchmark, together with a breakdown of the anticipated expense for each procurement option.

The analysis shows that if the contractor elects to procure the trade as a labour and materials package, the contractor's risk is low yet could result in a potential loss of £4,249, this being the difference between the estimated cost and budget. However, if the contractor splits the award and accepts greater risk by sourcing materials from suppliers and reverts to the estimator's decision at tender stage, it is possible the contractor could save £1,175.

When trade contractors' quotations are within budget, the analysis is straightforward and involves the quantity surveyor collating tenders in a summary together with a recommendation which is part of the negotiations towards the issue of an award. This is similar to the assessment of trade contract prices assessed during a contractor's initial tender period (see Chapter 3, Section 3.5.2 for details), with the exception being there is no budget at this time because it is created by the estimating process itself and confirmed once the project has been secured.

5.3.5 Negotiations

Once trade contractors' tenders are received and any cost analysis complete, the quantity surveyor would be wise to request a tender interview with each favourable trade contractor when considered a suitable contender for an award. A degree of common sense is required regarding the parties that should be interviewed, and preferably restricted to those that are serious with their efforts and presented offers in a professional manner. A face-to-face interview should be adopted where practical, and is one method of due diligence procedures a contractor can use as part of the selection process. The exception is with named and nominated subcontractors, as the contractor has limited choice with their selection.

The interview is part of the negotiations the quantity surveyor will need to conduct prior to making a recommendation for the issue of an award. This is a beneficial process for both contractor and would-be subcontractor, as it puts a face(s) to the name(s) and acts as an icebreaker to the start of a potential working relationship. It also provides insight into a subcontractor's capability for administering, managing and delivering the works. The quantity surveyor may have conducted research regarding a trade contractor's suitability prior to the issue of the tender consignment as part of the vetting process. However, an interview is beneficial in bringing the process up to speed in the event

Table 5.1 Cost analysis

BOQ Page 6/1	Description	Quantity	Labour	Plant	Materials	Total	Total budget (£)
	a. Block walls, 100 mm thick	950 m^2	25.00	*3.00	30.00	£55.00 m^2	52,250.00
	b. Form cavity 50 mm wide including ties	510 m^2	3.00	0	2.00	£5.00 m^2	2,550.00
	c. Partial cavity fill rigid insulation	510 m^2	10.00	0	10.00	£20.00 m^2	10,200.00
	d. Proprietary lintels 200 × 150 mm	56 m	35.00	0	65.00	£100.00 m	5,600.00
	e. Proprietary lintels 200 × 250 mm	16 m	40.00	0	100.00	£140.00 m	2,240.00
	f. DPC horizontal cavity tray	28 m^2	20.00	0	55.00	£75.00 m^2	2,100.00
	g. Adjustable stop ends to DPC	72 nr	5.00	0	2.50	£7.50 each	540.00
TOTAL BUDGET							**75,480.00**

Item	Description	ABC Blocklayers Ltd Quotation		Contractor's assessment - split package		
1.	Materials - block supply, 100 mm thick	950 m^2	£70.00	£66,500.00	1,093 m^2	£21.00	£22,953.00
2.	Materials - premixed mortar supplied	510 m^2	£1.88	£1,786.00	12 m^3	£148.50	£1,782.00
3.	Labour - block laying, 100 mm thick	950 m^2	Incl	0	950 m^2	£30.00	£28,500.00
4.	Plant - mortar mixers	950 m^2	Excl	0	6 wks	Excl	0
5.	Materials - cavity wall ties (4 nr/m^2)	510 m^2	Incl	0	2,500 nr	£0.30	£750.00
6.	Labour - form cavity 50 mm wide	510 m^2	Incl	0	510 m^2	Incl	0
7.	Materials - cavity fill rigid insulation	510 m^2	Incl	0	586 m^2	£7.00	£4,102.00
8.	Labour - fix cavity fill insulation	510 m^2	Incl	0	510 m^2	£5.00	£2,550.00
9.	Materials - lintel 200 × 150 mm	56 m	£100.00	£5,600.00	62 m	£57.50	£3,565.00

10.	Labour - install lintels 200 x 150 mm	56 m	Incl	0	56 m	£35.00	£1,960.00
11.	Materials - lintel 200 x 250 mm	16 m	£110.00	£1,760.00	18 m	£80.50	£1,449.00
12.	Labour - install lintels 200 x 250 mm	16 m	Incl	0	16 m	£35.00	£560.00
13.	Materials - DPC horizontal cavity tray	28 m²	£90.00	£2,520.00	34 m²	£54.00	£1,836.00
14.	Labour - DPC horizontal cavity tray	28 m²	Incl	0	28 m²	£20.00	£560.00
15.	Materials - DPC stop ends	72 nr	Incl	0	100 nr	£2.00	£200.00
16.	Labour - DPC stop ends	72 nr	Incl	0	72 nr	Incl	0
17.	Others/risk contingency	+ 2%		£1,563.00	+ 5%		£3,538.00
	TOTAL ESTIMATED COST			**£79,729.00**			**£74,305.00**

Notes:

- Material quantities in the contractor's assessment allow for waste.
- All rates for material supply include for delivery and off loading.
- Mortar is supplied and delivered @ £60.00 per tonne @ $1,650\,\mathrm{kg/m^3}$ + 50% shrinkage & waste for mixing with water = $£148.50\,\mathrm{m^3}$.
- The volume of mortar is $0.013\,\mathrm{m^3/m^2}$ of wall ×950 $\mathrm{m^2}$ wall area = $12.35\,\mathrm{m^3}$.
- Labour rates shown on the contractor's assessment are obtained from the quotation provided by ABC Blocklayers Ltd.
- ABC Blocklayers Ltd quotation is based upon fixed priced rates that include supervision, subcontract preliminaries, overheads and profit. An allowance of 2% is added for contractor's risk. There is a greater risk to the contractor with labour-only procurement, therefore 5% is added to the contractor's assessment.
- Scaffolding is excluded, part of the main contract preliminaries.
- Prices exclude VAT.

---

* Plant rate is for mortar mixers and not used for this assessment, as the cost for hire is in the preliminaries.

of changed circumstances and the fact the potential subcontractor will have become acquainted with the project when preparing the tender. Prior to the interview, the quantity surveyor must make it clear that by conducting the interview, it does not guarantee an award of subcontract. Furthermore, the potential subcontractor must be made aware that the interview is a process of ensuring the contractor and subcontractor have mutual understanding regarding the scope of works and services expected once an award is placed, and to confirm the price or payable schedule of rates for the works.

The purpose of the interview is to address and clarify any unresolved issues before entering into a contract, which to be effective should be recorded on a questionnaire. The questionnaire should be tabulated with a draft issued to the trade contractor prior to the interview, advising that a completed questionnaire would become part of any contract and will be used to settle any informal dispute that may arise post-contract. If information is requested by the contractor (e.g. insurance details), the trade contractor should be provided with an opportunity to provide the information prior to or at the interview and certainly before commencing works on site. The questionnaire should be straightforward with simple yes or no responses, and only elaborated where necessary with notes recorded on the form. The quantity surveyor will usually conduct the interview and may invite the contractor's project and site managers to attend. For reference, the questionnaire should commence with a heading stating the project particulars including title and address, type of trade package, name and business addresses of the contractor and trade contractor and the name(s) and title(s) of those attending the interview. The interview should aim to iron out any problems regarding discrepancies and expectations of the pending working relationship, and provide insight into the obligations and expectations of the parties. Figure 5.1 provides a suggested list of headings together with a theme of questioning to include in a questionnaire. Upon completion of the interview, parties should sign off the form and keep copies.

5.3.5.1 Lowest Price Guarantees

The quantity surveyor may wish to obtain a trade contractor's best price to potentially close a deal, and can ask for the fairest price on offer when selection of the company is of its choosing. While this bargaining power is haggling, it is no different than the buyer of anything in the commercial sense. However, what must be avoided is a Dutch style of auctioning where subcontractors compete with each other once an award is imminent and offer price reductions without request, with a revised price possibly arriving unexpectedly. Where applying, the quantity surveyor or any member of the contractor's business must not on moral or ethical grounds inform any trade contractor of their competitor's price, as those in competition may be eager to be issued an award and offer a lower price in an attempt to fend off competition. If a trade contractor offers a reduction that appears too generous it should be avoided as this may be a sign of desperation which, if accepted, may place the contractor with the risk of an insolvent subcontractor after an award is placed (a case of the buyer being aware). A small concession or reduction of 2.5% as a main contractor's discount (MCD) as a bargaining tool for monthly payments of invoices was at one time acceptable practice between contractors and subcontractors, which increased to 5% from suppliers. However, following the introduction of legislation with the Construction Act, this has placed tighter control on the flow of payments through the industry with the term now redundant in terms of a sale or contract. Notwithstanding this, a concession can be

General capability and status of the subcontractor

Questioning focus ......
It is necessary for the contractor to have an understanding of the subcontractor's business, even if engaged on any of the contractor's existing projects or is known to the contractor. Information a subcontractor is to provide includes:

1. Size of the existing workforce.
2. Proportion of the workforce to be committed to the project.
3. Name of contactable referees (not subcontractor employees).
4. Past projects for viewing by the contractor if required.

CDM compliance

Questioning focus ......
The subcontractor must confirm they will comply and cooperate with the requirements of CDM 2015 and the principal contractor's legal obligation to maintain the construction phase health and safety file. The subcontractor must confirm:

1. A willingness to provide safe working method statements (SWMS) regarding the site works, stating how the works will be performed, risks associated with them and how associated risks will be dealt with. These must be project specific and not generic.
2. A health and safety file (H&S) is to be provided which must comply with the contractor's requirements. The contractor must provide relevant criteria required for the file and the subcontractor must provide its file prior to commencement on site.
3. The H&S file must be project specific. The subcontractor must be aware that the contractor's H&S officer will check the file and the subcontractor must agree to comply with requests involving the subcontractor updating the file from time to time.

Subcontractor's capacity to deliver the works

Questioning focus ......
The subcontractor must demonstrate awareness of a commitment to the project by providing:

1. Start and end dates of the works or provide a specific programme of the works for the contractor's consideration.
2. Information regarding lead times for specified materials and production of shop drawings and material deliveries as well as any special attendances to be provided by the main contractor.
3. The main contractor must confirm any project-specific general attendances to be provided free of charge to the subcontractor, including methods of vertical and horizontal lifting of materials (e.g. crane, hoist and forklift), and the subcontractor must advise on a locality where the contractor will place materials for installation.
4. The name(s) and title(s) of supervisors the subcontractor will appoint to oversee the works.

Questioning leading to ·····

Acknowledgement by the subcontractor of receipt of tender documents

Questioning focus ......
The subcontractor must price the works in accordance with the contractor's enquiry and confirm the subcontractor's compliance with:

1. The scope of works and discuss clarifications and exclusions as well as compliance with the contractor's programme and agreement to the pending conditions of contract or provide notice of any modifications.
2. Any collaborative working methods, value improvements or environmental changes with suggestions noted.
3. Any special conditions of contract that affect the subcontract works (out of hours of working, parking arrangements, etc.).
4. A price for the works or schedule of rates and terms of payment.
5. A method of financial security (e.g. cash retention or bank guarantee).
6. At the contractor's discretion, the subcontractor will need to understand they must accept liquidated damages and the amount the subcontractor will accept as a capped sum if the root cause of a delay is solely due to the subcontractor when the contractor incurs liquidated damages.
7. Public liability and any other project insurance must be provided by the subcontractor, and may be issued after the interview.
8. The subcontractor must understand they will be liable to the contractor for payment of the value of any insurance excess included in the 'All Risk Insurance' policy in the event of the contractor making a claim for an event for which the subcontractor carries sole responsibility.

Figure 5.1 Interview questions.

used as a negotiating tool if the contractor agrees to rapid terms of payment (i.e. weekly or fortnightly) when gratuity is given for the concession. Prior to securing a discount for rapid payment terms the quantity surveyor must ensure such an arrangement is possible, because the contracting organisation may only permit monthly payments, meaning that the promise cannot be fulfilled.

5.3.5.2 Award Recommendations

Once trade contractor's tenders are compiled and the post-tender due diligence complete, the quantity surveyor in conjunction with the project manager can make a recommendation for an award. Depending on the contractor's procurement policy, this may require authorising by a commercial manager based at a head office if the value exceeds a certain threshold, with anything less authorised at the discretion of the project manager and quantity surveyor. Where applicable, the project teams' recommendation for authorisation is usually issued on a written form, stating:

- the project title;
- trade;
- names of tendering trade contractors;
- names of trade contractors who declined to submit offers after receiving the invitation to tender;
- prices received and amended prices following negotiations in comparison with budget;
- copies of quotations;
- completed tender interview and questionnaire form;
- name of the recommended subcontractor; and
- reasons for the recommendation.

The quantity surveyor should not make a recommendation based on lowest price alone and should consider a combination of price and service delivery; such a recommendation may be accepted or rejected by an authorising manager and, if overturned, may be for practical reasons. For example, a quantity surveyor/project manager may recommend a subcontractor that, unknown to the project team, has recently received an award for another project from the contractor. Here, the authorising manager may make a recommendation for the appointment of an alternative because there is a belief the subcontractor's business would be unable to meet the commitment if receiving an additional award. This may be contrary to the project team's understanding and subcontractor's belief, yet nonetheless a deciding matter. Any decision by an authorising manager to overturn a recommendation must be respected and the procedure for issuing the award to another implemented, which is usually to the second favourite. For this reason, the quantity surveyor should not commit an award verbally or in writing to any inquisitive trade contractor that submits an offer until the recommendation is approved. If any party that issued a tender makes enquires regarding the status of an award, where applicable, the quantity surveyor must state that the tender and negotiation procedure is complete with the decision pending authorisation. This approach should ensure any trade contractor understands it has not been given the green light to schedule materials, process orders and mobilise resources, as this may lead to disappointment.

5.3.5.3 Unsuccessful Tenders

After the quantity surveyor receives the authorisation to place an award in writing, the process of issuing the award may commence. At this stage, it is courteous to advise each unsuccessful trade contractor that their tender has been unsuccessful (without stating why), and thank them (in writing) for their interest. A comment in the notice may state that the trade contractor's information will be kept on file for future reference, with the company selected for involvement in any future projects the contractor may undertake in a tender. Advantages to the contractor for providing this feedback are that:

- it allows the project team to approach the trade contractor at a later date should the need arise;
- trade contractors usually respect feedback and will submit tenders for other projects; and
- it maintains an abundance of trade contractors interested in the contractor's business.

Unsuccessful trade contractors may be interested to know why they lost the project and communicate this via a telephone call/email. If solely because of price, and if asked, the quantity surveyor should not disclose the difference between their offer and the value of the award by stating an exact amount or the name of the successful party. The trade contractor may however be told of any legitimate reasons decided by the authorising manager.

5.3.6 Subcontractor Insurances

A contractor cannot expect a subcontractor to provide insurance to indemnify the contractor and/or project client from the personal injury or death of any persons that may occur while works are carried out on site. Neither can a subcontractor be expected to insure an existing property or provide indemnity in the event of an incident causing damage to the property even if caused by the subcontractor, as such policies may be too difficult or expensive for the subcontractor to provide. However, a contractor can seek individual policies from subcontractors to insure their installed works from damage, including the cost of replacing damaged unfixed materials stored on site appropriate to their works. This should extend to damage caused to the subcontractors' owned or hired plant and equipment when of its own creation. The supply of such insurance may prove an arduous task for the contractor to obtain, meaning the contractor would need to seek appropriate cover using its own devices. This can be achieved with an 'All Risks' insurance policy applicable to new works and/or works to/within existing structures including building extensions. This type of insurance and other types of cover is discussed further in Chapter 4 (Section 4.7.3), and is necessary in the event of a global claim. A global claim is an insurance claim for any event not restricted to an incident caused by a subcontractor, and has an advantage to the contractor because there is no need to obtain multiple insurance policies from subcontractors. A contractor will obviously wish to limit exposure to a global claim because of the amount of excess on a policy that would have to be paid in the event of a claim. If an incident occurs which gives rise to an insurance claim solely due to a subcontractor's default that is isolated and does not affect any other part of the works, it may not be treated as a global claim. Here, the contractor may rely on the subcontractor to provide insurance for its own works. However, if an incident has an impact on the performance under the main contract it would mean a global claim is unavoidable.

A contractor must rely on subcontractors to provide a degree of insurance protection in the event of a claim and, as a minimum, policies required from domestic and nominated subcontractors include:

- public and/or general liability insurance in the name of the subcontractor receiving the award in the event of a claim affecting the public, which can mean persons or the property of public persons;
- public and/or general liability insurance from any engaged contractor appointed by a subcontractor that becomes a sub-subcontractor and carries out works for the subcontractor, and applies to the likes of electrical contractors carrying out works for mechanical subcontractors; and
- worker's compensation insurance to cover amounts payable to workers on sick leave following the statutory provisions.

Public and/or general liability insurance policies may not be required from labour-only subcontractors if the contractor has employer's liability insurance for employees of the contractor where such insurance covers both entities.

Where a contractor agrees to pay a subcontractor for materials stored off site, the level of risk due to damage is increased and, subject to the conditions of contract, the subcontractor may need to provide appropriate insurance. Where applicable, a policy must be specifically obtained for the project showing the title and stating the type(s) of materials insured as well as the value of cover which must be adequate. To qualify for payment of the goods, a copy of the policy cover note is normally submitted by the subcontractor with a request for payment, and the contractor must be permitted access to the location to ascertain the quantities and quality of the goods. Where goods are stored in crates or containerised in a shipping port or similar facility, a Bill of Lading from the source of origin should be supplied regarding the type of products in store, as this usually states the contents of the consignment. Once goods are paid for, they become the ownership of the contractor and in turn the ownership of the project client once the client pays the contractor. In the unfortunate event of fire, flood, etc. damaging goods which are paid for, the subcontractor will need to replace them at no charge to the contractor, and in turn at no cost to the client by relying on the insurance provided.

When a subcontractor is providing design portions for a project, professional indemnity insurance must be provided. The requirement of this insurance will generally not apply if shop drawings are produced from a master design provided by consultants for a client or the contractor because indemnity insurance is provided by the designers. However, under a design-and-build arrangement, a contractor may list subcontractors as interested parties on a cover note to provide the project client with a single policy instead of multiples policies.

Any specific insurance that a contractor requires from a subcontractor for a project should be discussed during the negotiations for inclusion in an award of contract so the parties are aware of the cover required. While a project is under construction, it is important for the quantity surveyor to monitor subcontractor insurance regularly to ensure suitable cover is provided and current. Monitoring is possible with the creation of a register or a process management system of record keeping.

5.3.7 Bespoke Forms of Subcontract Agreement

A quantity surveyor administering a project must ensure each subcontractor receives an award in writing before commencing their works. With certain trades this can mean before starting on site, for example to prepare shop drawings. To demonstrate the contractor's commitment to a subcontractor, an award may be issued in any of the following ways:

- as a Letter of Intent outlining the contractor's intention to issue a form of contract once the parties agree to the terms and conditions of the agreement;
- as a Letter of Acceptance confirming the appointment and agreement to the terms and conditions, with the contract to be issued as soon as practically possible; or
- the issue of the contract, bypassing the need for either of the letters stated above.

When issuing a form of contract it may be a bespoke agreement, which is a purpose-made form drafted by the contractor's lawyers and which may be necessary due to the nature of the project with the wording revolving around the specific needs of a scheme. Alternatively, a generic/standard form of agreement for use on any project the contractor undertakes could be offered, and is a purpose-made in-house document. The latter form could be considered as 'subcontractor friendly', meaning the contractor loses a degree of control. The use of a bespoke form however, can be beneficial to parties where long-term working relationships exists or the works are repeat business, especially if the arrangement stands the test of time without the occurrence of disputes. The same of course could be said for a generic form, especially with those subcontractors familiar with the format. However, new subcontractors to a contractor may scrutinise either form and seek amendments which the contractor may be able to accommodate with generic forms, but not so much in the case of bespoke forms as any amendments would defeat the objective.

A contractor may adopt a two-tier policy for the issue of awards for trade packages by making a distinction between: what is considered a minor works agreement when the value does not exceed a certain threshold; and major works for anything in excess of a minor works amount. The clauses and brevity of minor and major work contracts will differ with both including sets of terms and conditions suitable for the works. Where this two-tier system exists, a contractor may delegate responsibility and authority for the issue of awards using a hierarchy and authorisation procedure, for example:

- minor works package not exceeding £5,000 only to be awarded by a competent person named within the project team;
- minor works packages in excess of £5,000 not exceeding £10,000 to be awarded by the quantity surveyor or project manager only;
- major works packages in excess of £10,000 not exceeding £50,000 to be awarded by the quantity surveyor and authorised by the project manager or the project manager alone may issue the award; or
- major works packages in excess of £50,000 require authorising by a company director.

This hierarchy approach is to be commended for large-sized projects as it places a level of trust on the project team and reduces bureaucracy that may otherwise stifle progression of the project.

Whatever form is used, it must cover aspects of the agreement and include the items covered in Sections 5.3.7.1–5.3.7.5 below.

5.3.7.1 Project Particulars and Parties to the Agreement
This is the introductory part of the form, and includes:

a) title and address of the project;
b) names and business addresses of the parties to the agreement;
c) contract sum or schedule of rates and terms of payment;
d) contractor's security from the subcontractor (e.g. cash retention or bank guarantee);
e) project commencement and completion dates; and
f) start and finish dates for the subcontractor's works.

The particulars may also name the project client and refer to industrial buzzwords used in the main contract for promoting a healthy working relationship. This includes collaboration, value improvement, health and safety and sustainable improvements that a subcontractor and contractor jointly elect to include in their agreement.

5.3.7.2 Documents Forming Part of the Agreement
This is the design and documentation usually included in the invitation to tender and subsequent negotiations and supplied as a document register. The format must also include a breakdown of the contract sum or schedule of rates for payment purposes. It should also include a schedule of insurances and daywork hourly rates, together with agreed percentage additions to the prime cost of the purchase of materials and hire of plant.

5.3.7.3 Collateral Warranties
Where required under the main contract, the contractor will advise the subcontractor of the names, titles and beneficiaries of collateral warranties as well as the parts of works that require warranties. The subcontractor must issue the warranties within a timeframe stated in the agreement, which is usually obtained from the main contract and advised by the contractor.

5.3.7.4 Terms and Conditions
The terms and conditions must focus on the rights, obligations and risks to the parties that form the body of the agreement under a series of headings that includes:

a) the contractor's access to subcontractor's representatives;
b) procedure the subcontractor must follow for claiming an adjustment of time;
c) the effect of site instructions issued by the contractor;
d) the parties adherence to health and safety obligations and CDM 2015 Regulations;
e) restriction of subletting of the works (e.g. a subcontractor may delegate their duties to another without assigning the rights of the subcontract agreement);
f) variation assessment procedure and authorisation process;
g) the contractor's right to instruct the removal of persons(s) under the control of the subcontractor from site for specified reasons;
h) termination of the agreement;
i) dispute resolution methods;
j) methods of addressing damage to the subcontractor's installed works and defective works; and
k) subcontractor's responsibility to maintain a clean working environment.

A dangerous and somewhat pointless exercise is for the contractor to include a term expressing that a subcontractor shall be liable for any of the terms and conditions written into the main contract. This type of ambiguous clause may be an attempt by a contractor to mirror its obligations to the subcontractor which, if sought, is best addressed with a back-to-back contract as discussed in Section 5.3.9 below. This drafting shortcut is not a credible catch-all clause, even if included in a signed contract that results in a dispute, because implications of liability to the subcontractor may be restricted by a court of law under the Unfair Contract Terms Act 1977.

5.3.7.5 Special Conditions

A detriment of some projects is when the contractor is not ready for a subcontractor to commence working operations with the subcontractor arriving on site as planned yet having no prior knowledge of the contractor's default. This can be frustrating to a subcontractor with the possibility of incurring costs due to abortive visits, without remedy in the agreement to recoup the loss. For this reason, subcontractors may request the inclusion of specific conditions and/or a list of attendances the contractor is to provide which is appended to the contract. Specific conditions can be suggested from a subcontractor that are published by a trade association affiliate which, if requested for inclusion in an agreement, should only be accepted if they are under the umbrella of a set of recommendations provided by the trade affiliate. For example, a piling contractor may be affiliated to the Federation of Piling Specialists (FPS), a trade association comprising members that carry out works for main contractors that are audited independently to ensure they maintain high standards of technical ability, quality management, safety, training and management systems. To reflect these qualities, the FPS produces a Schedule of Facilities and Attendances that a main contractor must provide for efficient and safe site operations (e.g. a suitable solid platform for manoeuvring a piling rig on site). If the quantity surveyor is requested by a subcontractor to include a trade affiliate's recommendations and is a member of the association, they should be considered reliable and appended to the agreement. Where a trade contractor does not have an affiliation and/or is insisting on a set of conditions of its own creation, the project team must consider the implications before agreeing to append such conditions to an agreement.

5.3.7.6 Conclusion of the Formality

It would be wise for the quantity surveyor to keep a register recording the dates of issuing awards and the date each award is formally signed off by the parties, which could be included on the letting schedule. Unfortunately, recording of this formality is not always adhered to with an assumption possibly made by the contractor that, once a subcontractor is issued an award, it is an end of the matter. This is incorrect conduct because the subcontractor might forget to sign and return the agreement and, should a dispute arise without recall on the executed contract, it may be hard to prove to a third party that a contract exists. If there is a failure to execute an agreement and the subcontractor commences works, the project manager must be made aware of the situation; if there is a dispute, the contractor may be exposed to limited remedy as the process towards formalising the agreement is incomplete. To conclude the formality, the quantity surveyor must issue two copies of the agreement to the subcontractor for execution with a proviso both are executed and returned to the contractor within a specified timeframe. Upon the contractor's receipt of both contracts and subcontractor's acceptance of the terms, the contractor executes the forms, keeps one copy and returns the other to the subcontractor.

5.3.8   Generic Forms of Subcontract Agreement

The JCT publishes a generic Short Form of Sub Contract (ShortSub), currently 2016 edition, which is ideal for use on a main contract where a JCT 2016 form is to be administered. Although not mandatory for use with a JCT main contract, it has the aims of maintaining consistency, and is suitable where the subcontract works involved poses low risk and is straightforward in nature. The adoption of this form has a distinct advantage for the parties as it is not prepared by the contractor; this means subcontractors should not be sceptical of using it as it is not drafted in the contractor's favour. JCT also provides a Sub-subcontract (SubSub) form (currently 2016 edition) for use between subcontractors contracted under the ShortSub for use down the supply chain, which again is not mandatory yet maintains the JCT consistency. Where a contractor elects to use a form of subcontract agreement provided by the same drafter of the main contract form, it does not mean a subcontractor has recourse to the main contract. Moreover, and by default, the project client will not have recourse to the subcontractor because the two are not contractually linked.

When a subcontractor is named in the Intermediate JCT forms of main contract, it has provisions stating the type of form a contractor must use for engaging subcontractors. For example, the JCT Intermediate Building Contract (IC 2016), clause 3.7 states the form to use is the Intermediate Named Sub-Contract Agreement ICSub/NAM/A. The issue of this form is a contractual requirement; the contractor must not issue an alternative as the execution of such form by the parties would create a breach of contract.

The current JCT forms of contract abstain from involvement with the nominated subcontract procedure. However, if a quantity surveyor is administering a project involving nominated subcontractors using alternative forms, it is necessary to understand the procedure; this would normally be found in appendices to the main contract and specific clauses.

5.3.9   Back-to-Back Forms of Subcontract Agreement

A back-to-back contract is a subcontract agreement that attempts to mirror the terms and conditions of the main contract into a subcontract form. In reality, these forms are hard to create and are of use when a contractor has repeat business with one or more clients that adopt the same main contract form and/or the project(s) is to run for some time. Where applying, the contractor's lawyers will review the main contract form akin to the project and edit particular clauses so they make commercial sense in a subcontract agreement.

Advantages of a back-to-back contract to the parties include the following.

- The contractor is in a position to declare to its subcontractors the form is not edited in any way that favours the contractor because it reflects the main contract requirements, parts of which can be provided for reference where a potential subcontractor has concern.
- It can be used for repeat business.
- With repeat business, execution is rapid as the parties are familiar with the form.
- It can reduce the number of disputes because it is specific, whereas standard forms are generic.

Disadvantages include the following.

- Subcontractors may be reluctant to be involved with such an extensive form, especially if creating an agreement with a contractor for the first time and unfamiliar with the process.
- A contractor may be disadvantaged if they enter into a main contract agreement using a standard form that is edited or modified at the client's request. In this case, and for the idea to apply perfectly, the adjusted main contract form would be used as a guide to create the back-to-back form which a lawyer will need to prepare, and the time may not be available.
- They are expensive to create due to the involvement of the contractor's management and lawyers.
- Where a contractor's business is diverse and carries out works under numerous standard forms, the back-to-back ideology is harder to create. This is because the lead time from the date of concluding the main contract to a date the first subcontractor is required to commence on site may be insufficient to draft the form.

Back-to-back forms of contract are ideal when they involve specific and/or technical situations because the contractor can mirror the criteria in an agreement in order for the subcontractor to perform exactly to the manner expected by the contractor under the main contract. This can be complex to arrange when the contractor wishes to reflect legal and commercial clauses that may be hard to conclude. In such a situation, the contractor may elect to leave such clauses intact by accepting the risk, yet be satisfied the technical aspects have been transferred which mitigates some project risk.

5.4 Material Supply Scheduling and Purchase Ordering

Where the contractor elects to procure works with labour-only subcontractors and purchase the materials, it is the contractor's responsibility to schedule such materials for the supply and delivery to site. The contractor may also schedule materials for certain preliminary temporary works items (e.g. timber for site hoarding for installation by labour-only subcontractors). The contractor does not need to schedule materials for domestic and nominated subcontractors, as they are usually responsible for this themselves and require no input from the contractor other than the supply of 'for construction' information. When a contractor is responsible for producing materials schedules and the ordering of materials, the information can be obtained from one or more of the following:

- builder's quantities prepared during the project tender period (as reference only);
- a contract bill of quantities (as reference only);
- updated builder's quantities or new material schedules;
- schedules in the specifications; or
- schedules on the drawings.

Builder's quantities or a contract bill of quantities used to assess an amount to tender or as a contractual document should only be referred to when it is known the works reciprocate the requirements at the time of tender. Even if this is the situation, the stated quantities should not be used for ordering purposes. This is because stated quantities reflect the net measured quantity installed in position, meaning they exclude additional quantities required for material waste created from cutting, working and shrinkage. Furthermore, where works are known to have altered since the contractor's tender was issued, the quantities will not be current unless they are updated, which is unlikely because of the time and cost involved in their preparation. This means it will be necessary to reschedule the actual requirements as new builder's quantities or as a schedule of material requirements. The quantity surveyor or buyer of the materials can carry out this exercise by adopting the original format and removing reference to labour for installation, while updating any product changes and quantities.

When the scheduling is complete, the project manager may elect to delegate responsibilities for the placement of awards to members of the project team. This will involve empowering individuals to seek chargeable rates from suppliers and placing purchase orders, which are usually to within financial thresholds with delegation to placing awards linked to job title, similar to the creation of minor and major work awards for subcontract packages as discussed in Section 5.3.7 above. Where an order value exceeds a certain financial threshold, those delegated to award orders must not exceed their authority and the quantity surveyor must adopt a commercial lead. This will involve the issue of invitations to tender to obtain best value following a similar procedure to procuring subcontractors. The post-tender procedures leading to the issue of material supply awards are not as robust as those used for placing subcontractor packages, as there is less risk to the contractor with the formality mitigated because:

- the value of awards is generally less than subcontractor packages;
- the contractor may have existing price agreements in place with certain suppliers of goods;
- there is no cash retention or financial security required;
- insurance provisions generally do not apply;
- liquidated damages do not apply;
- CDM requirements for health and safety are relaxed; or
- the contractor has more control with the timely supply of goods than with materials procured through domestic subcontractors.

The quantity surveyor must ensure project team members tasked with the responsibility of placing purchase orders are aware they must deal with suppliers that are approved by the contractor whose materials comply with the project specification and offer reasonable prices and/or discounts. In addition to obtaining best price, suppliers must be able to deliver products on time to suit the construction programme and the contractor's site requirements. It is equally important for project team members to be aware of budgets as they are the driving force that help secure projects for the contractor. In order for a contractor to remain competitive with tenders for construction works, the estimator will usually include the cost of materials that includes full commercial benefits. This commercial benefit may be lost if the contractor fails to manage the process and issues awards to businesses not offering discounts or are uncompetitive. To avoid this, a contractor's business may include a buying department based at the head office to procure materials, where prices are agreed by negotiation and/or with schedules of rates, and which liaises with the site and project managers for the delivery requirements. However, in the absence of a buying department, responsibility may rest with the quantity surveyor to be briefed by the commercial and/or project manager regarding purchasing arrangements and the companies they are to be with.

Where material procurement is a project team's responsibility, the method of issuing awards involves providing each supplier with a supply order that includes a unique reference or order number. The method of communicating supply orders varies between contractors, with some adopting a formal approach and an order numbering system and others using word-of-mouth agreements confirmed with site instructions. This latter approach cannot be recommended as it can lead to misunderstanding and errors. Orders of a small value for a project's preliminaries and minor material purchases for labour-only trade works may be placed over the telephone if the supplier will accept an oral order. When doing so, it is necessary to carry out the following.

- Advise of the unique purchase order number.
- Include a description of the goods, quantities and delivery dates.
- Complete a written order addressed to the supplier with project reference for identification. The written order must include the supplier's name and address; contractor's business and site delivery address; date of the order; date(s) required for the supply of the goods to site; a description and quantity of the goods; quoted rates(s) or a lump-sum price usually excluding VAT; and signature of the purchaser on behalf of the contractor.

To conclude an order a supplier may wish to include conditions of sale; if requested this is usually in accordance with the Sale of Goods Act 1979 (as amended). This legislation is aimed at reflecting commercial expectations in commonly arranged terms of agreement, and if requested should be accepted. When this legislation is not stated, the contractor's purchase order may include small print at the footnote of the order or on the reverse side referring to the Act. If the party's formalities are silent on the matter, there is no point in adding a clause to include the Act as it is a statutory right whether written into the agreement or not.

A supplier may request the buyer of the goods to complete a project registration form specifying the goods to be supplied as well as the quantities required. Suppliers adopt this approach when not wishing to be caught up in ambiguities, missing information or interpretations of specifications or drawings, and wish to rely on the contractor to state the requirements. For example, doors on a drawing and in a door specification may state a requirement for safety glass panels to be installed to BS6262 but nothing more. This is an obligation of the supplier to ensure the glazed panels comply with the standard. However, the supplier will need to know if the panels are to be clear or obscured, which may only be determined from viewing the plans to see if any doors are for rooms that require privacy.

When a verbal order is complete, the written purchase order should be sent as an attachment in an email, posted or uploaded to the supplier's host site (if applicable) as proof of the requirements and marked 'Confirmation'. It is necessary to note the order as confirmation because the recipient may not be the person accepting the telephone order and may duplicate the requirement in error.

If a contractor is a new account to a supplier, it may be necessary to complete a supplier's credit check arrangement which will require divulging the contractor's business details. Site managers and foremen delegated to place purchase orders may find this process a hindrance to project activities; if no existing account supplier is available, the quantity surveyor should complete the form and return it to the supplier for entry into their database to create the new account.

Once goods are delivered to site, each transaction is usually confirmed with a delivery note/docket issued by the supplier which is signed by the receiver as proof of delivery. The proof of delivery note should be kept in the site office for distribution to the contractor's accounts department at head office in order to process the supplier's payments based upon receipt of invoice and proof of the delivery.

5.4.1 Bulk Orders

A method of procuring large quantities of materials for delivery to site at sporadic intervals is with bulk orders. To fulfil the requirements of each bulk order it is necessary to ascertain a quantity required for the whole scope, which can be obtained from the bill of quantities or builder's schedules, and involves converting the unit of billed measurement into a unit of sold measurement plus an allowance for waste. For example, a bill of quantities may describe the external skin of hollow walls as 102.5 mm thick facing bricks to an area of $500\,m^2$ which, if verified as an accurate area, can be quantified as follows:

$$500\,m^2 \times 59\,nr/m^2 = 29,500 + 15\%\,\text{allowance for waste}$$
$$= 33,925\,\text{total number of facing bricks}$$

In this example, because the bricks are purchased by the thousand, the bulk order quantity must not exceed 33,925 if the budget is to be preserved, providing that the payable rate is comparable with the budget. The quantity surveyor must exercise skill when placing the bulk order and cap the quantity to the nearest whole unit advised by the supplier. To demonstrate, let us say a quotation is received from a brick supplier quoting a rate per thousand for the bricks including delivery to site and offloading based upon 10 packs of 550 bricks, that is, 5500 per delivery. What the supplier is saying is that it will need to supply this quantity per delivery to cover the cost of loading in the factory, hauling to site and offloading on site. If the quantities for each delivery are less, the price per thousand will increase to compensate for the loading and transportation costs that are fixed. Here, the quantity surveyor may place a bulk purchase order to a maximum of six full deliveries ($6 \times 5,500$) or 33,000 facing bricks, which acknowledges the suppliers terms of sale. This is also within an acceptable allowance for the budget to be safe and provides adequate allowance for waste, albeit slightly less than the budget assessment of 15%.

Bulk orders have a distinct advantage for running a project because only one order needs to be placed for each material, which reduces administration time. Imagine the chaos if a project requires 100 identical doors over a period of a year and one order was placed each time there was a site request for a door; that could mean up to 100 orders being placed!

A bulk order should include a proviso that it is for a maximum quantity which a contractor reserves the right to amend from time to time. The order should not be a fixed quantity or total lump-sum fixed price for the following reasons.

- The contractor's allowance for waste might be exaggerated resulting in the need to reduce the quantity required, which may only become apparent when the supply of materials on a bulk order is near completion.
- If an order is placed upon a fixed quantity that fails to transpire, the contractor may be exposed to a claim for loss of profit that a supplier would have gained if it had completed the order.

- The project may be terminated by the client, contractor or both.
- If the supplier defaults, the order can be terminated and a new order placed elsewhere.
- Prices may alter for the duration of the bulk order, which is a risk with this type of arrangement.

So, what happens if the final quantities are found to be in excess of the bulk order because the design has changed or errors are discovered in the bills or schedules in the specifications? If a variation occurs under the main contract altering the works or correcting the bill of quantities or schedules, the appropriate suppliers require to be issued with variation orders to adjust the quantities. This action is necessary because each supplier has an obligation to comply with their bulk order, and may not accept a request from a site manager for a further delivery if cumulative quantities exceed those stated on the order. This could have serious results if not communicated to the supplier, as a halt in deliveries may disrupt site production with the possibility labour-only subcontractors could seek reimbursement for lost time if they are delayed because of a shortage of materials.

A criticism from some site managers with bulk ordering based upon a bill of quantities is that they consider the quantities incorrect for ordering and production purposes. This may have a ring of truth about it, yet there could be compelling reasons for the discrepancy of which a site manager is unaware (e.g. billed quantities are as-built with product wastage allowed in the chargeable rate). Furthermore, if value engineering is carried out post-tender altering the design, the bills may not have been updated because the consensus is their production would cost time and money with the update considered unnecessary. The requirement for additional materials may also be because the contractor is wasting more materials than allowed in a bulk order; if so, information on the actual wastage should be fed back to the estimator for consideration on new works under tender. However, the project manager must be made aware of excessive wastage and be satisfied it is not site-specific or the result of site management methodology before altering the estimator, because the estimator will only allow wastage based upon efficiency and not something particular to a project. If excess waste is found to be the contractor's responsibility, it is generally of no concern to the supplier and the quantity on the bulk order must be adjusted with a variation or addendum/new order.

To manage the control of material supplies, it will be necessary for the site manger and foremen to record bulk order deliveries in a goods supply register to ensure each order is tracked. The management of these registers is important because they highlight any discrepancies between the contract design and documentation driving the contract sum and 'for construction' information showing the final requirements. For example, let us say under a construct-only contract, the steel bar reinforcement design to the poured concrete components was incomplete at the time of concluding a contract, with the design stating the requirement as a ratio of x kg/m^3 (i.e. a stated weight of steel in kilograms for the volume of poured concrete with varying ratios for beams, slabs, etc.). Here, the contractor has executed a contract based upon this information and assessed an appropriate allowance of steel to the volume of poured concrete. Once the works commence, a bulk order for steel reinforcement may be placed to include the supply of steel reinforcement to the maximum tonnage included in the contractor's assessment. Suitable management of the goods supply register will record the steel supplied for comparison with the bulk order allowance, and a problem can occur when:

- the builder's schedule allowance of steel bar reinforcement for the building is 500 tonnes;
- the bulk order for the supply is 500 tonnes with no variation deemed necessary; and
- the goods supply register totals 500 tonnes delivered with 50% of the works complete.

Here, the contractor is faced with a potential problem because the quantity on the bulk supply order has expired with half the works complete. This will require an investigation to assess the steel requirement for the first half of the project to see if a variation under the main contract is justified. Without these registers to track progress, a contractor will have no knowledge of material usage and the project manager and quantity surveyor must enforce their management.

5.5   Labour Hire Agreements

The hire of labour for the construction phase of a project falls into the following categories:

- general site operatives;
- semi-skilled operatives with an intermediate level of training;
- qualified skilled operatives as a labour workforce carrying out a range of site duties; and
- managers and other professionals engaged on a temporary basis providing a service not involving design.

General site operatives (often referred to as site labourers) are employed by a contractor with their cost to employ usually included in the preliminaries, and carry out duties that include (but not limited to) sweeping, cleaning, keeping the site in good order and general lifting and carrying. These operatives may also hold licenses which permit them to operate forklift trucks for transporting materials around site to position for installation by others.

Semi-skilled and skilled trade operatives refers to qualified personnel employed by subcontracting companies, labour-only subcontractors appointed by the main contractor and contractors engaged by the project client that are appointed to carry out specific tasks akin to their skill set. A general site operative does not usually assist these operatives directly, as the role is one of facilitator only to ensure areas are prepared and ready for skilled operatives use with the materials stored in close proximity. Moreover, and because of their training, facilitators of access and lifting equipment such as scaffolding, hoists and cranes are supplied by equipment installation companies subcontracted to the main contractor who are deemed skilled operatives. In isolation, skilled operatives may employ their own general operatives to assist with specific site operations. For example, a subcontractor engaged to apply a wet external wall rendering system over steel frames may engage operatives to place and install render boards under the watchful eye of skilled operatives prior to them applying the base and finishing coats; when doing so, these operatives are not deemed general site operatives for the contractor's use.

Occasionally, a contractor may need to hire additional general labour operatives to assist with site operations. Where required, recruitment is at the discretion of the site and/or project managers who may hire through a recruitment agency. When hiring in this way, a purchase order is placed with the agency and the operative(s) is hired at an agreed hourly/daily rate with the agency reimbursing the operative(s) as there is no financial arrangement between the operative(s) and contractor. The hours/days attended on site are recorded on hard copy timesheets which are authorised by the site manager, for invoicing in accordance with the authorisation and purchase order. Alternatively, time can be recorded online for electric authorisation by the site manager, which is a quicker sustainable system with the process spontaneous.

The quantity surveyor will need to monitor outsourced general site operative/semi-skilled labour hire under its direct control, as costs can escalate if left unchecked. A contractor's preliminaries will usually have a budget allowance for general site operatives per month and, if exceeded, must be reviewed to see if it is a short-term fix or something that is prolonged. A site and project manager will usually be aware of a project's preliminaries allowance and manage the process effectively by off-hiring the service when it is no longer required. However, continuous hire because of contractor complacency is not uncommon with an excessive general operative labour force not always justified which, if left uncontrolled, can cause financial harm to the contractor. This can be a contentious matter for the site and project managers as they might consider additional resources are required, yet understand there is no surplus budget. Where applicable, the quantity surveyor must endeavour to advise the project manager of the weekly/monthly costs versus the budget allowance to determine if the benefit of extra labour hire is worthwhile: will it assist production and recover lost time if works are behind programme?

5.5.1 Apprentices

Contractors are encouraged by trade unions, central government, educational boards and industry bodies to engage apprentices so they can combine college studies with practical on-the-job experience. Encouragement for contractor participation with training schemes comes via the Construction Industry Training Board (CITB), where contractors pay a levy for the service and benefit from the arrangement because of the productive work they receive. Further encouragement is endorsed by the British Government that in April 2017 introduced an apprenticeship scheme requiring all employers operating in the UK with a pay bill over £3 million per year to invest in apprenticeships. Applicable to all employers across a spectrum of industries with an annual pay bill of more than £3 million, each will need to spend 0.5% of their total pay bill on the apprenticeship levy. However, there is a 'levy allowance' of £15,000 per year, meaning that the total amount to spend is 0.5% of the pay bill minus £15,000.

When engaged on a construction site, apprentices are supervised by the contractor and are not expected to carry out general site operative duties as the intent is for them to learn the skills of the trade they are undertaking in an apprenticeship. The site manager inducts and briefs apprentices on the contractor's activities, including health and safety requirements, and the role the contractor plays as principal contractor to satisfy the requirements of CDM 2015. Thereafter, the apprentice commences training under the guidance of skilled trade operatives appropriate to their trade course for a specified duration before returning to studies.

5.5.2   Management Hire

Professional managers may be hired through a recruitment agency and paid on hourly/daily rates; where applicable, they are usually engaged on an *ad hoc* basis under a short-term agreement to aid a contractor's imminent needs without committing to long-term employment. Alternatively, a contractor may have a contact list of freelance persons for professional short-term appointments and elect to engage them directly instead of through a recruitment agency. For example, a project may be lagging with trade procurement and requires a freelance quantity surveyor on a temporary basis to assist with the placement of awards. Site supervision staff may also be hired as an auxiliary requirement, which is suitable when accelerating a works programme because of the number of operatives involved that warrants additional management. As these hire agreements are contracts of services, if they are sourced through a recruitment agency the agency incurs associated employment costs which is included in the chargeable rate. The agency is also usually responsible for withholding tax on payment to employees, as in effect they are employees of the recruitment agency. This is unless the supplier of the services is an independent contractor and self-employed who may elect to pay their own tax, which can also apply if the hire does not involve a recruitment agency.

5.6   Plant Hire Agreements

For any project, and particularly new build, a contractor will have a need to hire mechanical and non-mechanical plant to service the scheme undertaken. This is applicable to preliminaries items, labour-only subcontractor works, builders work in connection with domestic/nominated subcontractor and client-engaged contractor's works, as well as any trade works undertaken by the contractor which is not subcontracted (e.g. site clearance operations).

When there is a need for plant on a project the contractor may already own assets, and where suitable, are requested for delivery to site by the project manager from their place of store with the cost for delivery, hire and removal costed by the plant manager/accountant to the project through the cost purchase ledger. If outsourcing instead, the quantity surveyor must secure competitive rates from the equipment hire market with the cost based upon the length of time the plant will remain on site. Where a contractor has a buying department, the buyer(s) may negotiate prices upon schedules of rates applicable to fixed items (such as skips) and time-related items (such as forklift trucks and rubbish chutes) that may be hourly, daily, weekly or monthly rates plus delivery and collection charges which are issued as 'blanket' orders. This approach permits plant to be hired that can be later off-hired and rehired to suit the sequence of site operations without the need to issue additional purchase orders. For example, mortar mixers for use by bricklayers may be hired for the construction of a building which can be off-hired when the building is complete. The mixers can also be rehired to facilitate the construction of a brick boundary wall if it is not scheduled to be constructed concurrently with the building. Where a buying department does not exist in a contracting organisation, it is a matter for the quantity surveyor to ensure that 'blanket' orders are placed with competitive rates, which may be achieved by obtaining tenders or suitable predetermined rates if a contractor has repeat business with one or more hire companies.

When agreeing rates, it is important to determine where the responsibility rests for insurance in the event of plant failure or damage to the equipment while it is on hire. Such insurance is usually provided by the hire company, and should be confirmed as a matter of course.

Site mangers and foremen are usually authorised by the project manager to hire plant to suit the programme of works and are permitted to request delivery from any company in receipt of a blanket order to suit the site operations. Alternatively, these managers may place their own *ad hoc* orders if delegated with the commercial responsibility. When hiring items, it is necessary for the site manager and foremen to record the commencement and off-hire dates in a register to demonstrate the process is managed efficiently, and only applies to plant hired by the contractor for the contractor's use. Moreover, the register should distinguish between plant hired from a hire company and plant on site which is a contractor's asset, because there will usually be no invoices to reconcile for the latter. In the interests of cost management, the quantity surveyor should refer to the register regularly as it provides a snapshot of plant on a site at any time and is necessary for reconciling invoices from plant hire companies that refer to hire periods.

Any item of plant hired or owned by subcontractors to carry out their contracted works does not need to be recorded on the register as they are not a responsibility of the contractor. If the contractor hires plant including operators to aid a subcontractor's contracted works, it will in effect have taken over the works and, when doing so, is taking a risk. This is because a form of contract usually has provisions to deal with any event where a subcontractor fails to progress diligently. Usually, a contractor may only take over a subcontractor's works after the issue of a written notice to the subcontractor citing reasons for the conduct and stating a date the works will be taken over unless the subcontractor resumes operations beforehand. If the contractor decides to take over the works earlier than the expiry date of the notice, or omits to issue a notice and issue of a notice is a condition of contract, the contractor cannot recover expense from the subcontractor because the contract will have been breached. If this situation becomes apparent to the quantity surveyor, the project manager must be alerted because the contractor might incur expense it cannot recover.

The cost to mobilise, hire and demobilise cranes, hoists and scaffolding including the supply of qualified skilled operatives to facilitate their use is expensive. This involves the appointment of companies to provide the base plant as well as the operatives, maintenance, insurance, safety inspections and ancillary equipment required for the plant to function. In line with industrial customs, these hire companies are labour and materials subcontractors and procured in the same manner as domestic subcontractors, which involves a tender period, negotiations and issue of an award as discussed in Section 5.3 above. Some contractors however may own scaffolding equipment as tangible assets, and appoint operatives to erect, adjust and dismantle the scaffolding on rates or agreed prices on a case-by-case basis.

5.7 Consultant Appointments

The appointment of consultants is necessary when a contractor provides a design service for a project client or where the contractor requires a reporting service or range of supplementary services to fulfil its obligations under the contract.

5.7.1 With Design Input

With a design-and-build contract, the contractor appoints design consultants to carry out an array of services in order for the working design to become available in a timely manner for construction purposes. The selection process may be administered by the contractor's design manager or left to the devices of the project manager, which depends on the contractor's culture, the project itself and the number of consultant appointments required. As part of the appointment procedure, design consultants are invited to submit a fee proposal following the terms of an invitation to tender. Selection is by referral, experience or the continuation of services provided to the project client where re-appointment can take place by novation. The invitation to tender will usually outline the scope of services required from each consultant to furnish the contractor's needs, for which a fee proposal and defined scope of services is offered that usually comprises:

- initial briefing with the contractor;
- preparation of preconstruction design and documentation, including issue of the information to the contractor for approval;
- coordination with other consultants;
- issue of design and documentation for the construction phase including a predetermined number of site visits, attending meetings and a commitment to the contractor's requests for information (RFIs) on any inconsistencies, with the information supplied should the need arise;
- a review of as-built information provided by the contractor or its subcontractors; and
- supply of professional indemnity insurance.

Once a fee proposal is accepted, a contract of services comes into existence which creates a binding agreement. Ideally, the agreement should be a lump-sum fee the contractor agrees to pay the consultant for the services and applies where the scope is known with the client's brief and room data sheets at an advanced stage or complete. The agreement should also include a schedule of payable sums in accordance with the fee proposal for interim payment purposes and a schedule of rates for additional services. Where a subcontractor provides design portions, the contractor will not appoint the consultants and relies on the subcontractor to provide professional indemnity insurance to safeguard the parties to the agreement and the end user.

5.7.2 Without Design Input

The need for the engagement of these consultants includes:

- preparation of site accommodation layouts;
- where applicable, noise monitoring of site operations to BS 6472:1992 (2008 version), which is a standard with developed criteria relating to levels of building vibration that can give rise to an adverse comment if the noise is within a frequency range, applicable to an impact associated with construction activities if such monitoring is a requirement of planning approval;
- monitoring of dust air particles from site operations if a local authority request;
- field surveyor to set out the building and provide datums and levels where a contractor does not employ surveyors or engineers; and
- the creation of a construction programme if a contractor does not employ a construction planner.

When inviting tenders for supplementary services from consultants, a contractor will usually outline the objectives it wishes to achieve from the appointment to which the consultant submits a scope of services together with fee proposal for the contractor's consideration. If the fee proposal is accepted, it is included in an award for executing as a binding agreement. As with appointments including design, the contract should be in writing and include a schedule of payable sums by time and/or stage, and a schedule of rates for additional services by discipline (e.g. director, technician, project manager, etc.). Inclusion of these rates in an agreement will avoid any surprises should the additional services be required.

6

Running the Project

6.1 Document Control

The contractor's team tasked with delivering a project are usually issued contractual design and documentation for advice and, for practical purposes, must be provided working information as soon as the criteria become available. Thereafter, it is not unusual for the team to receive an influx of new or revised information prepared by the client, design consultants and manager in charge of the company's policies and procedures that may change from time to time. Separate to this, the team will generate documents associated with the running of the project derived from communications and instructions to and from the client, as well as consultants and the supply chain. Furthermore, information will be created in the form of progress reports, subcontractor and material supplier procurement, payment certificates and general administration. To deal with this, it is important for the project team to manage and store documentation while the project is active, as there will be a need to access any of these items during the construction phase and afterwards. Depending on the size of a project in terms of value, complexity and duration this can warrant a role of its own, and one which is best managed with the employment of a document controller. This role must not be considered general administration, and should be seen as ensuring the flow of information is consistent for the running of a project. To assist document control procedures, each team member must be responsible for storing and sharing suitable information with the distribution and storage of received information left to the devices of the document controller.

6.1.1 Design and Documentation Changes

At the commencement of a project, a hard copy of the contract drawings and documents are usually stored in files and racks within the site accommodation and backed up electronically on a computer server. The contractor also keeps a soft copy and at least one hard copy of 'for construction' information within the site accommodation for general use, which is updated as required with the updates stored electronically.

When procuring works, a contractor could engage scores of material suppliers and subcontractors, and it is important that appropriate information is supplied to each. An essential part of project administration involves the contractor issuing each supplier and subcontractor with a set of appropriate 'for construction' information, which is usually issued at the time of placing an award or soon after. Alternatively, an award may

Construction Quantity Surveying: A Practical Guide for the Contractor's QS, Second Edition. Donald Towey.
© 2018 John Wiley & Sons Ltd. Published 2018 by John Wiley & Sons Ltd.

state that the working information issued with the contractor's invitation to tender is 'for construction' purposes. However, if a client engages designers, material suppliers or contractors that have no contractual link to the contractor, the client's agent will normally issue information directly and provide the contractor with a copy of the consignment as a matter of course.

Design and specification changes can occur at any time during the construction phase (and possibly the defects liability/rectification period) that, when approved by the project client, must be issued to the contractor in a timely manner by either the client's agent on a construct-only contract or contractor's design manager on a design-and-construct project. There are various methods of supplying this information, for example: electronic collaboration system, email (subject to the size of the consignment), compact disc, flash drive and/plus hard copy (if hard copy supply is an accepted requirement by the parties). Each issue is usually accompanied with a transmittal notice addressed to the project manager who then issues the information for action by the quantity surveyor for commercial/contractual reasons and site manager for site operations.

There are reasons why new or revised information is issued that, when received by the quantity surveyor, must be understood in order to ensure appropriate supply chain members are issued with the correct information. To send changes blindly to all supply chain members would be irresponsible of the contractor meaning the contents to be supplied must be confined to the necessary parties. This is an important aspect of document control as the works on site may be awaiting the information in order to proceed, with any delay possibly disrupting site progress. If the quantity surveyor considers no expense will be incurred as a result of the changes, for example drawings that magnify certain details or confirm the design complies with the requirements of CDM 2015, the information must still be issued appropriately to ensure suitable supply chain members are in receipt of current and relevant information.

When a consignment involves revised information, it may state what the revision entails and, if involving a drawing, may identify changes with a symbol such as a cloud and describe in the revision section of the drawing what that change entails. However, consultants may avoid this for practical reasons, especially if numerous changes are involved, and may also wish to avoid liability by not clouding changes in the event any change not clouded is missed by the contractor. Here, a revision may state 'changes to suit the client's request' or 'changes following design coordination', leaving the contractor to spot the differences and decide which supply chain members should be supplied the information. In order to assess the impact of changes and the trade works affected, it is necessary for the quantity surveyor to understand the sequence of working operations as well as construction methodology and technology. For example, let us say revised architectural drawings are issued with an instruction to change the elevations of a building by amending the window dimensions which are not manufactured or the walls built. Figure 6.1 demonstrates this change in a diagram focusing on the subject matter and linking the effect to trades that require notifying of the change.

In the example in Figure 6.1, the glazing to the windows may be part of the window manufacturer's award which includes window installation. Here, the contractor would only issue details of the change to the window manufacturer, even if having knowledge of the glass supplier's and glazier's details. If there is a knock-on affect because of the window changes, say with the position of electrical power outlets that need relocating

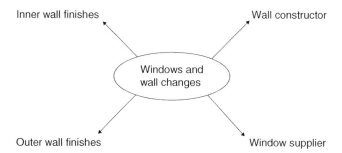

Figure 6.1 Work change affects.

because they clash with the revised window scheme, the party responsible for design coordination will coordinate the design with the electrical consultant. The electrical consultant will then amend and reissue electrical drawings and, once received by the contractor, requires to be issued to the electrical subcontractor for action. When issuing such information, the contractor must provide a document transmittal which may include a letter signed by the quantity surveyor if wishing to confirm a quoted price, which a subcontractor may already have provided. In the event of design and documentation changes applying to works where no subcontract award is in place, the quantity surveyor must recognise such changes and ensure that, when placing an award, the subcontractor is in receipt of the current information.

On a project of long duration, running for say a year or more, various subcontractors may be required for different periods of the construction phase and, in the process, may employ supervisors that may be based on site or provide regular visits. Here, distribution of new and revised information is usually via the contractor's site manager who places the information in pigeonhole message boxes located on site, such as in a general subcontractor's office, where each supervisor can retrieve the information. However, it is normally the quantity surveyor's responsibility to ensure the contents of each distribution is confirmed in writing, which can be achieved by including a document transmittal with each consignment. Failure to follow this procedure may lead to oversights by suppliers of goods and/or goods and services, as in effect they could be working to superseded information.

A wise contractor will not dispose of superseded information and ensure all details are retained in site files and racks. Where applicable, this information should be stored within the site accommodation and put out of sight to ensure they are not used as working information in error. If a project involves numerous revisions culminating in a large amount of superseded information, and a site office lacks storage space, the information can be kept as electronic copies instead (freeing up physical office space).

6.1.2 Contractor-Generated Documents

To enable collaborative working and effective communication between project team members, the creation of a shared reading file comprising written communications received in a working week for review by team members will aid a common understanding of colleagues' activities. This can be implemented by each team member placing received documents in a folder for circulation and review. A secretary would be the best person to arrange the process and involves circulating the folder for reading by project team members who sign off a sheet attached to the front to confirm the contents as read. When all team members have signed off the sheet, the folder's contents are removed for filing with new information inserted for the cycle to recommence. The use of a reading file is beneficial on large projects as each team member cannot be expected to know the daily activities of their colleagues. For example, a client's agent may have issued a number of written instructions to the site manager and, if the instructions are included in the file, the quantity surveyor will become aware of their contents and act as necessary by assessing any effect on the contract and alerting the project manager where necessary. Items to include in a reading file should be specific and not of a general nature, including:

- site instructions to the contractor;
- site instructions from the contractor to the supply chain;
- communication to and from the client/client's agent;
- document transmittals from the client's agent or consultants;
- minutes of meetings;
- health and safety notices;
- communications between the contractor and supply chain;
- communications between statutory authorities and other authorities involved with the project;
- communications from other project stakeholders; and
- contractor's requests for information (RFI) addressed to consultants to answer queries.

It does not need to include:

- delivery dockets from suppliers;
- quotations or payment certificates;
- variation prices;
- registers of hired plant and material deliveries;
- payment details;
- time sheets;
- general communications that are concluded and require no further input from anyone; or
- anything not considered relevant to the project team's roles.

Throughout the construction phase, each project team member has a responsibility to ensuring that electronic and hard copy information in relation to their role is accessible and maintained at all times. For example, a site manager is usually responsible for safekeeping site instructions issued from the client's agent, as it is a site management activity. Likewise, the quantity surveyor is responsible for the creation and management of documentation of a commercial and contractual nature. For security purposes, information retained in hard copies should be stored electronically and backed up regularly on a computer server located beyond the confinement of the site accommodation (in the event of a fire, flood, etc. that may destroy hard copies). The contents on the server may also be subject to risk from system failure and, for added security, such information can be uploaded and stored in the cloud for downloading when required.

6.2 Changes to the Works

The parties entering into a construction contract require a degree of flexibility with their agreement to reflect issues that can occur during the construction phase. For this reason, it is normal for a contract to include a clause(s) citing specific scenarios that, if eventuating, would be deemed variations, one of which may be to allow the client/employer to change the works under contract. To be effective the method of issuing a change to the works should be stated in the contract, and would normally be via a written instruction from the client's agent to the contractor. To determine if an instruction is in fact a variation at the same time, it should be called a variation and/or cite the appropriate variation clause(s) of the contract. However, there may be no requirement for the client's agent to call an instruction a variation or cite clauses of the contract if issued in a timely manner to allow the instructed works to proceed, leaving the contractor to decide if the instruction is also a variation. Upon receipt of such an instruction, the contractor's quantity surveyor must be first satisfied it is not merely enforcing the contract (e.g. changing installed doors because they are not in accordance with the approved design and specification). If recognised as a change at the client's request or a client responsibility, the quantity surveyor will need to assess the impact on two fronts: changes under the main contract; and changes to any agreements in place with the contractor's supply chain.

For clarity, it is wise for the quantity surveyor and project team to define variations as those listed under the terms and conditions of the main contract only. This is an important distinction as, through custom, contractors may use the term variations in general by associating them with additional works to any of the contractor's supply chain beyond the scope included in each agreement. To demonstrate, if a contractor executes an agreement with a plumbing contractor to install the free issue of sanitary ware with the supply part of the main contract, but the contractor later wishes the plumbing subcontractor to provide the sanitary ware, it is not a variation under the main contract. Here, it merely means that at the time of entering into the subcontract agreement, the contractor intended to place an award for the sanitary ware elsewhere; if later deciding to issue an instruction to the plumbing subcontractor to supply the sanitary ware, the supply is additional works to the plumbing subcontractor. However, if the client alters the type of sanitary ware that was part of the main contract, it becomes a variation under the main contract and in turn to the subcontract agreement if the plumbing subcontractor's scope includes for the supply of the original sanitary ware.

The quantity surveyor must understand what constitutes a variation because definitions stated in a main form of contract may be broad in context, as it is impractical to address all scenarios. If the quantity surveyor has doubts regarding the impact of an instruction, or if it is a variation, the project manager should be consulted before preparing a variation request or quotation because a client's agent may reject such a request if it has no grounds. As part of the research in determining this, the quantity surveyor will need to be aware of any latent conditions associated with the project as they may influence the impact of the instruction. Latent conditions are physical conditions that become exposed once works commence that a contractor could not have foreseen when preparing a tender by observation or by examining the site. This may relate to anything underlying or adjacent to the land when there is no specific mention in the invitation to tender documents which the contractor may have qualified as excluded from the tender (e.g. excavating in rock). Should any of these conditions manifest while works are in progress, they may be considered a variation by relying on the terms and conditions of the tender. However, this could be nullified if an Entire agreement clause exists, as it would mean the terms and conditions of the contract are exclusive and must prevail. There would be respite here for the contractor if the terms and conditions of the tender are appended to the contract as the contractor could then rely on such terms.

Once the need for a variation is verified, the quantity surveyor will prepare a variation request or quotation for the client's agent's approval. Driving factors influencing an amount to include in a variation request or quotation are if the variation changes quantities, quality or both.

6.2.1 Changes in Quantity

When a variation involves changes in quantities without altering the quality of the finished product, a straightforward approach in determining the value is to measure the original and revised works, quantify the variance and apply rates from the contract bill of quantities or builder's quantities to assess the price difference. So, let us say on a construct-only contract the concrete ground-floor slab to a new industrial unit is to be increased in thickness before construction commences with a bill rate of £120.00 m^3 including labour, plant and materials applying to the contract works. Assuming the bar reinforcement arrangement remains unchanged and no additional excavation is required to accommodate the extra thickness because the floor level will be raised, an assessment for the change in floor price could comprise:

> OMIT Original ground-floor details as per Contract drawing A1234 Revision 'A'
> 50.000 m × 10.000 m × 0.150 m thick = 75 m^3
> ADD Revised ground-floor details as per revised drawing A1234 Revision 'B'
> 50.000 m × 10.000 m × 0.250 m thick = 125 m^3
> Therefore, OMIT 75 m^3 ADD 125 m^3 = +50 m^3 @ £120.00 = £6,000.00
> Variation cost is an ADDITIONAL £6,000.00

There would be no additional charge for increasing the width of the formwork edge as it still falls under the category where it does not exceed 500 mm in width (NRM2, section 11.14.1). This assessment excludes an amount for the contractor's offsite overheads and profit which is usually charged at an agreed percentage to the cost of the variation, subject to the rate being exclusive of such charges. If the variation amount is a credit to the project client, only the cost of the variation is submitted. This is subject to there being no specific contract clause stating that a variation reducing the contract sum is to result in a proportionate reduction in the contractor's offsite overheads and profit. Value added tax (VAT) is normally excluded from the assessment and noted to that effect in the priced variation quotation.

6.2.2 Changes in Quality

Where there is a change in quality the assessment is more detailed, as it will involve assessing the chargeable rate as well as quantities if they also change. A method of assessing a price change for quality can be carried out by substituting the original material price and adding the new price together with any adjustment for a change in labour requirements. Alternatively, the change can be substantiated with verified quotations from material suppliers and trade contractors/subcontractors if appointed to demonstrate competition with prices. However, if the works are already awarded, only the subcontractor's quotation is required which can be negotiated; only in exceptional circumstances would others be appointed to carry out the change in works, with any other request to seek prices futile and a possible abuse of the supply chain.

Using the earlier example of altering the concrete floor to the industrial unit, let us consider a scenario where the structural engineer discovers the site investigation report indicates the presence of harmful chemicals in the subsoil and the concrete needs to be sulphate resistant. The engineer then wishes to change the Portland cement constituent of the concrete specification to blended Portland cement. In verifying a price for the material change, the quantity surveyor has obtained a quotation from the subcontractor in receipt of the award to carry out the concrete works. The quotation includes a letter from a ready-mix concrete supplier advising that the use of combined cements such as fly-ash or ground granulated blast furnace combined at the mixer with Portland cement will provide sulphate-resistant concrete and, if accepted, the subcontractor will be charged an additional £10.00 m^3. The subcontractor confirms there are no changes to the price for placing the material and requires a mark-up to the suppliers charge and wishes to charge the contractor £12.00 m^3. In light of this, the assessment can take the following form:

> OMIT Original ground-floor details as per Contract drawing A1234 Revision 'A' and Concrete specification S1234, 'concrete to include ordinary Portland cement' 50.000 m × 10.000 m × 0.150 m thick = 75 m^3@ £120.00 = £9,000.00
> ADD Revised ground-floor details as per revised drawing A1234 Revision 'B' and Concrete specification S1234A 'concrete to include blended Portland cement with suitable materials for the mix to be sulphate resistant' 50.000 m × 10.000 m × 0.250 m thick = 125 m^3@ £132.00 = £16,500.00
> Therefore, OMIT £9,000.00 ADD £16,500.00 = +£7,500.00
> Variation cost is an ADDITIONAL £7,500.00

A variation price should reflect the overall impact of the change where possible. For example, if window heights are altered on a new building prior to their manufacture, affecting the walls which are not built, the variation must address changes in quality to the windows as well as changes in quantities to the walls and finishes, as it is a waste of resources to submit one variation for windows and another for the wall changes.

When it is necessary to obtain subcontractors or suppliers quotations for pricing a variation, the quantity surveyor must be satisfied prices quoted are fair and reasonable. This means written quotations must reflect the changed scope as well as market prices for materials, plant hire and labour charges, which can be verified by the estimator. For example, and referring to the change in concrete specification discussed earlier, if the contractor's estimator advises concrete to the original specification will cost £100.00 m^3 to purchase, which a subcontractor advises is an extra £12.00 m^3 when the cement is blended to create sulphate-resistant concrete, the rate is reasonable. This is because the material price includes cement, sand, large coarse aggregate, water, blending materials, quality control and delivery to site with the blending part of the process. If a revised quotation states the extra as £75.00 m^3, representing a 75% increase merely because of the blending, it would be excessive and should be queried by the quantity surveyor before being included in a variation quotation.

On a construct-only contract, the contract may state the client's cost manager/quantity surveyor will assess the price of a variation with the client's agent having the final say. When this applies, the contractor's quantity surveyor must assist the process and not leave the client's cost manager/quantity surveyor to act alone as vital information may be required which, if not forthcoming, will lead to a fair assessment on merit without contractor input.

Where the quality change is radically different, it may be ideal to delete the original works by fully crediting the value and assessing a new price from scratch. This may mean introducing or calculating new unit rates if they are not provided in the bill of quantities that can be obtained from the estimator, and once known can form part of a composite rate. Where applicable, the contractor must drive the situation with or without subcontractor input as a project client will generally seek commitment from the contractor. To demonstrate, let us consider a project includes for 50 m of 1800-mm-high close board timber fence along a site boundary as stated in the drawings and bill of quantities. The fence is not erected or an award placed for the works, and the client's agent issues a price request for substituting the fence with a brick wall and there are no rates in the bill of quantities for the wall. What is the price of the variation? The options to calculate an amount to charge are to measure and quantify the works using NRM2 as a guide and applying market rates or create a composite rate applicable to the length of wall to create a job price. Table 6.1 demonstrates the latter process as a composite item and a method of calculating a suitable rate.

A contractor should endeavour to meet its obligations under the contract by issuing a variation quotation as a fixed price. However, if changes are impractical to measure, or the contractor considers the risk too great, a solution is to revert to dayworks which could be allowed under the contract. Here, the variation request must state the contractor's reasons for electing to carry out the works as dayworks and include a schedule of rates or affirm those in the contract. An item such as a boundary wall would generally not be considered acceptable as dayworks, as the price can be ascertained by a commitment from the contractor.

6.2.3 Changes in Sequence of Works

A suitable contract clause would allow the project client to seek a change in the sequence of working operations carried out by the contractor, which can be invaluable to a client because of the nature of the business (e.g. a retail store operator trading in a shopping complex). Furthermore, the clause could permit the imposition of limits on the working space available (such as dividing retail spaces due to newly occupied tenancies) and/or the rescheduling of working hours by the contractor which, by occurrence, would reduce the output of trades and/or prolong the fitting-out works and construction period. However, if a decision to alter the sequence of works is discussed early enough, operations could be reorganised in a new programme so they do not delay the date of practical completion, with the variation possibly resulting in no extra cost. Each assessment however is on a case-by-case basis, and a project client will usually be responsible for extra cost resulting from reduced labour output because of imposed restrictions (such as a lack of space) or changes in the sequence of operations that in combination

Table 6.1 Composite rate calculation

*** Boundary fencing change price request - Replace 1800 mm high timber screen with brick wall ***

OMIT - BOQ

Page 11/5 item g

Timber fencing 1800 mm high, close board feather edged 100 × 13, overlap 25; posts 75 × 75 × 2200 long at 2400 centres secured to and including 100 × 300 mm concrete foundation (1:2:4/40 agg) poured on earth.

50 m @ £75.00 = £3,750.00

ADD

Brick wall 215 mm thick, 1800 mm high overall above ground, in English bond facework (PC Sum £300.00 per thousand for brick supply) flush gauged mortar bed and joints pointed both sides as the work proceeds including brick on edge coping. Wall constructed of 600 × 200 mm reinforced concrete foundation (1:3:6/40 agg) with 300 mm deep substructure in common brickwork, including backfilling trench upon completion and disposal of surplus materials off site.

Analysis of new price based per metre run of wall

Foundation and substructure:

Excavate trench	1.00 m × 0.60 × 0.50	$0.30\,\text{m}^3$ @ £19.70	£5.91
Backfill with arisings	1.00 m × 0.60 × 0.50	$0.30\,\text{m}^3$ @ £26.00	£7.80
Earthwork support	2 × 1.00 m × 0.50	$1.00\,\text{m}^2$ @ £14.31	£14.31
Compact bottom	1.00 m × 0.60	$0.60\,\text{m}^2$ @ £1.00	£0.60
Reinforcement	1.00 m × 0.60	$0.60\,\text{m}^2$ @ £20.00	£12.00
Concrete foundation	1.00 m × 0.60 × 0.20	$0.12\,\text{m}^3$ @ £120.00	£14.40
Substructure, 215th	1.00 m × 0.30	$0.30\,\text{m}^2$ @ £45.00	£13.50
Damp proof course	1.00 m × 0.30	$0.30\,\text{m}^2$ @ £10.00	£3.00
Deduct backfill	1.00 m × 0.60 × 0.30	$0.18\,\text{m}^3$ @ £26.00	(£4.68)
Add dispose off site	1.00 m × 0.60 × 0.30	$0.18\,\text{m}^3$ @ £40.00	£7.20
Total			**£74.04 m**

Wall above ground

The wall height will be 1698 mm + 102 mm brick on edge coping to achieve 1800 mm overall. English bond comprises alternate rows of brickwork laid as stretcher and header bonds and there are 119 bricks per m^2. The contractor has agreed a bricklaying rate with a labour-only gang to construct the wall at £350.00 per thousand and £10.00 per metre for laying the coping course. This includes the use of the contractor's access equipment with subcontractor's quotations unnecessary. The contractor will need to provide the access equipment and mortar mixer and allows for this as consumables and attendance. The material cost for the mortar is assessed at £150.00 m^3.

Brick supply 1.00 m × 1.698 m × 119 nr + 15% waste	232 nr	@ £0.30	£69.60
Brick coping supply 1.00 m/75 mm wide + 1 + 15% waste	16 nr	@ £0.30	£4.80
Brick laying 1.00 m × 1.698 m × 119 nr	202 nr	@ £0.35	£70.70
Brick coping laying	1 m	@ £10.00	£10.00

(Continued)

Table 6.1 (Continued)

Mortar supply $1.00\,m \times 1.698 = 1.70\,m^2 \times 0.05\,m^3/m^2$	$0.09\,m^3$	@ £150.00	£13.50
Consumables and attendance, 15%			£25.29
Total			**£193.89 m**

Summary: £74.04 + £193.89 + Contractor's offsite overheads and profit of 15% = £308.12 m

OMIT	Timber fencing 1800 mm high	50 m @ £75.00	£3,750.00
ADD	Brick wall 1800 mm high	50 m @ £308.12	£15,406.00

Therefore, OMIT £3,750.00 ADD £15,406.00 = Variation price is an additional £11,656.00
Variation amount as a contract sum adjustment is an ADDITIONAL £11,656.00 (excluding VAT)

prolongs the works. If works are envisaged to be prolonged because of the client's request, this must be included in the contractor's assessment and will be the additional time a contractor will have as a site presence beyond the date of practical completion. These costs are ascertained on a daily or weekly basis and apply to time-related items (e.g. supervision, hired plant, site accommodation charges, etc.) which can be found within the preliminaries costing less the cost of any mitigating circumstances (e.g. off-hiring and re-hiring plant). Alternatively, and subject to an appropriate clause in the contract, the works could be accelerated in order to achieve the agreed completion date which, if cited in an instruction, could be assessed as 'crash costing' (i.e. the cost for additional resources and out-of-hours premiums). It is far better for the contractor to submit these costs as a variation quotation for acceptance or rejection by the client's agent, instead of permitting a situation to develop naturally when a contractor incurs expense or anticipates it will incur expense because of imposed restrictions or changes in the sequence of works without prior negotiation. Should this happen, it would become a claim under the main contract which is discussed under Sections 6.6 and 6.7 later in this chapter.

6.2.4 Abortive Works

Once a contractor is in receipt of a written instruction varying the works under contract, ideally it should not affect the works already carried out at the time of receiving the instruction. However, this is not always the case, and the contractor may have to deal with the situation the best it can. Unfortunately, this can mean carrying out deconstruction and corrective works to allow the changes to be implemented, resulting in abortive works. To demonstrate, and referring to the change in the ground-floor slab of the industrial unit as noted above, let us say the works are procured under a construct-only contract with part of the floor slab already poured on the date the contractor receives the instruction that states a new design and specification is to follow within a given timeframe. In such a scenario, time is of the essence and the contractor must comply with the instruction. However, the instruction may only state what the change involves and may fail to acknowledge the status of the constructed works and what the contractor is to do. Here, the contractor must act responsibly; assuming the instruction fails to acknowledge the status of the works, a suggestion for the quantity surveyor could be as follows.

- Advise the issuer of the instruction in writing of the status of the works within 24 hours of receiving the instruction or as soon as possible.
- Inform the issuer of the instruction that the contractor will carry out a site survey (stating the date and time) to record the status of constructed works and invite the issuer of the instruction to attend.
- Advise the issuer of the instruction that until such time the survey is carried out, the affected works will be placed on hold pending issue of the new design and documentation to mitigate abortive works which extends to any contractor's design portion, including the preparation of shop drawings and schedules.
- Advise the issuer of the instruction that the cessation of the works is isolated and not a suspension of the whole works (if applicable).
- As part of the survey, highlight appropriate drawings and make reference on these of the status of construction works, and take photographs using the camera date option. The drawings should be copied, signed off, dated and distributed to attendees of the survey before leaving site, and followed up in writing including the supply of the photographs.

After communicating the above with interested parties aware of the situation, the quantity surveyor in consultation with the project manager must agree a method of ascertaining the cost of abortive works. If this involves removing the existing slab and reconstructing a small area, it could be recorded and priced as dayworks. However, if a substantial area, a price should be sought from a demolition contractor to remove the area and a separate price calculated for reconstruction works using quantities and rates applicable to grading, filling, reinforcement, membrane, formwork and concrete works. (Note, using the example of the slab specification change discussed in Sections 6.2.1 and 6.2.2 above, the price to the client is the extra cost for the change which means the abortive works price must include for removal and reconstruction.) This approach should be seen as practical, as separating abortive works from similar unaffected concurrent works within the vicinity could be too hard to manage. Putting the choice aside, the site diary should record the duration of abortive works and the resources used as a part of usual site management that can be referred to at a later date should the need arise. Usually, the client's agent/issuer of the instruction will not interfere with the process, meaning the contractor must demonstrate they are taking sensible action. As soon as practically possible, the contractor should advise the issuer of the instruction of a timeframe for providing cost advice and the impact the change may have on the programme (time and money), which is not only a responsible action but may also be a condition of contract. In a scenario like this, the variation price will include a loss and expense claim comprising:

- where applicable, the price of dayworks as abortive works, which can only be determined upon completion of the affected works;
- trade contractor's verified quotations for demolition/removal if not priced as dayworks;
- priced schedules of reconstruction works where dayworks are not considered practical;

- the cost of products stored on site at the time of receiving the instruction that are of no use and considered lost (e.g. bar reinforcement); these should be set aside on site, as they would become the property of the client if agreeing to the price of the variation including abortive works which is subsequently paid;
- impact the delay has on the master programme, including costs;
- the cost of the variation excluding abortive works; and
- the cost of replacing designs, shop drawings and schedules that are a contractor's responsibility to procure (again, such items should be set aside for reclaiming as the client's property once the contractor receives payment, although this may not apply if such items are amended from the originals as it could result in a reduced cost compared to the price of full replacements).

If the change is so substantial it requires time and effort to collate the details, the effect on the programme excluding abortive works should be treated as an extension of time and financial claim. However, if such a claim is dealt with in isolation, it could delay any agreement to the value of the variation including abortive works. The subject of claims is discussed further under Sections 6.6 and 6.7 later in this chapter.

6.2.5 Variation Submissions

When issuing a variation quotation or price for approval to the client's agent, the quantity surveyor must state why a contractor considers an instruction confirming a variation is required. This can be substantiated with reference to written instructions and/or the supply of revised and new information. The issue should also include information used in determining the price such as subcontractor and/or material supplier's quotation(s) and the quantity surveyor's priced schedules.

A flaw of some contractors is they sometimes carry out works considered as variations without an instruction citing variation clauses, which poses the risk of not being paid. Strictly speaking, this may not be a contractual fault by a contractor because there may be a consensus to carry out variations based upon the issue of new or revised drawings considered an instruction. Ideally, such information should be accompanied by an instruction for there to be no doubt. In reality, if a contractor carries out works that it considers is a variation before issuing a variation quotation, it does not mean the request and price will be rejected, and will depend on the circumstances and responsibilities of the parties providing neither have breached the contract. If the quantity surveyor issues a variation quotation when the works are complete, it must state the works as complete because the contractor will be seeking payment in the next progress claim; ideally, any variation should be approved before making a request for payment.

Once a variation quotation is submitted, it is necessary to create a variation register to keep track of submissions which can be achieved with the use of an Excel spreadsheet. The register should list each variation in numerical order including a brief description of the changes, value, date of issue and status of the approval decision. It is wise to include a withdrawal notice on the status section to demonstrate where the contractor has no financial commitment when a variation is a quotation only with the changes communicated as not being required. Table 6.2 demonstrates a format of a suitable variation register.

Table 6.2 Variation register

Nr	Description	Date issued	Amount (£)	Status*	Approved amount (£)
	Client: Proprietary Holdings Ltd	**Project Nr: 1888** **Project Title: Construction of new industrial unit – Leeds** **Contract Sum: £1,588,390**			
V 1	Change of slab thickness and concrete specification		7,500.00	A	7,500.00
V 2	Boundary wall change		11,656.00	P	
V 3	Change external paving from concrete to stone		15,380.00	W	Not applicable
V 4	Reduce window W1 heights from 1200 mm to 900 mm		−4,700.00	A	−4,700.00
V 5	Abortive works associated with V 1		27,700.00	P	
V 6	Change colour of bathroom sanitary ware		NIL	A	NIL

*A: approved; P: pending; R: rejected; W: withdrawn

Variations are normally approved in a statement issued as a contract sum adjustment by the client's agent where a project is a fixed price with the adjusted sum equalling the original price plus/minus amounts shown on the variation register, excluding withdrawn items as agreed with the contractor. If a variation price is approved, it may be communicated to the contractor in a written instruction or as a contract sum adjustment and might include new or revised drawings and documentation. When the scope of changed works involves works by subcontractors or materials from suppliers that have agreements in place with the contractor, the changes must be communicated to the appropriate parties. This can be carried out by issuing a site instruction authorised by the quantity surveyor or in a letter from the contractor instructing the works to be carried out. Normally, site instructions exclude items of a financial nature and, if involving finance, should be issued as a letter or contract sum adjustment using a similar format to that under the main contract.

6.2.6 Rejection of a Variation

Once a variation quotation is issued, if the client/client's agent is of an opinion it should not exist and considers the works to be part of the contract and rejects the request, the contractor must act wisely with its response in a positive way. The same would apply if the price is rejected in part, for example there is disagreement to the value or scope of works covered by the variation quotation. If the contractor simply refuses to continue with the variation works because of a disagreement over the price and decides to issue a notice to that effect, it may trigger the client's agent to exercise clauses of the contract and terminate the agreement before the contractor. This would be an extreme situation and could occur if the sum involved is significant in relation to the value of the project or the financial turnover of the contractor's business, and could do harm to the business if the works were carried out and not paid. Such conduct suspending the works must be permitted by the contract or the contractor will be in breach of contract. However, if the contractor decides to commence the works, it is possible the decision could be perceived as acceptance of the client/client agent's position. Here, and after commencing works, the contractor must protest politely by issuing a letter to the client/client's agent explaining its grievance and grounds for the decision to proceed. The letter should be worded to the effect that the contractor respects the decision of the client/client's agent to reject the variation, yet reserves the right to leave the subject in abeyance pending future summary judgment under dispute resolution clauses of the contract or, if no dispute resolution clauses exist, seek remedy at common law. The letter must insist the client/client's agent acknowledges the contents of the letter and state that if not receiving an answer by a specified date the contractor will consider the contract repudiated. It is negative to view this approach as one of the contractor backing down or the client being blackmailed, and should be perceived as a method of recording events to place emphasis on the dispute resolution methods and common law rights available to the contractor.

6.3 Reimbursement

A contractor must be financially reimbursed for works under contract in a timely manner that follows the terms and conditions of the construction contract. Similarly, the contractor must reimburse its supply chain in a timely manner in order to fulfil the obligations of each agreement in place. Until the mid-1990s, contractors may have been excused for adopting a policy of 'pay when paid' meaning a contractor's creditors/supply chain would only be paid once the contractor is paid. This situation is generally no longer acceptable because of the Construction Act and associated legislative changes by way of the Local Democracy, Economic Development and Construction Act 2009 introduced in 2011. This Act changes the way construction contracts are entered into and includes remedies for non-payment using quick summary judgements through the process of adjudication, which is a type of alternative dispute resolution (ADR) to litigation (see Chapter 3, Section 3.7.9 for more details). However, section 113 (1) of the Construction Act relaxes the exclusion of pay when paid in the event of the insolvency of a third party. In effect, this means that if a project client becomes insolvent, there is no legal duty on the contractor to pay its supply chain engaged on a contract where the client is insolvent. Legislation regarding the flow of payments is not restricted to client and contractor agreements, and reverberates through the supply chain when 'construction operations' are carried out.

A delay in a payment to the contractor may be unintentional and could arise because of either the omission of a stated date in the contract for the issue of interim payment certificates, or poor time management by the contractor if forgetting to issue timely progress claim requests where such requests are a condition of contract. The contractor can take relief from this situation if the form of contract makes it a condition for the client's agent to assess the works in progress and issue interim payment certificates without contractor input. For example, clause 4.9 of JCT SBC/Q 2016 allows the architect/contract administrator to instruct the client's quantity surveyor/cost manager to prepare a valuation for the preparation of an interim payment certificate. Clause 4.10 of the same form permits the contractor to issue a Payment Application which, if submitted, would usually steer the process instead of the requirements of clause 4.9. This can benefit the contractor, as it releases time to focus on procurement activities and the carrying out of the works, and less so with the administrative requirements. As part of good practice, and where the issue of interim payments is a contractual obligation of the client, the contractor should ensure payment requests are made in a timely manner; if not, and the client's agent prepares certificates without requests (if permitted by the contract), the payments may differ from the contractor's expectations.

An important distinction regarding the flow of payments exists between those due to the contractor and those made by the contractor as, unlike the main contract, the contractor is not usually obliged to pay for works or goods delivered to site unless an invoice or application for payment is received. To manage the payment process, the contractor's accounts department will usually wish to pay creditors to preferred dates and may specify the latest date for receiving approved invoices which must be adhered to; if ignored, it could delay the flow of outgoing payments. The contractor's quantity surveyor is not responsible for the collection of payments owed to the contractor or for ensuring supply chain members are paid on time, as this is a responsibility of the contractor's credit controller. However, the quantity surveyor is responsible for putting the wheels in motion to generate these payments.

6.3.1 Client Interim Payments

To ensure interim payments are received on time, the contractor must be proactive and submit a payment application that becomes a payment notice if complying with the Construction Act (as amended), and to do this, a timeframe must be established. In order to understand this framework, it is easier to work backwards from the date the contractor is due to be paid, and using JCT SBC/Q 2016 as an example:

- the time a client has to make an interim payment after the Interim Valuation Date as stated in the Contract Particulars and clause 4.8, 7 days;

- the time for preparing an Interim Certificate for an Interim Valuation Date of say 25th (or nearest business day), as clause 4.9, 5 days

- the time to prepare and issue a Payment Application to the client's quantity surveyor/ cost manager as clause 4.10 (but not to surpass the Interim Valuation Date), say 3 to 5 days;

- the date the contractor's quantity surveyor should start preparing the application is therefore the 20th.

Each application is usually assessed by build stage or time. Where the preference is build stage, it is included in an application when the works reach physical milestones of completion as stated on Activity Schedules appended to the contract. This is a relatively straightforward process to follow where the quantity surveyor prepares an application based upon the completion of each milestone or part thereof as stated in the schedules. Where preference is by time, the quantity surveyor will need to drive the process by presenting a detailed progress claim schedule as an application for payment. The format can be a trade breakdown as discussed in Chapter 4 (Section 4.5.6) or as Activity Schedules that should be agreed with the client's agent prior to issuing a first application.

6.3.1.1 Assessing Measured Works

A method for assessing an amount to include in an interim application as measured works is to visit the site and inspect the works in progress. An assessment may take the form of copying each trade bill as hard copy and marking items in ink as complete or in progress at the time of the inspection. It is worth noting here that the assessment of works should be to that on the Interim Valuation Date and not a forecast of what may be complete on the date the interim payment becomes due. For example, the excavating and filling works can be assessed as follows:

Main heading on BQ: Excavating and filling
Subheading 1: Site preparation - 5 nr items 100% complete
Subheading 2: Excavations - 9 nr items 100% complete
Subheading 3: Support of faces to excavations - 4 nr items 50% complete
Subheading 4: Disposal - 3 nr items 50% complete
Subheading 5: Retaining excavated material on site - 2 nr items 50% complete
Subheading 6: Filling - 4 nr items 50% complete.

Upon returning to the contractor's office, percentages representing the works complete are valued against rates that form the breakdown of the contract sum, or schedule of rates where the contract is not a fixed sum. This involves identifying items under subheadings of the trade bill and applying a percentage of the financial allowance to arrive at a value for the trade. For example:

M/H: Excavating and filling
S/H1: Site preparation - 5 nr items - 100% of £7,000 = £7,000
S/H2: Excavations - 9 nr items - 100% of £25,000 = £25,000
S/H3: Support of faces to excavations - 4 nr items - 50% of £5,000 = £2,500
S/ H4 - Disposal - 3 nr items - 50% of £18,000 = £9,000
S/ H5 - Retaining excavated material on site - 2 nr items - 50% of £5,000 = £2,500
S/ H6 - Filling - 4 nr items - 50% of £60,000 = £30,000
Total value = £76,000
Therefore, the value completed is £76,000 out of a total £120,000, meaning the excavating and filling works are 63% complete.

A similar approach can be used for the value of variations for interim payment purposes, which is from physical observation and claimed up to the maximum amount of an approved variation.

6.3.1.2 Fluctuations

A fluctuations clause in a contract (sometimes known as a rise-and-fall allowance) is a provision allowing the contract sum to be adjusted in line with prevailing conditions that influence changes to the price of works during the construction phase. This includes changes to the cost of labour, fuel, materials and impact of anything caused by legislation. The inclusion of such a clause is more akin to a standard form of contract (e.g. NEC Option X1, JCT SBC/Q 2016 Schedule 7) and bespoke forms prepared by public authorities, and is a risk to a client as it favours the contractor. This is because the contractor takes limited responsibility for 'crystal ball gazing' at tender stage into changes of prices once the works are in progress that would otherwise be included in the tender. However, a fluctuations clause does not usually involve itself with changes in foreign currency and rates of exchange, which could be addressed under a separate clause (e.g. FIDIC CoC for Works of Civil Engineering Construction, cl 71.1 and 72.1–3).

The fluctuations clause is adopted by project clients where a lump-sum fixed price is considered too high a risk to the contractor and supply chain due to the anticipated length of the project, or if market conditions dictate the scarcity of obtaining a fixed price at a particular point in time (e.g. a period of high national inflation). A fluctuations clause in a contract can be described as being in full or in part and, when in part, the part(s) is identified and usually limited to changes in law (e.g. statutory changes altering employer's national insurance contributions and tax and levy adjustments, as found with JCT Schedules Fluctuations Option A). Where applicable, if such parts alter during the construction phase, the contractor is required to provide written proof of the date(s) the amendments become enforceable and include the price changes in each interim application from the date(s) they commence until the time they no longer apply. Where full fluctuations apply (e.g. labour and materials cost and tax fluctuations; JCT Fluctuations Option B (including Option A)), in order to qualify for payment, the contractor will need to provide proof of price amendments from sources such as material suppliers and wage-making bodies. However, if the contractor renegotiates terms of sale, this would be commercial bargaining and not fluctuations that, by default, would not trigger the appropriate clause of the contract.

An alternative to providing proof of fluctuations is to use the formulae rules adjustment, which to be effective must be part of the contract (e.g. JCT Schedules Fluctuations Option C, formula adjustment). In the UK, the origin of these rules dates back to the mid 1970s when inflation was running close to 30%. This saw the NEDO formula adopted as an interim assessment on rise-and-fall-style agreements in construction projects, while Baxter and Osborne adjustments were used for civil engineering works. Under a modern guise, the BCIS produces PAFI (price adjustment formulae indices), a comprehensive, detailed and easy-to-use method of measuring cost movement for building and civil engineering works, which is also used for pre-contract works to allocate risk between client and subcontract works. PAFI can also be used for clients as a model for deciding if a fluctuations clause is viable to include in a contract, with the information collected from BCIS resources and considered a reputable source for calculating price fluctuations.

Where formulae rules adjustments apply, work activities are categorised and assessed by an appropriate trade affiliation which divides the works into categories allocating each with an index alongside a base date with the indices reassessed at regular intervals (e.g. monthly or quarterly). Separate formulae also exist for other industrial disciplines (e.g. services industries, specialist engineering and lift installations) and, where 'formulae rules' apply, the appropriate affiliation may be stated in the contract. The use of formulae adjustments is advantageous over the laborious arrangements for providing proof of price changes with the amounts authentic because the indices for each category are obtained from a source which is free from contractor and client influence. The method of calculating a change in price levels for each category for inclusion in an interim payment application is:

$$RF = \frac{(IW - ID) \times VW}{ID}$$

where RF is the rise-and-fall value owed by/to the contractor; IW are indices applicable at the time of carrying out the works; ID are indices applicable at the base date; and VW is value of the works executed during the month of the works.

The variable IW may require an adjustment in line with valuation dates stated in the contract, as a date may be mid-month with the indices usually advised at the end of the month or each quarter. In general, formulae do not apply to the payment of materials stored on or off site or for loss and expense claims, as their assessments usually reflect prices prevailing at the time of the valuation which are at current levels. This would also apply to variations, except where the value of variations is defined in the contract as being derived from the contract bills of quantities with prices updated at the time the works are carried out because of the link to the formulae rules.

6.3.1.3   Materials Stored On or Off Site

If a contractor qualifies for the payment of stored materials, they can be assessed by any of the following means.

- A site inventory recording the quantities and types of materials purchased by the contractor and delivered for the works under contract which are in store at the time of assessing the interim application. This does not usually apply to temporary works and consumables that are part of the preliminaries.
- As above, applying to subcontractor's materials when it is a subcontractor's responsibility to create and manage an inventory. This applies to subcontractors that have agreements in place with the contractor for the reimbursement of suitably stored goods (i.e. mirroring the main contract).
- Physical observation of the storage area and count of materials which is recorded on a quantified schedule. This is also applicable to subcontractor's materials, which can be verified with photographs using the camera date option. The photographs should represent a broad sample of goods and be deemed sufficient evidence (e.g. pallets of bricks) without the need for detailed photographs that would be overwhelming.
- Proof of cost (i.e. invoices and purchase orders for goods procured by the contractor), and likewise with subcontractors from their suppliers with rates inserted into the quantified schedule. Normally, this would attract a mark-up for overheads and profit.
- Where materials are stored off site and if imported from overseas, a bill of lading confirming the consignment with dated photographs of the stored goods as proof.
- Where materials are stored off site and not imported, location of the store for inspection and the contents on the inventory.

Once reconciled, the details should be summarised as a total cost that is added to the value of the works in progress.

Confusion can sometimes arise regarding the inclusion of an amount for stored materials in an interim application, subsequently paid and still in store at the time of preparing the next interim application. The golden rule here is that the next application must include the price of the materials again despite them being paid, as the payment system under construction contracts is usually cumulative. This is best demonstrated with an example where, say, specialist equipment is stored on site for a number of months because the works programme is in time deficit or the production line warrants early procurement:

> Interim application Nr 1 (January): Value of construction works complete £5,000,000
> Materials stored on site - Specialist equipment £ 250,000
> Total: £5,250,000
> Less previously paid NIL
> Amount due £5,250,000
> Interim application Nr 2 (February): Value of construction works complete £7,000,000
> Materials stored on site - Specialist equipment £ 250,000
> Total £7,250,000
> Less previously paid: £5,250,000
> Amount due £2,000,000

Between interim applications 1 and 2 (during the month of February), the contractor carried out £2,000,000 worth of construction works. Should the identical materials included in the January payment be in store at the time of conducting the February application, they must be included again; if not, the contractor's request would be understated.

6.3.1.4 Assessing Preliminaries

When assessing preliminaries, the quantity surveyor should request payment for the full costs incurred if the project is running to the contract programme. For example, if by the end of the first month on a project the site accommodation is established with the foundation works in progress and is in accordance with the programme, the full cost should be recovered because the project is on target to complete on time. However, with the passage of time projects can be in time deficit. Let us say by the end of month four on the same project the site accommodation is established and the foundations have commenced, but the programme states by this time the building should be constructed with a roof. Here, it is still only possible to request payment based upon the project running to the contractually agreed programme. Should such a scenario be envisaged to delay the end date, the contractor must act wisely and consider where the burden of responsibility rests for prolonging the works. When it is a contractor's responsibility, any additional expense usually cannot be recovered and the contractor must mitigate the situation the best they can by accelerating the works programme. Until the acceleration commences, interim applications for payment for running costs can be assessed by charging the % value of the completed contract to the running costs budget. On the other hand, if the delay is a client's responsibility, the client may ask for an Acceleration Quotation if a condition of contract (e.g. as per Schedule 2 of JCT SBC/Q 2016). Where there is no provision for the contractor to accelerate the programme, the contractor must seek an extension of time to prolong the end date and issue advice of associated costs for the extended time (subject to there being no disagreement to the duration).

As works progress on site they may not go exactly to programme; assuming the agreed end date on a project is still achievable, and for assessing an amount to include in an interim payment application, it is necessary to forecast the preliminaries expense through to the end of the project. This can be assessed by identifying the components of the preliminaries budget and expressing a percentage of the allowance the contractor expects to pay for each month the project is to run until the programme expires. A determining factor here is that the sum of allocated percentages for each item must not exceed 100%, as this is the maximum the contractor is entitled to. Table 6.3 demonstrates this process on a construction programme to run for seven months with the assessment carried out on the first month.

From viewing the table, the first assessment (progress claim number one, PC1) represents an amount equalling the contractor's incurred expense plus overheads and profit, and will be included in an interim payment application as the project is running to the agreed programme. Thereafter, the forecast is based upon an assessment through to completion. The forecast should be considered flexible to enable future anticipated expense to be modified in line with each progress claim. For example, and using code P117 (minor mechanical plant) as a sample, the requirement for half the budget amount has been allocated to the first two months of the programme with the remainder proportioned equally over the remaining five months. If on the second month the requirements are less and result in reduced expenditure, the residual amount can be allocated to month three or later month(s).

6.3.1.5 Interim Application Submissions

The value of each progress claim as a payment application is assessed as the cumulative value of works complete including stored materials, preliminaries, overheads and profit (builder's margin), plus/minus entitlements, less cash retention (if applicable), less an amount already paid to produce an amount due excluding VAT where applicable. The quantity surveyor should prepare and issue each progress claim on this basis for advice to the project manager and contractor's financial controller, as it acts as a guide to cash flow forecasts for the project. Depending on the terms and conditions of contract, a client may appoint a cost manager/quantity surveyor to assess the application and make a recommendation to the payment certifying officer who may be the client's agent. Once the application is submitted, the client's cost manager/quantity surveyor will normally make arrangements with the contractor's quantity surveyor or project manager to visit site to check the application. At each meeting, the client's cost manager/quantity surveyor may benefit from the contractor's quantity surveyor's knowledge of the project, as items included in the application may not be immediately apparent from the site visit and may need discussion. For example, a number of subcontractors may have issued shop drawings and are seeking payment for their preparation, which is included in the application. The cost manager/quantity surveyor may not be able to see this by visiting any structures in progress, and will need to verify inclusion in

Table 6.3 Preliminaries: progress claims forecast

		BQ allowance (£)	Cumul. claimed (%)	Cumul. amount claimed (£)	Progress claim PC1		Forecast expenditure (%)					
						PC2	PC3	PC4	PC5	PC6	PC7	
P100	Pre-commencement	25,000	100	25,000	100% £25,000	0	0	0	0	0	0	
P101	Management and staff	140,000	15	21,000	15% £21,000	14	14	14	14	14	15	
P102	Site establishment	20,000	90	18,000	90% £18,000	0	0	0	0	0	10	
P103	Running costs	60,000	10	6,000	10% £6,000	15	15	15	15	15	15	
P104	Temporary services	10,000	75	7,500	75% £7,500	0	0	0	0	0	25	
P105	Temporary works	15,000	0	0	0% £0	100	0	0	0	0	0	
P106	Security	21,000	14	2,940	14% £2,940	14	14	14	14	14	16	
P107	Employer's requirements	5,000	90	4,500	90% £4,500	0	0	0	0	0	10	
P108	Contract conditions	5,000	0	0	0% £0	0	0	0	0	0	100	
P109	Fees and charges	10,000	60	6,000	60% £6,000	10	0	10	0	0	20	
P110	Insurances, bonds, etc.	8,000	60	4,800	60% £4,800	40	0	0	0	0	0	
P111	Health and safety	5,000	80	4,000	80% £4,000	10	0	10	0	0	0	
P112	Control and protection	10,000	20	2,000	20% £2,000	10	10	10	20	20	10	
P113	Site records	5,000	0	0	0% £0	15	15	15	15	15	25	
P114	Cleaning	6,000	0	0	0% £0	10	10	10	10	10	50	
P115	Non-mechanical plant	15,000	10	1,500	10% £1,500	35	10	10	10	10	15	
P116	Mechanical plant: major	50,000	10	5,000	10% £5,000	35	10	10	10	10	15	
P117	Mechanical plant: minor	20,000	25	5,000	25% £5,000	25	10	10	10	10	10	
P118	Post-completion requirements	5,000	0	0	0% £0	0	0	0	0	0	100	
Total		**£435,000**	**26%**	**£113,240**	**26% £113,240**							

a payment recommendation by confirming sight of the shop drawings in the contractor's possession.

If the cost manager/quantity surveyor disagrees over the value of a progress claim request because it is a genuine calculation error by the contractor, it is normal practice to issue a notice to the contractor advising of an amended amount which the contractor should accept. If there is a discrepancy which is not a genuine error, the cost manager/quantity surveyor may advise the contractor with reasons for the change, with supply of this information possibly a condition of contract. However, this is not always the situation, and the cost manager/quantity surveyor may only be obliged to make a recommendation to the payment certifying officer/client's agent without advising the contractor. If receiving advice of a change, the contractor's project manager must be informed who may accept the situation or raise a query with the client's agent/payment certifying officer. If any payment certificate fails to materialise, a Payment Notice (e.g. clause 4.10, JCT SBC/Q 2016) stating an amount due can be issued by the contractor, but only if a Payment Application is not submitted. Once a payment certificate is received, and if there is disagreement, only then can it be disputed.

6.3.2   Subcontractor's Payments

Once a subcontractor is in receipt of a trade package award, the quantity surveyor must initiate the subcontractor's business details in a computer database and recognise the financial commitment on the cost management system (CMS). This can be to fixed amounts in line with the award or, where a schedule of rates applies, the commitment can be an estimate based upon approximate quantities to create a cost target. This process is required to enable the system to recognise the value of the award and generate payment certificates. A subcontractor may already be engaged by the contractor on existing projects or have had prior dealings with the details stored in the CMS database under a unique code. Once the details are retrieved, this code is migrated to the project reference set-up in the CMS for it to be recognised as a project vendor, with the financial value of the award committed to the system. Once initiated, the committed sum requires distributing to cost codes (or cost centres) within the CMS that also records the budget which may have been created at the pre-commencement stage of the project. To demonstrate, Table 6.4 shows details of a committed order issued to a subcontractor to carry out excavating and filling works in comparison with the budget.

A distinct advantage of this arrangement to a business is that the CMS is usually designed to ensure the value of each award is not exceeded by successive payments, meaning a subcontractor cannot be overpaid.

The quantity surveyor is responsible for assessing the value of works in progress, which in effect is a payment withdrawal from the value of a committed award. However, there may be occasions when the value requires adjusting, for example:

- if there was an error in the original amount input into the CMS that differs from the award
- changes in VAT/legislation;
- client's variations to the contractor to be carried out by the subcontractor;
- additional works authorised by the contractor; or
- builders' work in connection with trades that the contractor wishes the subcontractor to carry out.

Table 6.4 Committed order input

Cost centre	Standard Description	Budget (£)	Order value XYZ Earthworks (£)
A100-1	Remove overgrowth	0	0
A100-2	Topsoil strip	7,000	6,000
A100-3	Reduced level excavations to 1 m deep	3,000	2,000
A100-4	Reduced level excavations over 1 m deep	0	0
A100-5	Filling to level up to 1 m deep	65,000	62,000
A100-6	Filling to level over 1 m deep	0	0
A100-7	Basement excavations	0	0
A100-8	Detailed trench excavations	20,000	19,000
A100-9	Detailed pad excavations	0	0
A100-10	Detailed lift pit excavations	0	0
A100-11	Earthwork support	2,000	2,000
A100-12	Deposit spoil on site	0	0
A100-13	Deposit spoil off site	18,000	19,0000
Total	**A100 - Excavating and filling**	**115,000**	**110,000**

Where applicable, financial adjustments may require authorising by the project or commercial manager; once approved, they are entered into the CMS to recognise the contractor's liabilities.

6.3.2.1 Interaction with Accounts Department

A feature of a CMS is that it is also an accounting system which links committed order values to accounts payable. In order for this to run smoothly, the quantity surveyor must consider timing and the latest dates the accounts department must be in receipt of payment certificates. Where this applies, the quantity surveyor must consider the length of time needed to assess the works, prepare the payment certificates and obtain authorisation to enable the deadline to be met.

A cause of a dispute between a contractor and subcontractor can arise because of the contractor's failure to pay on time as a result of a failure to comply with their own procedures. It is therefore important for the quantity surveyor to adhere to the accounts department request for deadlines; missing a date and submitting late payment certificates may result in payments not being paid on time and possibly create a breach of contract with the subcontractor. If a subcontractor is not paid on time, in general the client will have no interest in the process because the project client has no binding agreement with the subcontractor. However, failure to pay a nominated subcontractor on time because of a fault by the contractor may trigger mechanisms of the contract for breach that possibly enables the client to pay the subcontractor direct and recover the sum paid from the contractor by reducing the payment of the main contract works.

6.3.2.2 Subcontractor's Responsibilities with the Payment Process

Time management is necessary to avoid delays in payment, which requires the coopera-tion of subcontractors. To initiate and sustain this ideology, the quantity surveyor must instruct subcontractors to adhere to the latest date(s) invoices are to be received, which is driven by any deadline set by the accounts department. However, this can create a problem if the contractor has a policy of logging receipt of invoices by accounts person-nel to record liabilities in the event of a subcontractor activating a dispute and making a claim under the Construction Act. Where applicable, the quantity surveyor must understand the arrangement and make due allowances for time. To mitigate any delay, a suggestion would be to instruct subcontractors to issue payment requests by email to the quantity surveyor and send hard copies as follow-up by post.

6.3.2.3 Assessment of Subcontractor's Works for Payment

The normal credit terms of subcontractors are monthly payments which, to be effec-tive, must be stated in each agreement. However, some labour-only subcontractors and domestic subcontractors with unique agreements may be paid weekly or fort-nightly. Where applicable, the quantity surveyor is normally jointly responsible with the bonus surveyor or site manager in conjunction with the project manager to authorise rapid payments. Where normal monthly credit terms apply the amount invoiced in any given month can be considerable, and those in receipt of major works packages where the works span several months (or even years) should be requested to issue timely sequential progress claims as interim applications instead of invoices. This is to permit the quantity surveyor to assess the works and advise the subcontrac-tor of an amount to include on an invoice. Payments cannot normally be processed without an invoice unless the contractor uses a self-billing system. This is a recipro-cated system where a tax invoice is generated by the contractor equal to the payment with the invoice generated at the same time and sent to the subcontractor, meaning any VAT component is equal to the payment. The arrangement is suitable for refur-bishment works and some aspects of new construction that attract VAT. Where self-billing does not apply, a subcontractor will usually issue a progress claim or invoice to a timeframe stated in the agreement. If the works are under a minor contract that commences and finishes within one month, a single invoice to the maximum value of the award as one full and final payment should suffice. For ongoing works lasting more than one month, the value of each progress claim must be a fair representation of the works complete, including any entitlements permitted by the agreement(s). In order to assess an amount due, the quantity surveyor must view the works in progress from physical observation in a similar manner to preparing a progress claim under the main contract. Alternatively, the quantity surveyor can prepare an Activity Schedule and update progress on a monthly basis, which acts as a snapshot of the status at any time for all works.

A progress claim (possibly called an interim/payment application in each agreement) differs from an invoice in that it is a request to assess the value of works for payment purposes, whereas an invoice is an automatic debt. However, it may be called a 'progress claim/invoice', in which case it must be considered an invoice. Once an invoice is received, and if agreeing to the value, the amount can be processed without the need to communicate the agreement to the subcontractor. If it is a progress claim only with the amount subsequently agreed, the subcontractor must submit an invoice to qualify for payment. However, there may be reasons why the quantity surveyor may disagree with the value on an invoice or progress claim, including:

- incorrect calculations;
- errors involving VAT;
- inappropriate value of the works completed;
- claims for materials stored on or off the site that are not part of the agreement;
- disagreement over the value of materials stored on or off the site when part of the agreement;
- no insurance provided with the application if requesting payment for materials stored off site;
- the wrong project;
- dayworks not authorised by the contractor;
- amounts already in dispute and not resolved which are still included; or
- additional works or variations not authorised by the contractor.

Where applicable, the quantity surveyor must act responsibly and formally reject the amount requested stating grounds for the rejection, and communicate the matter in writing to the subcontractor. The rejection must be called a 'pay less notice' and include a statement of a recommended amount together with breakdown (it also applies if the amount to be paid is nil), and must be issued no more than five days before the payment is due. When adopting this manner, the contractor will have complied with legislation and the terms and conditions of the subcontract agreement. However, if the process is not stated in the agreement, it must still be carried out as the legislation addressing the subject is supreme. If a subcontractor issues a progress claim and later agrees with the quantity surveyor to adjust the amount to include on an invoice, there is no point in communicating agreement to the value of the invoice once it is received. Moreover, awaiting the issue of a credit note for an invoice is not an option to delay payment, meaning the quantity surveyor must still certify the amount stated as due in the 'pay less notice'. If the contractor fails to issue a 'pay less notice' with the payment amount a surprise to the subcontractor, the subcontractor may issue a default payment notice. The final date for payment is then extended by the period between when the contractor should have issued the 'pay less notice' and when the subcontractor issued the default payment notice. Such procedures mirror the requirements of the main contract works, which a client also has a duty to acknowledge.

During the construction phase the value of variations and additional works can be substantial, and the creation of a variation register will aid the process of managing payments. This is possible by creating an Excel file for each subcontractor, listing variations arising from the main contract in chronological order together with a description of the works/items and amount of each. The list can be extended to include additional works instructed by the contractor. Alternatively, this can be on a separate register with the format for either being similar to the variation register created for administering the main contract as demonstrated in Table 6.2 earlier. The value of any approved variations and additional works amounts requires to be entered into the CMS, which will automatically adjust the contractor's commitment to the subcontractor. It should be noted here that details entered in the CMS are those creating payments as they represent the contractor's commitment for works under the subcontract agreement, plus or minus any changes brought about by approved variations and additional works. The Excel spreadsheet register is for reference only, for which the contractor's accounts personnel will have no knowledge (or need to have knowledge) as they will be relying on the CMS for information regarding the contractor's liabilities.

The payment process itself is dictated by the level of automated information the CMS permits. A problem with some systems is that they have a one-line approach, that is, an amount to draw off the contract sum and client variations and another for additional works, which may produce a variety of payment certificate formats. To demonstrate, Table 6.5 shows an interim payment certificate for a subcontractor carrying excavating and filling works. Once the certificate is authorised, it is necessary to append the invoice to the certificate which is usually required by the contractor's accounts department for auditing purposes and matters associated with VAT.

The retention fund noted on each payment certificate is held by the contractor and released in accordance with the subcontract agreement. This is usually half the amount held at practical completion with the balance due at the end of the defects liability/rectification period (i.e. usually mirroring the requirements of the main contract). Any accrued interest on retention sums are usually retained by the contractor. This is because the cost to establish, manage and close a trust account for each subcontractor is not practical, as amounts are usually low in comparison with any retention sum held under the main contract.

Table 6.5 Progress works payment certificate

CONTRACTOR:	BUILDER CO LTD	
SUBCONTRACTOR:	XYZ EARTHWORKS CO LTD	
PROJECT:	Construction of new industrial unit - Leeds (Project Nr 1888)	
Contract works award value		£110,000.00
Agreed additional works		£1,000.00
Agreed client variations		£0.00
Builder Co Ltd commitment to XYZ Earthworks Co Ltd (Excluding Value Added Tax)		£111,000.00
Value of works in progress **(63%)**	£69,300.00	
Value of additional works in progress **(100%)**	£1,000.00	
Value of client variations in progress **(0%)**	£0.00	
Total value of works	£70,300.00	
LESS Retention (5%)	(£3,515.00)	
Net amount certified (excluding VAT)	£66,785.00	
LESS previously net amount certified	(£28,500.00)	
Amount due (excluding VAT)	**£38,285.00**	
Signed:	**Approved:**	
Quantity Surveyor	**Project Manager**	
Date:	**Date:**	
Accounts entry on CMS		
Date:		

6.3.2.4 Effect of Site Instructions with Subcontractors Payments

A subcontractor may request payment from the contractor if in receipt of a written instruction authorised by the contractor that alters, changes, enforces or reschedules the contract requirements or includes the supply of information pertaining to the works under contract. This may be instigated by an instruction from the client's agent, as the agent cannot normally issue instructions to subcontractors appointed by the contractor or their subcontractors. The contractor's personnel can also issue instructions for the contractor's purpose only, and there is nothing stopping a contractor issuing an instruction to a client-engaged contractor (although the contractor should be aware of any ramifications when doing so). Occasionally, the contractor's personnel may be unaware of a subcontractor's contractual responsibilities and may issue an instruction in any event. Where applying, the instruction may include the notation 'for record purposes' or initials FRP with the contractor remaining elusive to the fact such an instruction constitutes additional works and/or a variation. Once a site instruction is issued it is enforceable, and the contract may be worded to this effect; this means site personnel must be aware that, by including the reference FRP on an instruction, the issue is committing the contractor.

If a subcontractor requests payment for additional works authorised by the contractor, the original written instruction or copy must accompany the invoice/progress claim. It is not the contractor's quantity surveyor's responsibility to search for these instructions even though the contractor issues them, and a subcontractor must ensure such information is attached to a payment request to qualify for payment. If a subcontractor requests payment for additional works stated as being carried out upon an oral instruction given by the contractor's site personnel, the amount should be rejected as it is hearsay and would generally go against the conditions of the agreement.

6.3.3 Material Suppliers and Hire Company Payments

Companies supplying materials to a contractor normally issue invoices for payment after they have been delivered to site, less any amounts paid in advance. This may also apply to hired plant including operatives, whereas the supply of general and semi/fully-skilled site operatives and other services from recruitment agencies are invoiced progressively as the services are hired. Depending on the functions of the CMS, a supplier of goods or services may need an order value entering on the system in the same manner as a subcontractor to facilitate the processing of payments. Here, it may only be possible to input an approximate value because the price is not fixed, which can be achieved by using priced schedules stated on the purchase order charged at approximate quantities. For example, a bulk order for the supply and delivery of bricks may be to an uncertain quantity. Furthermore, a site cabin may be hired for an estimated period of time or general site operatives hired for an unspecified duration. To drive the commitment certain criteria must be answered and, using the abovementioned examples, answers to a range of questions such as the following could fulfil the requirements.

- What is the maximum quantity included on the bulk order for the brick supply?
- How long will the site cabin be required on site?
- How long will the general site operatives be required, and is this to be a frequent request?

Of course, the calculated order value is not a full commitment to the vendor and the quantity surveyor may generate future savings with value management once the maximum quantity or hire duration crystallises. A less desirable CMS is one without an order value commitment facility where invoices are processed and allocated to cost codes or cost centres. This has a disadvantage compared to an order committed system: it is prone to human error because supplier's invoices can be duplicated, meaning there is the possibly of overpayment. To avoid this, a CMS may have a facility allowing the invoice number to be logged and tagged to the supplier's identification code, with the system not accepting a duplicate invoice number for the same supplier.

6.3.3.1 Assessment and Authorisation Process

The quantity surveyor is normally responsible for authorising material supplies and hire invoices for payment, which involves signing off the amounts as well as validating supplier and project codes and cost codes/centres. To authorise an amount the goods delivered or items hired must be to the satisfaction of the contractor, which is usually substantiated with a delivery note signed by the receiver of the goods (who ideally should be an authorised signatory of the contractor). If there are queries regarding the quality and quantity of goods delivered, and noted to that effect on the delivery note or communicated by the contractor after delivery, they should be addressed formally. This will involve contacting the supplier and requesting a credit note if goods are to be, or have already been, returned. The same applies if goods are rejected upon delivery or are not the same as any written notation on the delivery note. For example, a brick supplier's invoice may state that it delivered ten pallets of identical bricks as per a written order which corresponds with the delivery note. However, what the invoice may fail to state is that the receiver of the goods signed the delivery note with a caveat that five pallets were not acceptable because they contained the wrong bricks and were left on the vehicle for return by the driver. Here, a credit note will be required for a proportion of the invoiced amount, and it is equally important for the supplier to fulfil the order by delivering the replacement materials and issuing a separate invoice in the aftermath.

As a matter of course, invoiced amounts must be in accordance with the purchase, bulk or blanket order issued by the contractor. This can be validated if the invoice cites reference to the order number, a copy of which will be in the contractor's possession. Suppliers of goods to subcontractors in receipt of awards for the supply of labour and materials may issue invoices to the contractor in error and, if received, should be rejected. Even if a representative of the contractor signs for receipt of goods on behalf of a subcontractor in any given scenario (e.g. an evening delivery), the credit arrangement is not altered. The inverse of such an arrangement would only exist on agreement with the contractor and could apply if the subcontractor has temporary business problems.

When plant is hired, a delivery note for mobilisation is normally issued and attached to an invoice which confirms the date for commencement of the hire period. Ongoing hire invoices do not usually require attachments, and the hire period can be verified by referring to the contractor's plant hire register that should also record the date of demobilisation. In general, if plant is hired with an operator, it becomes a goods and services supply order and invoices should be treated in the same manner as a subcontractor; this means the quantity surveyor must issue a 'pay less notice' advising of an amount due if wishing to authorise a lesser amount than stated on the invoice(s). Labour hire invoices issued by recruitment companies should include the days and number of labour hours worked as authorised by the contractor's representative, and does not normally fall under the category of goods and services.

A contractor's accounts department will usually create a list of ongoing projects including the names of quantity surveyors or project managers responsible for authorising payments. However, an accounts department may seek approval of invoices by a quantity surveyor assigned to one project that are for other projects undertaken by the contractor, and care needs to be taken while authorising invoices when the same suppliers are serving a number of projects. To avoid this type of error, invoice(s) must correspond to the appropriate purchase order number and delivery note(s) stating the project title and address, including a signatory confirming receipt of the goods which is recognisable. If the quantity surveyor is in receipt of an invoice addressed to the correct contractor and issued to the wrong quantity surveyor and is for another project, it should be returned to the accounts personnel for redistribution and not simply forwarded on to the correct quantity surveyor as it could cause havoc with the system (e.g. lost invoices).

Suppliers may also make errors on their invoices. For example, a concrete supplier in receipt of numerous awards from a contractor for a number of projects under construction at the same time may confuse project titles and/or phases on an invoice, which can be clarified by observing the delivery note that must relate to the invoice. In the event of an error(s), the supplier must issue a credit note(s) and raise a replacement invoice(s) to correct the mistake(s). In an extreme situation, a contractor with works in progress on two different sites at the same address may appoint the same supplier and, with invoices unchecked, either may receive the wrong invoice. In any situation where the contractor is not liable for the debt, the invoice must be returned to the supplier and not entered on to the CMS.

Payment terms with suppliers of goods or services are usually monthly, and the quantity surveyor must acknowledge the time frame of the accounts department for receipt of authorised invoices to ensure payments are made on time.

6.3.3.2 Factoring Company Payments

Some suppliers of goods or services may use factoring companies to assist with their cash flow, and the quantity surveyor should be aware of the arrangement where it applies. When factoring is involved, the supplier appoints a factoring company and advises the company of the value of the debt the contractor owes. The factoring company then pays the supplier the debt within a shorter timeframe than under the agreement with the contractor, less a commission, and raises an invoice for the full debt to the contractor in its name with the contractor paying the factoring company. A flaw of the arrangement is that the contractor may misjudge the situation and pay the supplier, thus ignoring the factoring company arrangement. At worse, both supplier and factoring company may be paid, which is unusual and more likely an error on the contractor's part (as there is usually only one invoice per consignment with invoice duplications rare). Factoring is not assignment in the legal sense because the supplier remains responsible to the contractor for the quantity and quality of goods and continues to comply with the terms and conditions of the agreement. Where a supplier factors its business dealings and is in receipt of an award from the contractor, it would be wise for the contractor to request a letter from the supplier confirming the arrangement together with the name, address and bank details of the factoring company.

6.3.4 Consultant's Payments

When consultants issue progress claim requests or invoices for payment, they are usually in accordance with a stage payment breakdown that forms part of an executed agreement. For the purpose of the CMS, an order value for each award needs to be input into the system to represent the contractor's commitment and to generate payments in a similar manner to preparing payments for other vendors.

With construct-only contracts, an invoice may be issued by a consultant for a single payment based upon an *ad hoc* arrangement such as the issue of a report as advice on a specific matter, with receipt of the report generally considered entitlement for payment. With a design-and-build project however, numerous consultants are engaged and stage payment requests may be issued for works carried out 'behind the scenes' without any information being supplied. This is because the design and documentation may be undergoing development that requires peer review and approval prior to being issued to the contractor for working purposes. Here, the consultant will have incurred expense for project initiation and pre-commencement activities, and is usually entitled to payments as per the agreement. As supporting evidence, the appropriate consultant(s) requesting payment could supply a draft document register listing the design and documentation undergoing development to substantiate the payment request. If supplied, the register is solely for demonstration and must not be used on the project as nothing is approved, with the details subject to possible change. Understandably, a design consultant may be reluctant to release any information in the event of it being interpreted as working details, which of course it is not; the quantity surveyor must act responsibly if supplied with such information.

In general, consultants providing a service are treated in the same manner as goods and services suppliers. This means the quantity surveyor must ensure invoices are processed and issued to the accounts department in a timely manner for payments to be made on time, and provide any relevant consultant with a 'pay less notice' with supporting information if payment is to be less than the invoiced amount.

6.3.5 Project Banking

Project banking, or Project Bank Accounts (PBAs), is an exclusive arrangement involving the setting up of a project account from which payments are made to parties in the supply chain. Funds in the account can only be paid to project beneficiaries, that is, members of the supply chain named in the account such as the contractor and certain supply chain members. As a consequence, supply chain members do not have to wait for the contractor to process payments and benefit from receiving the monies directly. This ensures beneficiaries have certainty and security with the supply of speedy payments, allowing a reduction in the need for borrowing or financing credit. Furthermore, the arrangement facilitates a reduction in the need to chase payments as well as a reduction in the number of disputes that can otherwise occur with construction projects that do not adopt the arrangement. The use of a PBA for a scheme does not affect procedures for valuing and certifying payments, and does not remove the contractor's responsibility for selecting and managing the supply chain.

The private sector has not fully embraced PBAs; however, they are used widely with government construction works contracts on any project where a significant amount of subcontracting is required. The process involves the project client depositing advance funds into the PBA, whereupon each beneficiary is able to gain access and transfer amounts into their own account via a bank transfer restricted to the value of the appropriate payment certificate. Each advance deposit made by the project client represents the anticipated expense a client expects to pay over an agreed duration, with the expense determined by the client's agent or quantity surveyor/cost manager. However, due to the nature of construction projects, changes to programmed activities can mean the forecast expenditure is modified from time to time. The preparation of each forecast involves creating a payment schedule and breaking the sum down into amounts theoretically due to each beneficiary by time, linked to stages of physical completion. This is similar to a cash flow forecast but is more detailed; a cash flow forecast is usually for a single drawdown, whereas the payment schedule involves an assessment of each beneficiary's award. To be effective, all payments from a PBA require authorising by the client's agent and main contractor; this means the contractor's quantity surveyor prepares a payment certificate for each beneficiary, citing it as for a beneficiary of the PBA to distinguish it from other subcontractor's payments. The selection of beneficiaries must stem from an invitation to tender for the works/services which is mirrored in the terms and conditions of each agreement. Prices within supply chain agreements are deemed to include any concession for the commercial privilege, which may also be a deciding factor for the contractor when submitting a tender for the works. The client's agent is responsible for managing the PBA, meaning there is a requirement to:

- monitor the PBA to check lodgements and payments occur as authorised;
- identify charges and accrued interest on a PBA and ensure charges and interest are appropriately passed on to the client and recorded in the client's accounts; and
- close the PBA on completion of the project.

A contractor may be concerned that the use of a PBA could expose the profit margin built into the contract sum. However, as PBA deposits are for beneficiaries only, the amounts represent a proportion of the value of the project meaning it will not be possible or there be a need to determine the profit margin if not a contractual requirement. The use of PBAs is included in a number of standard forms of contract, for example NEC3, PPC 2000 and JCT (which also produces the Project Bank Account Documentation (PBA) as an auxiliary supplement).

6.4 Cost Centres and Financial Reporting

The commercial aspect of running a project usually involves producing monthly or bi-monthly cost reports on the financial status of the scheme to the contractor's senior management. Each report is normally produced from the CMS that tracks financial activities, possibly in a variety of formats. Formats can include a record of the expense incurred allocated to cost codes/centres and a record of the value of awards placed as commitments. A condensed progress report will also display the values of package awards against each cost code/centre in comparison with the budget, as well as expense incurred and the anticipated final account, to help streamline the process into a single document as a frame of reference. Suitable management of the CMS is vital for the appropriateness of the report and commences with setting the budget as discussed in Chapter 4 (Section 4.5.7). Once the budget is committed and an order for the value of awards allocated to cost codes/cost centres, the CMS creates a financial residue or deficit on each cost code/centre (i.e. the difference between the budget and value of an award). To demonstrate, Table 6.6 shows a financial progress report for the excavating and filling works for a project, together with notes for managing the cost centres/codes. This table is a sample of one part of a project, with the sum of all cost centres/codes driving the projected final account.

The format of the report produced from the CMS may be in text format, and for presentation it may be necessary to reproduce summaries of the cost code/centres in an MS Excel spreadsheet or table in Word format. This permits the information to be embedded in a broader cost report that includes commentary for electronic and hard copy distribution.

6.5   Tracking Expenditure

The contractor's quantity surveyor can be involved in projects worth millions of pounds running for long durations. With any project on this grand scale, a number of project team members may be delegated with the task of placing awards and authorising payments. However, the quantity surveyor as cost manager must be aware of all project expenditure for which the use of a CMS is an invaluable asset; its use means all commitments are enclosed within the system, effectively ring fencing the overall expenditure to compare with budgets and/or cost limits. Project expenditure is usually shown on a cost ledger that breaks down the constituents of cost incurred, which is integrated with the CMS to show summary expense by cost code/centre. A periodic check of the contents of the cost ledger provides the following advantages:

- it provides a snapshot of the progression of the project (e.g. £5.0 million spent of a £10.0 million project indicates the project should be *c.* 50% complete);
- it acts as an audit check; if left unchecked there may be times when the cost exceeds the anticipated final account, which may highlight errors in the payment process; and
- it verifies the amounts stated on CMS reports.

The quantity surveyor would be wise to track project expenditure regularly, a suggestion being monthly, in order to avoid detailed investigations at later dates if errors become exposed that may have occurred months prior that went unnoticed. By tracking expenditure, it is possible to identify mistakes that may expose duplicate cost allocation or incorrect coding, for example a brick supplier that issues two invoices for two sites with one invoice incorrectly allocated to a project identification code, or concrete purchase costs allocated as brick supply costs because of errors in material coding. In the event of exposing such errors, the accounts department usually has a method of cost code reallocation. This is carried out by entering the details on a form and crediting an amount from one code and debiting the other.

6.5.1   Accruals

An accrual is an accounting term used to describe the collective forecast of payments in the process of being paid that represents a business's liabilities to its creditors. Accruals in the plural sense may be allocated to cost codes that are assessed automatically or

Table 6.6 Financial progress report

Client: Proprietary Holdings Ltd
Progress Report Date:

Project Nr: 1888
Project Title: Construction of new industrial unit - Leeds

A100 - Excavating and filling

A	B	C	D	E	F	G	H	I	J
Cost centre	Description	Budget (£)	Committed contracts (£)	Contracts to commit (£)	Committed additional works (£)	QS judgment (£)	Projected final account (D+E+F+G) (£)	Variance (C-H) (£)	Expenditure to date (£)
A100-2	Topsoil strip	7,000	6,000	–	–	+1,000	7,000	0	7,000
A100-3	RL excavations	3,000	2,000	–	–	–	2,000	+1,000	2,000
A100-5	Filling to level	65,000	62,000	–	+1,000	–	63,000	+2,000	31,000
A100-8	Trench excavations	20,000	19,000	–	–	–	19,000	+1,000	19,000
A100-11	Earthwork support	2,000	2,000	–	–	–	2,000	0	2,000
A100-13	Dispose spoil off site	18,000	19,000	–	–	–	19,000	–1,000	9,300
TOTAL		115,000	110,000	0	+1,000	+1,000	112,000	+3,000	70,300

(A–C) These are created during the pre-commencement stage with information supplied by the estimator and input into the CMS. (D) Order values or an estimate based upon approximate quantities charged at applicable rates. (E) Activated upon initiation of the cost centre and represents part of the budget that is uncommitted. (F) Authorised additional works instructed by the contractor extra to the committed contract value. If 'additional works' are carried out as client's variations, they may be coded separately and issued in a separate report as the contractor will be reimbursed. (G) A flexible assessment which is manually inserted to recognise works or scope not included in Columns D–F that the quantity surveyor considers is a contractor's responsibility. With this column, the CMS may have a facility where it is possible to create a brief description of the works and calculations in a back screen that drives the amount. (H) Calculation. (I) The contractor's gain/loss is shown as +/–. (J) Must not exceed the projected final account (H).

manually. The latter of these can be an arduous task as it means tracking and assessing works in progress, possibly without receipt of invoices or payment applications, which can lead to wild amounts being recognised as liabilities. However, an effective CMS provides more accuracy if it includes an accrual facility. With this facility, the CMS lists cumulative expense which is allocated to each cost code/centre plus an accrual amount to produce the total expense at a particular point in time. Each accrual includes debts owed to supply chain members for the supply of materials, plant and labour hire and works in progress to the value of pending invoices and amounts in question on received invoices that will probably be paid. Accruals are usually assessed at the end of each month, which are automatically reversed at the end of the following month. At the time of reversal, processed payments are recognised as an affirmed cost with new accruals added to update the liabilities.

An accrual facility on a CMS has a clear advantage over a system lacking the option, as it means the quantity surveyor is able to recognise current expense and unclaimed amounts that acts as a reliable source for tracking expenditure. For example, a project may have a cost allocation of £50,000 for the supply of bricks which the quantity surveyor assesses to be the final cost as the supply is complete and a bulk order exists on the system. With all bricks delivered, a CMS with an accrual facility may show the expense incurred as £35,000 plus an accrual of £15,000 for unauthorised or pending receipt of invoices from suppliers. However, without the accrual facility, the CMS would show the cost as £35,000 and only recognise the remaining £15,000 once unauthorised and/or pending invoices are processed for payment. Without an accrual facility, project cost reporting in comparison with progress payment requests to a project client will always be one month in arrears. This is because the progress claim is current, yet the expense for comparison is lagging one month. In the example with the brick supply above, any project expense report must therefore include the actual cost plus accruals, as the progress payment application to the project client will be for the value of works including the price of all supplied materials. If the accrual is not recognised in the report, a false amount would show that would dissipate once the costs catch up.

The logging of accruals involves allocating amounts to a cost code/centre which includes a vendor identification code obtained from the CMS database. To demonstrate, Table 6.7 shows a theoretical accounts ledger report for the cost of brick supplies using a CMS that has an accrual facility.

In addition to vendor identification, an accrued amount may be allocated by the contractor's accounts department for items deemed onsite costs or preliminaries items (e.g. rental of site cabins if a contractor's asset and the cost to employ the project team), as the expense is recoverable under the main contract agreement.

Information used to create accruals is derived from:

- logging an invoice amount on the CMS once received;
- assessing the value from delivery notes pending receipt of invoices;
- an estimate calculated by the quantity surveyor based upon work in progress; and
- use of a valuation system where a percentage of a committed order is ascertained as works in progress.

A degree of accuracy is required with accruals to avoid significant swings in the stated amounts from one month to the next. For this reason, a CMS with an accrual facility provides a hands-off approach and is the most reliable. However, where *ad hoc*

Table 6.7 Accounts ledger with accruals

B100-12: brickwork supply. Project nr 1888			
Supplier code	Date	Cost code	Amount (£)
Month 1: opening balance			**0.00**
B34877	Month 1	BL01	1,483.00
S13778	Month 1	BR55	6,350.45
Accrual	Month 1	BL01	2,700.00
Accrual	Month 1	CC01	7,300.00
Month 1 closing balance			**17,833.45**
Monthly movement			**17,833.45**
Month 2 opening balance			**17,833.45**
Reverse accruals	Month 1	–	−10,000.00
B32566	Month 2	BL01	2,700.00
456678	Month 2	CC01	7,300.00
Credit note 1	Month 2	CC01	−450.00
244556	Month 2	BR55	200.00
Accrual	Month 2	BL01	1,400.00
Accrual	Month 2	DT01	3,046.00
Accrual	Month 2	BK12	8,400.00
Month 2 closing balance			**30,429.45**
Monthly movement			**12,596.00**

situations occur meaning subcontractors can claim additional compensation because of events allowed under the contract (e.g. a change in law on employee national insurance contributions), it is wise to create a register of forecast expense to include as an accrual. This is to recognise liabilities and avoid embarrassment should late invoices be submitted without any contractual provisions to reject them, even if a result of poor administration on the subcontractors' part.

6.5.2 Cost Value Reconciliations (CVRs)

A cost value reconciliation (CVR) is a reporting exercise carried out at intervals of the construction phase of a project to reconcile incurred expense with the value of works assessed from the budget, and applies to the works in full or in part and the preliminaries. The accuracy of any variance in the reconciliation is determined by the relationship between the cost (including accruals) and the value of works in progress, which on occasion may be stark in contrast because of accrual inconsistencies. A disadvantage of logging, assessing and estimating accruals is they are prone to human error which can produce inaccuracies with the cost. However, a CMS with an integrated valuation and accrual system is not prone to such errors, as accruals are generated from the progress of the works. For example, if a subcontract order value for ceilings and partitions works is committed to the CMS for £100,000 without any prior payments made and valued at 50% complete, an accrual of £50,000 will be generated. Conversely, if the budget for the same works is £110,000 and valued as 50% complete, the value will be £55,000 reconciled to a cost of £50,000, meaning the contractor has produced a £5,000 gain. However, scenarios can occur where an order value is not committed with the expense permitted to run out of control if not checked in a CVR (e.g. preliminaries items). To demonstrate, Table 6.8 shows a CVR that a CMS may produce for the preliminaries.

By tracking expense against the budget and values on the CVR, strategic action can be taken if there is evidence of cost overruns. This is evident when reviewing the information provided in Table 6.8, in particular with 'management and staff (P100-2)' and 'mechanical plant - minor (P100-18)', as the expense is high in comparison to the value. Questions to be asked about this CVR could be:

- P100-2: As this is a time-related item, has 15% of the project duration surpassed and, if so, why has more expense been incurred than envisaged, and is the cost allocation correct?
- P100-18: What is the basis of the cost and is the accrual overstated or incorrectly allocated? Is it possible any invoices are duplicated? Are there any items of plant on hire that can be off-hired?

Once answers to question are ascertained, the quantity surveyor must alert the project manager to the findings as they might impact the overall budget if the trend continues; the objective of a CVR is to provide a realistic view of the current financial status in comparison with the budget. Furthermore, an objective of cost reporting is to use foresight to predict whether the budget can be preserved and address any underlying issues that could be managed differently by the project team.

The allocation of expense for management and staff within the preliminaries is usually input by the contractor's accountant or business manager on a monthly basis and shown on a cost ledger. Here, the quantity surveyor usually has no control over the matter because no award or order value exists. However, the quantity surveyor has a duty to ensure that accruals and allocation of expense for these items reflects the level of attendance on a project. For example if, during a given month, attendance was provided by a project manager, quantity surveyor, site manager and two foremen in line with the staffing programme, the expense should be comparable with the budget. However, because management and staff attendances on a project can fluctuate over time, the expense incurred should be monitored to ensure the charges reflect the actual attendances; this is because the accountant or business manager is not usually based on site and will proportion expense based upon information provided, which may be from time sheets issued by the project manager. Such criteria must therefore be checked for accuracy to ensure there are no 'phantom' attendances, and that the expense is allocated to the correct project and cost codes.

6.6 Extension of Time (EOT) Claims

For business efficacy, a suitably drafted form of construction/building contract will include clauses placing responsibility on the contractor to issue a notice of delay when the contractor considers the project will not complete by the date stated in the contract

Table 6.8 Cost value reconciliation

Project nr 5678

Cost centre	Standard Description	BQ allowance (£)	Cumulative valued (%)	Cumulative value (£)	Cumulative cost* (£)	Variance (£)
P100-1	Pre-commencement	25,000	100	25,000	24,000	+1,000
P100-2	Management and staff	140,000	15	21,000	27,595	−6,595
P100-3	Site establishment etc	20,000	90	18,000	18,750	−750
P100-4	Running costs	60,000	10	6,000	5,980	+20
P100-5	Temporary services	10,000	75	7,500	7,180	+320
P100-6	Temporary works	15,000	0	0	0	0
P100-7	Security	21,000	14	2,940	2,610	+330
P100-8	Employer's requirements	5,000	90	4,500	4,380	+120
P100-9	Contract conditions	5,000	0	0	0	0
P100-10	Fees and charges	10,000	60	6,000	5,205	+795
P100-11	Insurances, bonds etc	8,000	60	4,800	4,200	+600
P100-12	Health and safety	5,000	80	4,000	3,780	+220
P100-13	Control and protection	10,000	20	2,000	1,500	+500
P100-14	Site records	5,000	0	0	350	−350
P100-15	Cleaning	6,000	0	0	0	0
P100-16	Non-mechanical plant	15,000	10	1,500	1,305	+195
P100-17	Mechanical plant: major	50,000	10	5,000	5,000	0
P100-18	Mechanical plant: minor	20,000	25	5,000	12,780	−7,780
P100-19	Post-completion requirements	5,000	0	0	0	0
TOTAL		**435,000**	**26%**	**113,240**	**124,615**	**−11,375**

* Includes accruals

or agreed date after the contract was executed. The requirement to give such a notice is usually within a stated timeframe expressed in days, and can be the number of days from the event, circumstances giving rise to the event, or the awareness of an incident triggering the event. If a notice of delay procedure does not apply because the contract is silent on the matter, the contractor must submit an extension of time (EOT) claim instead. Here, the contractor could also raise a time at large argument (a term used to describe disagreement to the end date) under common law if delayed by the client, which would render the contractual completion date void. If time does become at large, the contractor is only obliged to finish the contract within a reasonable period of time.

The term claim suggests it is a request for additional money where, in reality, an EOT claim is one arising from a notice of delay seeking to prolong the construction period and is different from a financial claim. However, a notice of delay or EOT claim may give rise to a financial claim; if applicable, it is necessary to keep the two separate to avoid confusion. In essence, a notice of delay or EOT claim is the contractor's method of mitigating exposure to liquidated damages when the completion date is considered unachievable. If the contractor is of an opinion the end date cannot be met because of their own default, they must accelerate the programme at their own cost and should not issue an EOT claim. Whether a notice of delay should be issued when the contractor is at fault will depend upon the requirements of the contract; in most situations, a contractor would have an ethical duty to inform the client's agent of the situation together with mitigating actions, even if not contractually obliged to do so.

If the contractor is of an opinion the delay to the completion date is a responsibility of the client, the notice of the delay or EOT claim must cite reasons. These may be unique, in which case the circumstances must be explained, or may be relevant to the contract. For example, clause 2.29 of JCT SBC/Q 2016 lists Relevant Events that could give grounds for a notice of delay (e.g. variations altering the works so considerably the original scope cannot be completed on time, exceptional adverse weather, etc.). Here, clause 2.27.1 requires the contractor to issue a notice of delay citing the Relevant Event(s) where applicable. This is advantageous, as it helps to avoid disputes that could occur where a contract is silent regarding what constitutes a delay.

A confusing matter regarding the validity of a notice of delay or EOT claim is with the subject of client-authorised variations. Here, the contractor must act responsibly if of the opinion the end date will be delayed because of variations, as the client's agent can only usually issue an instruction to prolong the end date and not anything on the critical path. The value of authorised variations that increases the financial worth of the works under contract may not be a reason for the contractor to issue a notice of delay or pursue an EOT claim as they may be disproportionate to time (e.g. an increase in the contract sum by 10% may not delay the project duration by 10%). Furthermore, the issue of timely approved variations that alter the quality may not be a valid reason for a delay. For example, if a contract programme of 50 weeks includes floor tiling to commence laying on week 40 with product selection to be received by week 20 which is on a 3-week delivery, it may still be possible to achieve the end date if more expensive tiles on a 6-week delivery are selected on week 30, despite the increase in cost. This is subject to the tile-laying procedure being identical that would not prolong the works. By comparison, if a contractor on a construct-only contract is building a 10-storey apartment block comprising 200 mm thick in-situ concrete floors has the thickness increased to 300 mm by a variation, it may constitute a valid claim to delay the end date because the quantities will substantially increase the works. In this scenario, a client's agent's initial response to a notice of delay or EOT claim may be to request additional labour, which is a valid point. However, situations like this are considered on a case-by-case basis by the contractor, and could involve:

- observing any float available in the programme;
- considering risks and space for storing additional materials; or
- assessing availability of additional labour by communicating with subcontractors.

Here, it would be unacceptable for the contractor to advise the price of the works variation, receive an approval and let the end date surpass because of the prolonged works. In this scenario, it would reasonable for a client to assume the project will complete on time as nothing to the contrary has been advised, meaning the contractor would be liable for the delay. However, if the contractor issues the variation with a notice of delay or EOT claim, the client may consider options and could elect to use precast concrete instead to save time.

Normally, the quantity surveyor will not issue a notice of delay or EOT claim, as it is a project manager's responsibility. However, in assessing the impact of a delay, the project manager may seek input from a planning manager to reassess the programme and the quantity surveyor to provide substantiating documents as evidence. This includes citing details of the Relevant Event(s) (where applicable), including date(s), letters, instructions, document transmittals, etc. The quantity surveyor may also be required to provide an estimate of approximate cost associated with the delay. For example, if there is a 10-week delay which the contractor is of the opinion is a client responsibility, the cost will be based upon time-related expenses assessed from the preliminaries budget or known running costs that will continue to be incurred for 10 weeks commencing from the current date of practical completion. Once the costs are assessed, the project manager will usually decide if the delay is manageable, and may elect to commit the contractor to a length of delay and advise the client's agent of the decision.

When issuing a notice of delay or EOT claim, it must relate specifically to the clause(s) of the contract, be clear, concise and free from personality to represent authenticity. As soon as practically possible after a notice of delay or EOT claim is issued, it is in the contractor's interest to update the client's agent on the status of any associated financial claim, which may be dependent on the supply of information from third parties such as subcontractors. It is therefore good practice to record delay claims in a register that includes:

- reason(s) for the delay;
- clause(s) of the contract supporting the reasons for the notice or EOT claim and entitlements;
- date of issuing the notice or EOT claim;
- length of the delay and whether it is committed or to be negotiated; and
- date of response and what extension is granted.

The date of the response from the client's agent is important as it sets the basis for assessing any financial claim, often referred to as a loss and expense claim in the contract where the contractor will be out of pocket because of a delay it has no control over.

A contractor may also face a notice of delay or EOT claim from any of its supply chain and, in particular, subcontractors that carry out most of the works. Where applicable, the delay may have a link to a claim under the main contract. However, if it is a separate matter with the contractor having no recall from the main contract, the contractor should refer to the subcontract agreement and consider options which may include:

- accepting the request if the delay can be accommodated by the programme;
- reject the request and advise the supply chain member to reconsider the situation with solutions; or
- seek an alternative subcontractor by terminating the agreement in accordance with the terms and conditions of the agreement.

If a subcontractor's notice of delay or EOT claim is valid and not the result of its default, the contractor must explore options to settle the matter as there may be no grounds to terminate the agreement. Even if the subcontractor is at fault and hindering progress, the situation might be temporary and voluntary termination may do actual harm to the project. The subject of voluntary termination of subcontract agreements by the contractor is discussed further in Section 6.8 later in this chapter.

6.7   Financial Claims

It is possible for the contractor to be faced with a financial or loss and expense claim on two fronts: firstly, under the main contract; and secondly, if receiving a claim from a member of the supply chain which may or may not be associated with a main contract claim. The quantity surveyor plays a vital role in dealing with the preparation and defence of such claims by ensuring project administration is impeccable at all times (e.g. keeping communications and transmittals, ensuring design and documentation is suitably stored, registers are up to date, 'pay less notices' issued on time, etc.). This is an important aspect for running a project because the reasons a client's agent may reject a claim under the main contract can be due to flaws in administration, when information is either missing or substituted with assumptions which are exaggerated due to a lack of evidence (e.g. a delay in the works caused by statutory undertakers, where the confirmation of start and finish dates is missing). Likewise, a contractor may reject a supply chain member's claim if the information provided lacks credibility (e.g. if relying on verbal instructions).

A claim must be differentiated from a variation even though it may be linked to one or more variations. For example, if on a construct-only contract the contract drawings define weights of reinforcement in kilograms per cubic metre applicable to the profiles of in-situ concrete slabs, beams and columns because the reinforcement is not designed with the actual requirement in excess then, subject to the terms and conditions of the contract, the difference could be a claim because of the inconsistency and not a variation. This is because variations are usually restricted to changes described under the variations clause of the contract with anything else being a claim.

A method of preparing and issuing a claim can be with a Position Paper. This is a document representing an opinion about an issue, which is usually authored by or for the entity expressing the opinion that is often published in law and politics papers and can apply with claims in the construction industry. However, this is more for advice when one party to a construction contract issues the other with the Paper to avoid a scenario(s) from manifesting that, if left uncontrolled, could give rise to a claim. Position Papers range from the simplest format of a letter through to complex presentations.

The Paper should clearly analyse the party's positions by setting out the facts in clear and unambiguous terms in order to arrive at a sound conclusion regarding the various principles concerning a scenario(s) that can or does give rise to a claim. In effect, a Position Paper can form part of a claim restricted to facts and at the same time allow the claim to elaborate on details of the loss and expense account. The structure of a Position Paper when also part of a claim should comprise:

- introduction (project overview);
- executive summary including the identification of issues citing supporting documents, e.g. bills of quantities, drawings, minutes, etc.;
- conclusions;
- appendices, including copies of supporting documents cited in the executive summary; and
- schedules of expense (incurred and to be managed) to substantiate the loss and expense claim.

Once in receipt of the claim, and as part of the assessment, the client/client's agent may engage a cost consultant/quantity surveyor with experience in claims to evaluate the contents who may ask the contractor for further information, which of course the contractor must endeavour to provide.

In order to be successful with a financial claim, the injured party must prove that payment of the loss and expense would put the injured party back into the position it was in before the event and/or chain of events took place, which in effect means the party making the claim will be no worse off once paid. However, the party issuing the claim must prove it did not contribute to the event and/or chain of events, and is satisfied there is nothing in the contract apportioning risk that has not gone in their favour.

The rejection of a claim can be in full or in part, which in any case is still a rejection. When rejected in full, the notice can be blunt if the claim is considered to have no grounds whatsoever. However, a rejection may have a theme of negotiation if the recipient is of an opinion the claimant is not entirely responsible for the event(s) giving rise to the claim. Where applicable, this may be because the recipient considers the claim has grounds in full or in part yet is missing vital evidence for it to be conclusive. In essence this can mean further information is required and, to save time, the recipient may suggest the sum be negotiated to a different amount, creating a 'win-win' scenario for the parties in the process.

The settlement of a financial claim can be included in a contract sum adjustment which, to be instrumental, must be based upon the terms and conditions of contract. The contract sum will not usually be adjusted if settlement of the claim is subject to an insurance policy pay-out, which may apply for the likes of repairs after a storm or fire damage that affects the works under contract. Here, an extension to the end date would normally be provided without costs, as any financial claim arising from the granting of an EOT is usually part of the insurance claim.

6.7.1 Delay and Prolongation

The procedure for issuing a notice of delay is instrumental as it ensures the contractor remains responsible, which in the process dismisses a culture of permitting a claim to be pursued as and when the contractor sees fit. It is not in the contractor's interest to issue a notice of delay and financial claim at the same time, because the length of time requested by the notice may alter following client input for mitigation. Neither is it in the contractor's interest to issue a financial claim after the date of practical completion while incurring liquidated damages in the hope the claim will be granted with costs, including the refund of any liquidated damages, as it will probably not happen. Where prior notice and a procedure for the issue of a financial claim is lacking in a contract, the EOT claim can include time and money at the same time. However, the contractor should be pragmatic in such a situation by advising of the delay first and at the same time provide an indication of the time it will take to prepare the loss and expense claim.

The format of a claim must be accurate and not involve remote matters that have little or no influence on the facts; the intention is to identify, clarify, quantify and value the direct loss and expense without overwhelming or providing inadmissible information. The format should comprise a chronological list of main events that either contribute to or cause the delay, including descriptive contents, together with their effect on the works cross-referenced to the appropriate clause(s) of the contract. Each main event should include back-up information (e.g. drawing register, site diary pages, photographs, site instructions, etc.) as proof. Where a chain of events are linked to a main event, these should be listed separately and cross-referenced to the main event including back-up information. This information should not be duplicated with the main event as it would create repetition (which is unnecessary). Once each main event and time delay is substantiated, the cost of the effect of the delay requires to be calculated. This will include:

- contractor's preliminaries (e.g. idle plant, equipment and supervision);
- subcontractors charges due to the delay;
- loss of profit;
- financing costs (e.g. interest on prolonged loans to fund the project); and
- inflation costs on remaining works under the contract which are halted due to the delay.

If the quantity surveyor has sought assistance from a consultant to prepare a claim, the cost of engaging the consultant should be included as it is a direct cost. As some parts of the claim may be a forecast, it will be necessary to separate actual cost from the estimated expense the contractor elects to manage for the length of the delay. Once the costs are compiled, any plausible savings to mitigate the situation are credited (e.g. off-hiring plant, relocating site staff to other projects, etc.). The total cost is then divided by the number of days requested in the notice/EOT claim (or subsequently negotiated) to equate to a cost per day that attracts no profit, as the intention is to recover the actual loss and expense without profiteering. However, the daily cost will attract an amount for offsite overheads which may be somewhat difficult to calculate. One approach is to express the contractor's margin used in the tender as a percentage of the contract sum/forecast total price of the works, minus profit, charged at the value of the loss and expense claim. Another approach is to use Hudson's formula derived from information published in the Hudson's Building and Engineering Contracts manual:

$$Dl = \frac{Os}{T} \times \frac{Cs}{Cp} \times Dp$$

where Dl is daily loss; Os is contractor's offsite overheads; T is contractor's business turnover; Cs is contract sum; Cp is current contract period before the event; and Dp is delay period.

Using this formula, if a contractor has an annual turnover of £50 million and incurs offsite overheads of £4 million in a relevant 12-month period and undertakes a contract worth £10 million to run one year and a 10-week extension applies, the daily loss per calendar day is:

$$\frac{£4,000,000}{£50,000,000} \times \frac{£10,000,000}{365} = £2,192 \text{ per day,}$$

implying a loss and expense for offsite overheads over 70 days of $70 \times £2,192$ per day $= £153,440$.

If a client wishes to retain the date of practical completion by not wanting to go down the path of a prolongation claim, and there is an acceleration clause in the contract, the contractor can withdraw the notice of delay and loss and expense/EOT claim. Where applicable, this should be substituted with a loss and expense claim caused by disruption of the programme, which includes acceleration costs to recover lost time. Here, withdrawal of the delay notice and financial claim can only be instrumental if the disruption and acceleration claim is approved as, in effect, the approval would systematically cancel the prolongation claim.

6.7.2 Disruption and Acceleration

Disruption is a term used to describe an intervening event causing a loss of production part way through a project caused by an event that delays a part of the works and lengthens the critical path. The domino effect here is that the disruption may delay the end date which, if applicable, would need to be recognised in a delay notice or EOT claim as discussed in Section 6.7.1 above. However, disruption is a fundamentally different matter as it may not delay the date of practical completion if the works affected by the disruption are accelerated, which would become a client's responsibility if the contractor is in no way responsible for the disruptive influence.

For a disruptive claim to be valid, the contractor must demonstrate that a specific disruptive influence hinders the *actual* progress and not *planned* progress. Moreover, the claim must identify the event and record it as being either a breach of contract by the client or the result of an event (or sequence of events) written into the contract that gives rise to a notice of delay (or EOT claim). Isolated variations alone of a minor value in comparison with the value of the project do not usually hinder planned progress because they can be mitigated by supplying additional resources. However, if a client's agent issues a series of urgent variations requiring operatives to frequently postpone their contract works to attend to the urgencies, the contractor may be able to claim for loss of productivity on the originally contracted works because of disruptions. However, the contractor must take care to examine all of the facts first before issuing a disruption claim because of responsibilities. For example, disruption can occur due to the late issue of design information. In this situation, a client may avoid liability if the contractor fails to give notice of an intention to suspend the works because of the lacking information or fails to make a request for issue of the information if such obligations are written into the contract.

The presentation of a disruptive claim is similar to a prolongation delay (or EOT) claim. In preparing the claim, the quantity surveyor is not expected to observe or record disruptive works as this is normally a responsibility of the site and project managers. However, the quantity surveyor may be expected to assess the expense incurred because of the disruption. To permit this, it is necessary to be provided with the following:

- information on the type of works disrupted;
- reasons for the disruption, confirming it is not the contractor's responsibility;
- attempts adopted by the contractor to mitigate the disruption; and
- delay to the programme as a result of the disruption.

For example, let us say a project manager advises the quantity surveyor that formwork gangs are being instructed to carry out variations elsewhere on a project at the request of the client. As a result, formwork operations under the contract are sporadically abandoned to carry out variation works as there is no additional labour available to take over the abandoned works or the variation works in isolation. The scope of works under contract affected is formwork to the soffits of slabs totalling $500\,m^2$ that the project manager advises was on an 8 week critical path that will now take 12 weeks. Here, a starting point is to recognise the financial allowance in the contract to pay for labour and plant for the affected works, which may be found in the project bill of quantities; see Table 6.9 for an example.

As this amount represents labour, plant and materials, it is necessary to ascertain the labour component affected by the delay. In order to assess the labour amount, the contractor's estimator may be asked to give advice if unavailable in the estimator's handover file issued at the commencement of the project. If the split is found to be 60% labour and 40% materials, this means payment for labour for an 8-week period without disruption would be 60% of the total costs (i.e. £46,500). If the labour component for 4 weeks of variations accrues £25,000.00, the disruption claim is the loss of productivity and would be £21,500.00, which represents the shortfall that would have been earned if the works had not been disrupted. Furthermore, props used for strutting on a prolonged hire period should also be included in the assessment, providing they were on site at the time the first variation was issued and were not off-hired.

With a disruption claim, the contractor must include and demonstrate any methods used to mitigate the event such as off-hiring plant during a redundant period when practical to do so, as well as methods used for obtaining additional labour and any

Table 6.9 Priced bill of quantities

Item	Description	Quantity	Unit	Rate	Total
	In-situ concrete works				
	Formwork				
A	Plain formwork, soffits of horizontal work for concrete not exceeding 300 mm thick, propping over 3 m but not exceeding 4.5 m high	400	m^2	150.00	60,000.00
B	(As above), sides and soffits of isolated beams, regular shape 600 × 300 mm	100	m^2	175.00	17,500.00

concessions. In the formwork calculation example above, concessions might be required because the amount is assessed from billed rates without conceding ineffi-ciencies in construction management or acknowledging the payable rate to the subcon-tractor, which may be higher or lower. When compiling the claim, it must be a true reflection of the disruption and the costs incurred as a result, and presented as a loss and expense claim because the exercise is to recover out-of-pocket expenses only which excludes profit.

Acceleration is the antidote to disruption that aims to put the programme back on track to achieve the end date without the need for a prolongation (or EOT) claim. Disruption costs and acceleration costs are two different matters, and require to be addressed differently. If the contractor is not responsible for acceleration, with the responsibility resting with the client, the acceleration costs represent the extra expense for utilising additional resources which must be submitted as a variation quotation in lieu of a claim for approval from the client/client's agent. The quantity surveyor assists the process by scheduling additional requirements advised by the project manager or obtained from a revised construction programme, which are charged at suitable rates. To be effective, if the client wishes the programme to be accelerated, an instruction must be received before commencing the works; until such time, the prolongation (or EOT) claim is sustained. As with all claims under a contract, it is important to recognise submission periods which are generally before the date of issue of the final payment certificate. This is to ensure claims are recognised and included in the final account or, if inconclusive, negotiated or referred to dispute resolution methods as per the terms and conditions of contract.

6.7.3 Common Law

A contractor can make a claim under common law through the courts as a statutory right, normally when there are no provisions within the contract for remedy, and usu-ally applies when there is a breach of expressed terms. Common law claims are claims for lump-sum payments that, if successful, are awarded as damages, which can be due to negligence, a breach of contract or statutory duty or as tort (a civil wrong). A contrac-tor may choose to make a common law claim instead of a contractual claim based upon the same facts if there are grounds for remedy at common law. However, the outcome of a common law claim through litigation may not be as swift as the outcome of a claim under the contract using alternative dispute resolution (ADR). If the contractor initiates a common law claim instead of ADR, the client and client's agent must be informed that the claim is not being addressed as a contractual matter because the contractor cannot claim the same amount twice. If a court awards damages in favour of the plaintiff (con-tractor), the client's payment is separate to the contract and the contract sum is not adjusted. The inverse would occur if the client is the plaintiff and awarded damages payable by the contractor that, when paid, would also mean the contract sum is not adjusted.

6.7.4 *Ex Gratia*

This term is a request for payment in good faith that has no basis for a claim under the contract or common law. A contractor may submit an *ex gratia* payment request due to financial hardship in relation to a business problem that can be beyond the contract.

It is inaccurate to perceive the request as compensation to avoid issuing a loss and expense claim or as an attempt to settle a dispute prior to involving a third party such as a court, as such suggestions would be arbitrary.

The preparation of an *ex gratia* claim request should not be exhaustive as the intention is to sell nothing and demonstrate the situation in a genuine manner. It should also be in writing, citing grounds for the favour and include back-up documentation where available. A client can refuse to pay such a claim and is entitled to do so, and can stand by the decision usually without triggering any dispute resolution process. However, if wishing to pay something in full or in part, it will usually involve a benefit in return. For example, a contractor may be on the brink of insolvency because of an insolvent debtor and seeks financial relief from other debtors. Here, the other contracting party (i.e. client) may not wish to risk the possibility of an insolvent contractor and, in the spirit of the request and along with other debtors, may issue a payment in return for expected performance which is not enforceable or conditional. The payment is usually final and not considered a loan or repeat process, with the contract sum not usually adjusted.

6.7.5 Set-Off

'Set-off' is a term used in a contract to describe the temporary or permanent withholding of money from a stated amount on an interim or final payment certificate for specific reasons. The term must not be confused with the temporary withholding of cash retention as security for the payer, which is covered by a separate clause. When applying set-off, the process for issuing a 'pay less notice' as required by legislation applies; this is usually included in the contract as a matter of course, and means the payee must be informed 5 days or more of the reduction before the payment is due. The amount to set-off is not usually included in the value on a payment certificate and is deducted by the payer which can be a one-off or ongoing if applying to circumstances mutually agreed by the parties to the contract. For example, a main contract agreement may exist where the cost for the engineer/architect's onsite supervisory overtime is described as being included in the construction contract and part of the lump-sum price for the works with an amount included in the preliminaries. In a situation like this, the project client can make a payment to the engineer/architect and seek reimbursement from the contractor using the set-off clause, which the contractor can include as an amount in a payment application to balance off the sum. This way, the contractor should be no worse off once payment is received, providing that the budgeted amount is not exceeded by the amount included in the set-off.

6.7.6 Claims from the Supply Chain to the Contractor

Any member of the contractor's supply chain can issue a claim to the contractor using the methods in their respective agreement. Moreover, any member can issue an *ex gratia* claim or access the legal system by activating a common law claim against the contractor, which must be instead of a contractual claim if it manifests into a dispute as the same dispute cannot be settled through litigation and a type of ADR. To be effective, a condition of the agreement between the contractor and each supply chain member may compel the supplier to forewarn the contractor of a pending claim by providing a notice, and can apply to an EOT or financial claim (i.e. similar to the contractor's obligations under the main contract). Any claim is usually addressed to the project manager in the first instance and, upon receipt, action rests with the contractor for review that when concluded must consider where the burden of responsibility for action rests (i.e. under the main contract or solely with the contractor).

6.7.6.1 Main Contract Claims

If the contractor has prepared or is preparing a claim under the main contract they may include claims from one or more supply chain members and, where applicable, will usually involve subcontracted works. Here, the contractor must take care when handling subcontractor (and other supply chain member) claims that form part of a claim under the main contract as the rights, risks and obligations under the main contract may differ from each agreement the contractor has with their supply chain. This can be as obvious as a subcontractor's claim for a delay due to adverse weather, that may trigger clauses of the main contract or a subcontractor's claim for a delay in the sequence of works because of the late issue of variations considered the main contractor's responsibility. To mitigate the contractor's exposure, the contractor can create a back-to-back form of subcontract agreement mirroring the main contract (as far as possible, and making commercial sense) that when executed by the parties may insulate the contractor from the full burden of responsibilities under a contract without the arrangement. However, if the contractor is in dispute with the client because of a claim that involves one or more subcontractor's claims, it does not mean to say any subcontractor is bound to the pending outcome; the quantity surveyor will need to look at individual agreements for clarification. Here, the general exception is if the client is insolvent and the contractor's claim is against the client's estate that includes supply chain members' claims, as it may mean the contractor and supply chain members may be unsuccessful.

If the contractor decides to commence legal proceedings with a project client which is to include one (or more) supply chain member(s) as co-plaintiff(s), the contractor must notify the appropriate supply chain member(s) and seek participation. Prior to this, the contractor is in no position to state to any supply chain member that payment will only be made upon an award in damages as such a statement is gallant. The same could be said if the contractor is in dispute and seeks remedy from an ADR method permitted under the main contract, and asks supply chain members to act as co-claimants. Here, such action may constitute breach of their agreement, leaving any aggrieved supply chain member to take action against the contractor using an ADR method possibly permitted under the agreement or under a common law claim.

6.7.6.2 Contractor's Liability

A circumstantial claim from a supply chain member (usually subcontractor) may relate to an event or sequence of events that disrupts their works under contract, and may be due to either the contractor's style of site management or default of another subcontractor(s). If a claim is the result of another subcontractor(s) default, the claim must be presented to the contractor and not the other subcontractor(s), as there is rarely a tripartite contractual link comprising of two subcontractors and the main contractor. However, freedom of contract legally permits parties to contract as they wish on the proviso it is legal, and if two subcontractors form a legal agreement pertaining to a common project, it is generally of no interest to the contractor (e.g. a low voltage subcontractor to the contractor appointed to install power and lighting that is also a sub-sub contractor to the mechanical subcontractor to hard wire equipment). If a sub-subcontractor has a claim against a subcontractor, it is usually of no interest to the contractor and any such claim presented to the contractor should be rejected. These types of claims would be hearsay or may in fact be misdirected common law claims; even if redirected to the contractor, rarely will a quantity surveyor be involved with a common law claim as the contractor will usually refer the matter to their lawyers. Once the quantity surveyor is satisfied receipt of a claim is the contractor's responsibility to review in detail, the task involved is to understand if the contractor has liabilities and, once deciphered, take the appropriate action.

Other supply chain members to a contract that may submit claims can include design consultants engaged on a design-and-build project if of an opinion the scope of services provided has gone beyond those included in the fee proposal and the contractor's subsequent acceptance. This may be presented in the form of a pleading, and can include:

- additional visits to site during the construction phase at the request of the contractor;
- abnormal number of drawing and specification revisions, possibly due to variations in the works;
- excessive reviews of as-built information, again due to variations in the works;
- new and/or supplementary reports prepared and issued at the contractor's request; or
- peer review of materials stored on or off site for including in interim applications for payment, and can mean physical inspection of the goods or review of technical and consignment documents (this would normally apply for specialist products in unique locations, e.g. mechanical primary equipment imported from overseas and stored in a port).

The contractor's quantity surveyor involvement with an EOT claim from a subcontractor is usually limited, as it would normally be the site and project manager that would assess the impact on the master programme. Upon receipt of a financial claim however, the quantity surveyor must review the evidence submitted which should include site instructions, photographs, etc. to understand the facts and the amount claimed, and would apply to an EOT claim with costs and any other type of financial claim. After reviewing, it may be possible to recognise a sum of say £5,000 out of £10,000 claimed as the contractor's liability because of credible evidence provided. However, as the balance is in doubt, the quantity surveyor is not in a position to acknowledge if this is the contractor's liability and must under no circumstances process a payment of any amount until authorised by the project manager.

The process towards reaching settlement of a financial claim can commence with a meeting involving the project and site managers and claim issuer; the quantity surveyor should be proactive in promoting this culture, instead of leaving matters to chance. A meeting may expose the claimant's reliance on verbal discussions or 'understandings', and it may be wise for the parties to attempt to settle on an amount in the spirit of the meeting while the claim is fresh in people's minds. It would certainly be in the interests of the contractor and claim issuer to address the situation at the earliest opportunity and seek a win-win scenario instead of procrastinating the matter. The worst case is one where there is no agreement with the meeting adjourned when, months after the project is completed, the claim resurfaces meaning the reopening of a situation that will take time and effort to conclude to avoid a dispute.

Ex gratia requests tend to be more commonly requested from the supply chain to the contractor rather than from the contractor to the project client. This is because of the number of supply chain members engaged by the contractor on a project where the risk is disproportionate, even if written into an agreement, when the party accepting the risk realises in hindsight it is too heavy to shoulder. If an *ex gratia* request is received, the contractor should consider what effect rejection of the claim could have regarding responsibilities under the main contract and what benefit any payment could have if wishing to gain something in return. A subcontractor's *ex gratia* claim may be submitted when their profits are diminished or the business is operating at a loss, possibly because of an oversight of the terms and conditions of the invitation to tender that were subsequently included in an agreement (meaning what started out as a lucrative scheme has failed to transpire). Here, a contractor would have every right to reject such a claim on contractual grounds. However, and before rejecting, the contractor should consider the situation, as it may have an impact on the status of the project at the time of receiving the claim. For example, what should a contractor do if a block-laying subcontractor under a supply and construct contract with works 90% complete issues an *ex gratia* claim request for £25,000 as a contribution to a £50,000 increase in the cost of materials that cannot be recovered under the subcontract agreement, as it is a fixed price? If the subcontractor provides proof of the increase with letters from suppliers, the contractor may make a decision based upon the subcontractor's performance to date. In other words, when considering payment, a decision may be influenced by the subcontractor's work completed because it can be difficult to confirm if the subcontractor has contributed to a delay resulting in a prolonged block-laying programme. If deciding to pay anything, the contractor may request the subcontractor to complete the remainder of the subcontract works, exclude further additional work payment requests unless expressly instructed by the contractor, and include the *ex gratia* claim in a final account which the contractor will pay. This promotes collaboration and may be an effective negotiating tool to avoid a dispute regarding delay.

6.7.7 Claims from the Contractor to the Supply Chain

A contractor can issue a claim to any member of its supply chain (more so with subcontractors, who usually carry out most of the works) and might wish to deduct the amount from an interim payment as full and final settlement. Alternatively, a claim may be issued separately and left in abeyance for agreeing as part of a final account, which may be appropriate if a counterclaim against a claim from the supply chain member. If the contractor wishes to reduce a payment in order to settle a claim, the reduction is usually permanent and a contractor may refer to this as a contra charge or recharge which, to be effective, should be agreed with the payee beforehand. This must be administered with the issue of a 'pay less notice' where works are carried out as 'construction operations' as recognised by the Construction Act. Alternatively, the contractor may issue an invoice for the amount; however, this does not guarantee the contractor will be paid.

The permanent withholding of money is not to be confused with the temporary withholding of cash retention in accordance with a subcontract agreement as security for the contractor. Neither should permanent withholding be confused with reducing an amount on a progress claim (or invoice) when the quantity surveyor amends a sum and issues a 'pay less notice' (with reasons), advising it is a temporary reduction (e.g. trade works under contract being 50% complete and not 100% as invoiced or claimed). This is because the contractor has a liability under the agreement to pay an amount upon completion of the works, unless reimbursement is on a schedule of rates when the price is not a fixed sum.

A claim made by the contractor to any supply chain member should reciprocate the merits expected from the supply chain if issuing a claim to the contractor (i.e. substantiated with proof such as photographs and site instructions). In order for the quantity surveyor to identify the party the claim is to be made against, there must be prior communication in writing from site management to the appropriate supply chain member. Where contra charges apply, works executed by others must be authorised by the contractor's personnel using either a site instruction or daywork sheet(s) recording the cost of labour, plant and materials. To be effective, the instruction/daywork sheet(s) must state the works are a contra charge and specifically name the supply chain member(s) to incur the charges. For example, dayworks may be authorised to the painting and floor finishes subcontractors to make good walls and floors due to damage caused from a burst water supply pipe that occurred during the final commissioning stage. Here, daywork sheets may state that the cost of the painting and floor finish repairs the subcontractors will charge the contractor are to be a 'Contra charge to 123 Plumbing Contractors Ltd'. This advice is of no interest to the painting and floor finishes subcontractors, and is for the quantity surveyor's information when in receipt of the subcontractor's payment requests. The exception here is if the cost of the repairs is substantial enough to be recovered under the main contract works insurance policy.

The quantity surveyor must ensure a contra charge is genuine and not a result of the contractor's inefficiency or management style, which must be ascertained prior to advising the appropriate supply chain member of the intention to deduct an amount from a future payment. When communicating such intent, the quantity surveyor should reserve the right to apply a proportional administrative charge for preparing the claim and request a credit note or reduction on a subsequent progress claim to acknowledge the contra charge. If any supply chain member in receipt of such communication contests the validity of the contra charge, they must be referred to the person authorising the instruction/dayworks and the quantity surveyor must not agree to mediate. Of course, the supply chain member will have access to dispute resolution options if wishing to commence the path which, to be instrumental, will apply once the amount is deducted from their payment when it is a permanent deduction. However, where a contra charge is a minor amount and objected to by the supply chain member, third party intervention is not usually warranted; in this case it is best resolved with the contractor through negotiation, or the contractor must stand by their decision. Alternatively, the recipient of a contra charge may involve their insurance company if wishing to make a claim against their policy. Here, the contractor should permit access to the works by any insurance representative if wishing to investigate the circumstances.

6.8   Voluntary and Involuntary Contract Terminations

During the construction phase, the contractor's quantity surveyor will administer the main contract as well as supply chain agreements and, in the process, may have to deal with contract terminations that are voluntary or involuntary. A termination is voluntary when one party ends an agreement or is mutual to the parties. Involuntary termination means the reluctance of the parties to terminate the agreement that must happen because one (or both) parties cannot perform under the contract despite best intentions. This can occur with insolvency, which is the inability of a business to pay its debts when they become due. Here, the business is wound up under a process of liquidation by an insolvency practitioner acting as receiver, rendering the entity obsolete and naturally unable to perform under contract with the contract terminated as a matter of course. To avoid insolvency, a business can be placed in administration for a defined period (usually up to 12 months) with the administrator having a dominant role in the business and accounting activities of the company. This can involve chasing debts and making suggestions to stakeholders for restructuring the business, allowing it to rekindle itself to be in a better position to trade. During administration, the struggling business may seek to acquire assets or source a loan from an investor or lending institution to obtain a debenture, which is a document evidencing a secured debt. However, seeking such equity is not a duty of the administrator, and any investor or lender must be confident the business is in official administration, as the process provides a neutral overview of the potential borrower's business arrangements and how it would operate once in receipt of sufficient equity and/or assets. From a contractor's perspective, a struggling project client or supply chain member contracted to the main contractor may elect to be open in discussions regarding its situation and inform the contractor of pending administration. In such a situation, and once appointed, the administrator must abide by the contract and may suggest assignment be initiated if allowed under the agreement. Where permitted, this may be to a parent or side company that must inherit the same obligations, rights and responsibilities, including price and deliverables stated in the agreement, to permit the works under contract to continue.

A contract can also be terminated involuntary by frustration where an intervening event occurs making the remaining obligations under the contract so radically different that the performance each party intended to carry out is not considered possible. Before a party considers frustration has occurred and the contract is to be terminated, the facts require examining. This is to confirm if in fact frustration exists because the hindrance of frustration may be contrary to the parties' belief. For example, a natural flood causing damage might delay a project; however, the contract may not be frustrated if it can still be completed after the repairs if it is a desire of the parties. Each event causing harm to a project therefore requires to be assessed on a case-by-case basis, meaning the parties should resort to legal advice if a contract is considered frustrated.

6.8.1 Main Contract Termination

The reasons giving rise to voluntary or involuntary terminations may be stated in the contract and, where applicable, must be cited in any termination notice. Failure by the terminating party to cite such reasons or state other reasons when performance under the contract is still possible will constitute a breach of contract, meaning a degree of care must be taken before issuing an official termination.

6.8.1.1 Voluntary Termination

Probably the most humiliating situation a contractor can endure is the project client's voluntary termination of the contract. The client's agent alone cannot terminate the contract, although this entity can issue a termination on behalf of the client citing reasons that may also refer to any prior written instructions or communications used for arriving at the decision. Once in receipt of this type of termination, the contractor's quantity surveyor must follow the procedures of the contract that cannot be departed from as there may be rights available for possible breach. Once the notice is received, the procedure usually involves the contractor issuing a statement of what it considers is due, and also applies if the contractor voluntarily terminates the contract. If the contractor is of an opinion the termination is a breach of contract, the statement should include a loss and expense claim identifying the risks and opportunities the contractor would have endured if fulfilling its obligations, which may be referred to in the event of dispute resolution. In the meantime, the client may or may not choose to continue with the works and could counterclaim a request for payment and refuse to release any cash retention fund until settlement of the account.

6.8.1.2 Involuntary Termination

If termination of the contract is involuntary because of a client's insolvency, the contractor must act practically and officially terminate the contract. Where involving a company, such official termination should only be issued following receipt of confirmation from the insolvency practitioner (or receiver) that the company has been wound up under the Insolvency Act 1986. The same applies if the defunct entity is an individual, except that the term used is bankruptcy. If the contractor is in a fortunate enough position to negotiate a new contract with a new party to complete the works, it may be wise to sustain the site establishment subject to the works recommencing swiftly. If negotiations are not possible, the contractor must mitigate the situation for their own benefit by possibly demobilising and removing its presence from site. The contract may state the contractor's duties in this regard and will usually involve the removal of unfixed materials from site providing they are not paid for by the insolvent client; if this is the case, they become the property of the receiver. Although not usually a contractual obligation, the contractor should inform subcontractors to do the same, with the exception of any unfixed materials paid for by the contractor.

After the termination, and to within any stated timeframe written into the contract, the quantity surveyor must issue a statement to the client's trustee or receiver stating the value of completed works. The statement should also include a loss and expense claim as a consequence of the termination in order to create a final account. The final account will demonstrate the contractor's version of the liability to the contractor less an amount already paid to produce a balance due to the contractor or, in a rare event, amount due to the receiver. The final account should also request release of any cash retention fund which belongs to the contractor, which is usually kept in a separate fund and theoretically immune from the client's insolvency. However, the contract may not distinguish voluntary termination from insolvency when considering cash retention with the funds possibly not readily available that require some effort to obtain through the receiver (or trustee).

If the final account shows the contractor is owed money, the balance due is claimed from the estate. Thereafter, the contractor may be invited to attend meetings chaired by the receiver who will provide information on the distribution of any available funds derived from sell-offs once assets are melted down and prioritised payments are made. The driving force regarding the hierarchy of prioritised payments is the Insolvency Act and, assuming a floating or fixed charge (a security interest over a fund of changing assets which floats until it is converted into a fixed charge) applies, comprises:

- secured creditors with fixed charges (i.e. banks);
- insolvency practitioner's fees;
- preferential creditors (e.g. Inland Revenue, VAT payments, etc.);
- secured creditors without fixed charges; and
- unsecured creditors.

The contractor would probably be an unsecured creditor, and because of the hierarchy arrangement would only receive payment after prioritised payments are made that unfortunately may take some time.

As the main contract has ceased and is officially terminated, subcontractors generally have no recall on the contractor for loss and expense claims, and a wise contractor will include this as a condition of each agreement irrespective of whether termination of the main contract is voluntary or involuntary. However, the contractor is usually responsible for payments to subcontractors for their completed works and expenses in connection with completed works. In isolation, nominated subcontractors may have a claim on the estate if there is an insolvency clause written into their contract, which may mitigate a degree of risk to the contractor. Needless to say, if the contractor becomes insolvent it is the client and/or client's agent that addresses the situation.

6.8.2 Supply Chain Terminations

6.8.2.1 Voluntary Termination

Rarely will a supply chain member seek termination of the agreement it has with the contractor, and if voluntary termination is to take place, is usually the contractor's decision and not the result of any instruction from the client's agent. However, a form of contract may include provisions permitting the client's agent to issue an instruction requesting the exclusion of a person(s) from the site, and may apply if there is disregard to health and safety or gross negligence by an individual or group of individuals regarding the works. The issue of this type of instruction and subsequent compliance by the contractor should not be seen as a full reflection on a supply chain member's performance engaging the person(s) to be removed, with the removal not necessarily grounds for the contractor terminating the agreement.

A contractor must perform under the contract it has entered into and will rely on its supply chain to do the same, as grounds for a client's voluntary termination of a contract usually include the contractor's failure to perform regularly or diligently with the progression of the works and/or design if the contractor has design responsibilities. If a supply chain member is considered not to be performing it may be due to pending administration or, if a subcontractor, a reluctance to admit it has an overstretched workforce and is unable to service the contractor. A flaw of some contractors is that supply chain members, and particularly subcontractors, may be expected to perform 'at the drop of a hat' because a programme or arrangement has not been communicated properly, with the contractor and project in a temporary state of disarray. If there is genuine concern regarding a supply chain members' performance, with the project team of the opinion the default is impeding the works, the contractor must exercise its rights under the agreement and issue a letter expressing the contractor's concern with the performance. The letter should confirm any verbal requests already made or attempts made by the contractor to communicate the concern and include reference to the supply chains' obligations to perform under the contract, citing the appropriate clauses. The letter should state that a written response is required within a stated time frame, and that such a response must state the intentions to rectify the shortcomings. A proactive response is one where the supply chain member acknowledges the situation and commits to an immediate improvement, even if the contractor has some responsibility (e.g. a failure by the contractor to provide storage space for a subcontractor's materials in a timely manner and what the contractor is to do). However, should the supply chain member not respond or issues a defensive response leaving the contractor none the wiser, the project team must consider the contractor's position before deciding to terminate the agreement with the quantity surveyor consulted regarding the impact of any termination. A method of recognising the consequences of termination could be considered by answering the following questions.

- Is the defensive response justified, and does it highlight a shortfall in the contractor's performance?
- Has the supply chain member been paid in accordance with the agreement if it is claiming the slowdown is because of delayed payments?
- If involving a subcontractor, are areas ready for the subcontracted works to commence?
- If the contractor is at fault, is it possible it could lose in any dispute resolution process?
- If termination is the only option, what is the contractual process?
- What time will be lost by terminating the agreement and seeking a new supply chain member?
- What costs are involved with a replacement?
- If seeking to terminate a named, specialist or nominated agreement, has the client been informed?

Quantity surveyors generally dislike dealing with this type of termination because the contractor's decision may be hasty without exploring the risks involved. Moreover, if answers to the above questions are inconclusive, termination must be treated as the last option.

In the event of termination by the contractor, the quantity surveyor must ensure it is contractual and served in writing. Thereafter, a final account should be issued stating what (if anything) is due to the supply chain member from the contractor or vice versa. If the supply chain member disputes the termination notice and is a subcontractor, it must accept the contractor's decision and remove itself from site. In the aftermath of this type of termination, the contractor might be issued with a loss and expense claim which must be reviewed and either formally accepted or rejected. If rejected, and the subcontractor objects, it is up to the subcontractor to commence any dispute resolution process.

Depending on the circumstances of termination, it may be necessary for the quantity surveyor to prepare a Scott schedule. This is a schedule listing items of defective workmanship and fixed or unfixed materials appearing defective that are paid for by the contractor. The estimated cost of repairing the works or replacing the goods is listed against each scheduled item together with comments which are signed off by both parties that can be later independently assessed by a third party such as a tribunal if necessary. This approach would apply when considered practical to do so, and does not need to be considered final. This is because oversights or too zealous an approach regarding the severity of a defect observed when creating the schedule is possible, and could have a different impact closer to the date of sectional or practical completion.

6.8.2.2 Involuntary Termination

Termination may be considered involuntary when a supply chain member succumbs to insolvency. A contractor cannot voluntary terminate a contract because the other party is in administration thinking it is about to become insolvent, and an astute contractor will observe signs when a supply chain member may be facing administration or insolvency. This is with particular reference to subcontractors that may have substantially performed under the contract whose mode of conduct may be different from usual. Indicators to be aware of include:

- the contractor receiving complaints from subcontractor's material suppliers that are not paid;
- the failure of labour and/or materials to arrive on site at agreed times;
- the subcontractor reducing labour supply not instructed by the contractor;
- contractor's receipt of *ex gratia* payment request(s); or
- contractor's receipt of a request for revised payment terms which are to shorter periods than stated in the agreement.

Upon receiving notice of a supply chain members' insolvency from an insolvency practitioner, and where applicable, the contractor must reappoint an alternative as soon as possible to maintain the programme and mitigate the possibility of a delay in the progress on site. This will usually be restricted to labour and material subcontractors and requires the quantity surveyor to do the following.

- Ensure no further payments are issued in the name of the insolvent business.
- Attend to any requests issued from the receiver in writing that may involve a request for the payment of outstanding debts, which would normally be denied and stated in writing with reasons. This is usually because the contractor may need to place an award(s) elsewhere with the expense to incur unknown until a new award(s) is placed, possibly resulting in a difference that would need to be recovered from the insolvent business's estate.
- Ensure unfixed materials brought to site by the insolvent subcontractor and paid for by the contractor are stored in a location known to the contractor.
- Ensure surplus or unwanted materials not paid for by the contractor and plant owned or hired by the insolvent subcontractor are put aside for claiming by the receiver. If these are to be collected, the receiver must state in writing to the contractor the capacity of the collector and the contents for collection. At the time of insolvency, the subcontractor may have been hiring plant and the hire company may wish to recover their assets, which the contractor cannot permit without authority from the receiver.

In the aftermath of a supply chain members' insolvency the contractor may decide to appoint a lawyer to deal with the matter, and the quantity surveyor will assist the situation by preparing a notional final account. This is required to assess the contractor's liability to the receiver or vice versa and the implications the insolvency has on the contractor's performance under the main contract. The notional final account recognises the cost incurred by the contractor at the time of the insolvency, plus a cost to complete using whatever means needed. From this amount the sum committed in an award(s) is deducted to produce a liability, with the difference between the value of the award to the insolvent business and the liability a balance due to or owed by the contractor. To demonstrate, let us say a formwork subcontractor has an award valued at £500,000 for supports to in-situ concrete floor slabs and beams and becomes insolvent. At the time of insolvency the formwork is substantially complete with interim payments made, but one invoice for the works in progress is not paid and is being sought by the receiver. Table 6.10 demonstrates this possible scenario for issuing as a notional final account to the receiver or contractor's lawyer.

The notional final account must put the contractor back to a position it would be in if it was not for the other party's insolvency. In the example used in Table 6.10, the liability of Builder Co. Ltd is based upon payments already made at the time of insolvency plus the estimated expense that will be incurred for completing the works using rates from

Table 6.10 Notional final account

NOTIONAL FINAL ACCOUNT SUMMARY	
PROJECT Nr 5678: **NEW MULTISTOREY CAR PARK: SALISBURY**	
CONTRACTOR: BUILDER CO. LTD	
SUBCONTRACTOR: FORMWORK CO. LTD	
Agreed contract works value	£500,000.00
Agreed additional works	£ 25,000.00
Builder Co. Ltd commitment to Formwork Co Ltd upon completion of the works	£525,000.00
Builder Co. Ltd liability to carry out the works	
Nett payments to Formwork Co Ltd as works in progress	£360,300.00
Plus the cost for works to be completed by others:	
Plain formwork, soffits of horizontal work for concrete not exceeding 300 mm thick, propping over 3 m but not exceeding 4.5 m high	
660 m² @ £150.00	£99,000.00
Plain formwork, sides and soffits of isolated beams, regular shape 600 × 300 mm	
180 m² @ £175.00	£31,500.00
Making good and preparing surfaces incomplete by	
Formwork Co. Ltd 150 hrs @ £75.00 (Gang rate)	£11,250.00
A. Total cost of the works	**£502,050.00**
B. Builder Co. Ltd commitment to Formwork Co Ltd	**£525,000.00**
C. Builder Co Ltd Gain/~~Loss~~ (A−B)	**£ 22,950.00**
D. Cash retention refund to Formwork Co Ltd	**£ 18,963.00**
E. Amount ~~due~~ to/owed by Builder Co. Ltd upon works completion	**£ 3,987.00**
All figures exclude VAT	

the bill of quantities. This example demonstrates the benefits of withholding cash retention from payments as it mitigates the contractor's exposure. Where no cash retention is withheld and a bank guarantee is supplied instead, the contractor must contact the issuer and present the document.

When issuing a notional final account, and if demonstrating an amount as due to the estate of the insolvent business, the contractor must make its position clear that any payment due to the receiver will only be issued upon completion of the works. Furthermore, the contractor must reserve the right to amend the account by providing periodic notices, including an estimate of probable expense to be incurred until the cost to complete is affirmed. This is because the cost to complete is risk to the contractor and based upon an estimate that can only be confirmed once commitments are made with others to complete the works. Once the works are complete or the remaining works committed in new contracts, the notional final account is replaced with a final account, with any amount due to the insolvent business paid to the receiver. If the final account demonstrates an amount due to the contractor which cannot be recovered through the cash retention fund or bank guarantee, the contractor will need to register itself as a creditor. In this situation, the contractor will probably be an unsecured creditor and may only receive an amount from any remaining finances after payments have been made to secured creditors, the liquidator and preferential creditors.

Extreme and unfortunate situations can occur when the insolvency of a supply chain member prevents a contractor from achieving the date of sectional or practical completion, resulting in the contractor incurring liquidated damages. With this scenario, the contractor's final account will need to include a loss and expense claim to recover the damages which must prove the insolvency as the root and only cause of the delay without any mitigating factors being available.

The quantity surveyor will need to educate the project team that insolvency of a supply chain member does not mean the contractor is certain to be reimbursed by the receiver, with the likelihood of being paid anything remote. Any payment that may be forthcoming will take time and effort to acquire and, if receiving anything, it will probably not be what the contractor considers they are entitled to. For this reason, the project team should not become complacent where supply chain insolvency occurs and works are incomplete, and must endeavour to mitigate the situation the best they can.

When a contract names or nominates a subcontractor(s) that succumbs to insolvency, the project client will be involved and should instruct the contractor what to do. This will involve either renaming or nominating a new subcontractor(s) or instructing the contractor to take over the trade works or, at worst, possibly omit the works to be completed. In any case, the contractor may be entitled to an extension of time and reimbursement for loss and expense if the client delays the renaming or nominating process. When preparing a notional final account involving renaming/nominating new subcontractors, it may be a complex issue. This is because if the value of any provisional or prime-cost sums included in the contract price is in relation to a named/nominated subcontractor's works that is substituted with a variation as a committed cost which cannot be honoured because of the insolvency, a further variation would be required. However, if the contractor agrees to carry out the works and receives the appropriate instruction, it must mitigate the expense as it no longer becomes the client's responsibility. Naturally, if the contractor succumbs to insolvency, it is the client and supply chain members who must act and deal with the receiver.

6.9 Project Reporting

The senior management and directors of businesses operating in the construction industry may seldom see the activity of their ongoing projects. A reason for this may be the distance of the head office from the projects, as well as dedicating time for running of the business which includes securing new work and dealing with recruitment. In the interest of a main contractor's business, shareholders and project stakeholders will be interested in the progress of every project on their order books and the impact it is having on the business. Senior management and/or company directors may also have been involved with the initial tender and subsequent negotiations for securing the project and wish to see if their business intent for obtaining the work is proving worthwhile, which can be provided with project reporting. Normally, executives of large contracting businesses are not involved with the day-to-day running of projects and delegate the responsibility for deliverables and reporting to the project team, with the project team periodically presenting a report on the status of the project for advice on the direction in which it is going.

Contractors vary in the contents to be included in a progress report, with some interested in the bottom line restricted to finance, health and safety issues and completion dates with others adopting a thorough approach including a broad range of topics (see below for more details). Contractors requiring a large amount of information may have reporting templates as part of project administration, so that the reporting system for all projects follows an identical format. From the executive's perspective, the use of a standard reporting format makes sense as it creates an orthodox system and avoids *ad hoc* reporting.

A progress report must provide suitable information and should comprise a cover page stating the project title, date of issue, description of the project and commentary on the status of works at the time of reporting. It should also include the project team member's names and position titles, together with a list of topics and contents. A suggested list of topics and contents to include is as follows.

1) *Contractual*:
   - contract sum and the amount claimed for payment at the time of reporting;
   - value of variation quotations/approvals and amount claimed for payment at the time of reporting;
   - start and finish dates of the project;
   - current construction programme indicating build status by group element (i.e. substructure, superstructure frame, services, etc.) with anticipated completion dates for each;
   - any EOT claims and the likelihood of achieving the agreed completion date;
   - statement of claims and disputes under the main contract and with the contractor's supply chain;
   - status of the supply of 'for construction' design and documentation;
   - the contractor's industrial relations with the client's team and/or subcontractors;
   - status of project insurance and contract securities; and
   - status of the supply of warranties.
2) *Health and safety*:
   - number and type of accidents and record of near-misses;

- contractor's relations with the CDM principal designer;
- comments on amendments to the construction phase health and safety file; and
- status of the supply of subcontractors' health and safety files, with comments on any shortfalls and actions in place to rectify.

3) *Commercial and finance*:
- status of awards issued to the supply chain and those yet to be awarded;
- cash flow report;
- summary of additional work payments to the supply chain; and
- cost value reconciliations through the cost management system, with comments on overspends and gains and the anticipated final account.

4) *General*:
- neighbour relations;
- environmental issues;
- risks and opportunities for the contractor, which may involve finance and time;
- total number of hours worked by operatives recorded from site attendance registers in the health and safety records or site diary; and
- photographs of works in progress.

The quantity surveyor is normally responsible for the contractual, commercial and financial contents of each report as well as risk and opportunities in the general section. Time management is therefore important to ensure information is collated and correct for inclusion, which must be coordinated with other team members. This is because failure to issue reports on time is not an option, as it serves no purpose to the project team's pride on the job if dates are missed. Each report should be informative, complete without missing parts, concise, clear and accurate to give advice on the true status of the project, and must be developed from available data. Comments on each report from a gut instinct which are either too optimistic or pessimistic should be avoided. Furthermore, and where possible, effects and outcomes that are unknown should be identified, clarified and quantified for including in the risk and opportunities section, as this helps to identify potential scenarios for discussion and action. For example, on a design-and-construct contract, a contractor may wish to change a building's basement perimeter walling from reinforced in-situ concrete to precast concrete in order to save time. Although this may cost more to build, it could produce a benefit of saving time that may be important if a programme is lagging.

An intention of a progress report is to provide executives with a snapshot of the status of the scheme at any time, as well as to gain an understanding of current liabilities and responsibilities and what may lie ahead. In general, each periodic report is initially issued electronically to enable stakeholders to understand the contents before requesting project team members to attend a meeting to discuss the details. Such a meeting will be an experience for those involved and may initiate some interesting discussions and debates with participants encouraged to contribute and be proactive and not reactive to any points of concern.

In addition to the contractor's internal report, the contractor may be required to issue regular progress reports to the client and client's agent, the format of which may be included in the invitation to tender documents. In the aftermath of issuing such a report, meetings are usually held at the client or client's agent's offices to a series of scheduled dates (usually monthly) with a minimum of one person representing the contractor (usually the project manager). However, the project manager may request the site manager and/or quantity surveyor to also attend on an *ad hoc* basis depending on the contents of the report. Naturally, the contents issued to the client/client's agent will be the topic of discussion, with the contents of the contractor's concurrent internal report beyond the meeting (although some information such as the contract sum, date for practical completion, etc. will be identical).

7

Project Completion

7.1 Sectional and Practical Completion

The period leading up to the agreed date of sectional or practical completion of a construction project is a busy time for the project team who will be finalising works in anticipation of receiving a written statement of completion. As part of quality control, the finalisation procedure involves the contractor's supervisors preparing a pre-handover checklist where 'snagging' takes place. Snagging is a generic term used by builders to describe a method of spot-checking items of works considered complete or near-complete to identify missing items, flaws, errors and unacceptable finishes, and applies to newly constructed and refurbished works. It involves the physical examination of private areas, such as rooms and offices, and common areas such as landings and stairs from which a checklist of works items is created, describing them as complete or incomplete. Where works are incomplete the checklist includes comments noting the actions required to make them complete, which can be as simple as repainting a wall because the background is visible to more extensive works that may involve replacing warped doors. The contractor's supervisors will normally advise supply chain members (mostly subcontractors) that must endeavour to remedy incomplete works in accordance with the contractor's request. The contractor is responsible for snagging works under the contract and only the contractor can enforce the terms and conditions of agreements it has with subcontractors and suppliers. However, it may be a contractual duty for the contractor to also snag items installed by the client or client-engaged contractors in order for the completion statement to be issued.

Snagging cannot raise new items; if it does, it would be a request for additional works and usually applies to subcontractors and client-engaged contractors that might seek reimbursement for completing the works. This is something the contractor's quantity surveyor must understand, as it could involve additional expense at a late stage of the project that may not be envisaged. Once a supply chain member advises the contractor it has completed its works, the supervisor(s) will normally confirm the works as acceptable or reject the advice; if rejecting, it will require a revisit by the supply chain member with the cycle continuing until the works are acceptable.

The satisfactory rectification of snagged items is in anticipation of a handover meeting between the contractor and client's agent (and possibly the client) prior to the agreed date of sectional or practical completion. At the handover meeting, the client may visually accept the condition of the works which may be to the elation of the contractor.

Construction Quantity Surveying: A Practical Guide for the Contractor's QS, Second Edition. Donald Towey.
© 2018 John Wiley & Sons Ltd. Published 2018 by John Wiley & Sons Ltd.

However, this does not constitute completion in the real sense, as only the client's agent or other title named in the contract (e.g. architect/contract administrator) has the fiduciary powers to issue the certificate. This means that a notice from a clerk of works or even written advice from the client will have no standing if they are not named in the contract as the certifier. When works are in sections and statements of sectional completion are issued for partial possession, it will be necessary for the contractor to obtain a practical completion statement once the final section is certified to formally complete the project.

7.1.1 Final Certification

In order for the contractor to receive a completion certificate, the contract usually requires the contractor to provide evidence that the physically complete works complies with the contract, which can be achieved through certification and satisfactory test results. There may also be a requirement for the design to be evidenced as being compliant with the Building and CDM 2015 Regulations which would be supplied via design consultants. This is usually supplied as a matter of course prior to issuing the first 'for construction' consignment that may be affirmed for procedural purposes following supply of the last 'for construction' consignment. The process involves designers and installers separately providing self-certification of the designed and installed components confirming compliance. If the main contract works are procured under a construct-only contract then, subject to the contract, the project client may seek certification of the design portion directly from the consultants. However, with a design-and-build contract, this is the contractor's responsibility. For example, a completed sports stadium procured under a design-and-build contract may look physically excellent, but how will the certifying officer be ensured that the piled foundations are suitably constructed or that the installed spectator seats pose no risk to health and safety? In this situation, the piling subcontractor would submit test results as required by the contract to the contractor, that in turn issues them to the design consultant for approval. If the design consultant is satisfied with the results, normally a certificate is issued to the contractor. With the spectator seating arrangements, the subcontractor that installed the seats may provide self-certification which must be in accordance with the approved design and specification. In isolation, where the contractor carries out some permanent works themselves, self-certification is also required in addition to any test results required by the contract.

Any certificate issued to the contractor or prepared by the contractor must be titled 'Certificate' and state 'I/We Certify…' It must then state the scope of works and what the certificate stands for, that is, workmanship and materials carried out to British Standards, citing the relevant number and part of the standard (if applicable) and year the standard came into effect. The certificate must include the project title and name and registered business address of the company issuing the certificate. The certificate must be dated and signed by a qualified person who is competent to certify the completed works or services as being compliant, and not a notary public alone who may not possess the required qualifications to make the certificate valid. Once the contractor is in receipt of the certificates and/or has prepared certificates for the design and/or works they have carried out, the originals should be sent to the certifying officer with copies kept on file for the contractor's records.

The contractor's quantity surveyor plays an important role in the certification process and must ensure the certificates and test results of subcontractors (and client-engaged contractors where applicable) are issued to the contractor in a timely manner, which also applies to temporary works. However, as this information is applicable for a brief duration, the final certification requirements are usually restricted to permanent works. In order to meet a date of practical completion and obtain appropriate works certification in a timely manner, it is important to liaise with the certifying officer for the requirements that can commence three months or more prior to the completion date, with this timeframe considered flexible to suit the type of project. Here, the certifying officer may assist the process by supplying a list of information required from the contractor in order for the completion certificate to be prepared.

In addition to contractor-driven certification the client may seek independent certification of some works, and in the process may nominate a certifier(s) to carry out inspections and issue the certification. This can be a minefield for the contractor to deal with, especially if the process is to be carried out prior to the agreed completion date which, to be instrumental, is only effective if a condition of contract meaning the client cannot impose the requirement on the contractor as an afterthought. However, and where applying, the contractual requirement to supply independent certification may not prevent the issue of a completion certificate. Here, a suitable clause might state that any discovered flaws in the works highlighted by the independent certifier before or after the completion date will become a defect and included as a rectification item, to be rectified before the end of the rectification period. This clause would be effective if for example an item of mechanical plant installed on a flat roof is found to be in a different location to that shown on the plans where in fact it could be left in the same position at the end of the rectification period if there is a genuine reason for the change (e.g. a clash of ducting with the roof members with the relocation not detrimental). However, if part of any mechanical plant is missing, it may be accepted as a defect or, at worst, could result in the issue of a non-completion certificate.

To comply with CDM 2015, the contractor must acknowledge its responsibilities under law and provide the client/end-user with a health and safety file which includes:

- design and specification criteria;
- information needed for the health and safety of operatives that will maintain or clean areas inside and outside of the building; and
- details on products, substances and manufactured components used to construct the building.

In order to obtain a completion certificate, the contractor's project team should progressively collect certification and create the health and safety file during the construction phase, and prompt subcontractors not to be tardy with the supply of the information. This is to ensure the file is complete prior to the date(s) of sectional/practical completion for as smooth a handover as possible.

7.1.2   Definition and Effect of Practical Completion

To avoid any doubt of what constitutes completion, an expressed term in the contract will define the requirements. For example, clause 10.1 of the FIDIC Red Book (for Building and Engineering Works designed by the Employer) entitled 'Employer's Taking

Over', the works are considered complete when 'Works have been completed in accordance with the Contract...and a Taking Over Certificate for the Works has been issued'. Here, the green light permitting the Employer to take possession of the project (or taking over) will occur after the issue of the Taking Over Certificate, which in other forms of contract such as JCT SBC/Q 2016 is referred to as 'the Practical Completion Certificate'. The form of contract may be silent regarding any specifics the certifying officer may seek in arriving at a decision to issue a completion certificate, meaning the only reliance may be with the clause seeking the conditions of the contract to be met. With JCT forms, the word practical is a driving force for the decision to issue a completion certificate, which may be decided by the physical condition of the building on the date of practical completion as well as the status of the relevant paperwork in place at the time. Here, a minor outstanding item preventing the function of a component might mean the building is impractical for use and a certificate of non-completion would be issued. On the other hand, works noticeably incomplete which do not restrict or impair the primary use of the building can mean the completion certificate could be issued. For example, a building may have an impeccable physical presence without a water supply that might categorise the building impractical for habitation, because water is an essential service. By comparison, a building may have all its essential services operating with the soft landscaping outstanding, meaning the building is practical for the intended use with the landscaping not considered essential because it is ornamental.

The project team will show interest in the handover procedure as the goal is for the contractor to receive a completion certificate that triggers contractual mechanisms and relief from certain obligations. To the contractor, the effect of the Taking Over/Practical Completion Certificate means the following:

- the liability for liquidated damages cease;
- the effect of any outstanding claim for an extension of time can be ascertained that influences the amount of any liquidated damages;
- the defects liability/rectification period commences;
- the contractor's contract works insurance liabilities cease;
- the defects liability insurance commences;
- the final account can be prepared;
- half of the cash retention is due for release and/or the return of one bank guarantee; and
- the local authority is able to issue an occupational certificate, usually applicable to large public buildings, which in the process can discharge some conditions of planning approval.

It can therefore be seen that obtaining the completion certificate is a massive relief for the project team, constituting a job well done.

7.2 Operating Manuals and As-Built Information

Operating manuals describe how key components of installed items and systems in a building are operated and maintained to ensure the building meets its full potential. Buildings may be subdivided into public parts, such as front and back of house areas to a hotel, and private parts for use by tenants or owners as found with apartments in residential construction. For ease of maintaining, it is a usual contract requirement for the contractor to provide operating and maintenance manuals for each different part. These manuals generally comprise:

- a contractor's cover letter stating the project title;
- subcontractor/installers names and contact details;
- product manufacturer's specifications and warranties;
- product name(s), codes and illustrations for operating parts; and
- help guides including general maintenance and cleaning responsibilities of owners.

Any completed areas of a building for common use such as lifts, corridors, basements, facilitating plant, landscaping, etc. are normally under the control of facilities or property managers, for which the manuals are invaluable. These managers are responsible for the ongoing maintenance of the building or works after practical completion, and are provided with a unique manual that includes operational and maintenance information. Where products require operator training as found with the likes of a building management system (BMS) that controls and monitors a building's mechanical and electrical equipment, the installer may provide demonstrations, possibly in real-time mode showing how the system is managed for the purpose of facilities and property management. With other types of sophisticated systems such as sanitisers for cleaning medical equipment in hospitals, the installer may conduct in-house training shortly after a building is occupied via a workshop which is attended by end-users, with participants issued a certificate of attendance upon completion. The same installer may also provide services after the completion date by entering into a comprehensive maintenance contract with the building owner or facilities management contractor to ensure the products are suitably maintained.

It is a usual condition of contract that manuals are examined by consultants that designed and specified the works and approved the shop drawings. After examination, each consultant normally signs off the manuals to confirm they contain sufficient information for the operation of the parts and are an accurate guide of the designed and installed works. The contractor will rely on its subcontractors to provide these manuals, which should be an expressed term of each subcontract agreement where the supply of manuals is a requirement. Each agreement should also include a clause that the subcontractor has a duty to amend the details of any submitted manual as requested by the consultant, including resubmittal, until such time it is approved at no cost to the contractor. Subcontractors may develop their manuals by coordinating directly with the contractor and design consultants to confirm the final requirements, the idea being for each manual to be approved on its first submission. The preparation of these manuals is usually progressive and carried out during the construction phase. This is to allow time for them to become available as close as possible to the appropriate completion date, with the contract usually stating if manuals are required for each section or if they can be provided at practical completion. The quantity surveyor is normally a coordinator in the process to ensure subcontractors provide their manuals in a timely manner.

Manuals for facilities or property managers usually include details of specific components and technical information of installed items such fan coil units, electrical distribution boards, etc., together with a maintenance schedule of the parts and responsibility for servicing the parts (i.e. owner or trade contractor qualified to service the parts). The schedule should also include periodic maintenance requirements for maintaining or servicing each component or part of a component. This is planned maintenance of a major component that has the aims of preventing the systematic breakdown of the component, known as preventative maintenance. The facilities or property manager must comply with this requirement, as it may be a condition of warranty for the works and a statutory requirement (e.g. maintenance of fire/smoke detection and fire-fighting systems installed in a building). The manual may also state that certain products must be maintained under a maintenance contract provided by the installer during a warranty period, with the owner or facilities management contractor reserving the right to negotiate a service contract thereafter. The manual must also include copies of any collateral warranties from subcontractors as required under the terms and conditions of the main contract.

A well-prepared contract will also include a condition that the contractor provides the facilities/property manager with two sets of keys for lockable components and a schedule of spares in addition to the supply of operating manuals. A schedule of spares is a defined quantity of requirements needed in the event of breakage or damage and applies to items such as tiles or carpets. Materials included on a schedule of parts are usually items that would be needed in an emergency or are purpose-made that would be otherwise on long delivery, meaning that managers can have a limited supply of spares in the event of an accident or wear and tear.

In addition to the supply of operating and maintenance manuals, it is a usual contractual requirement for the contractor to provide as-built information, which may be referred to as Work as Executed Plans. This information is provided in print and soft copies on a compact disc or USB, with the soft copies produced in AutoCAD exchange such as *.dwg or *.dxf files and also supplied in *.pdf. The drawings must represent the built aspect of the completed building by discipline (i.e. mechanical, electrical, etc.) that may differ from the contract design because of variations, and are needed to acquire knowledge of the whereabouts of installed components. Assignment of any copyright of the drawings is not usually required if the intention of the issue is for operating and maintenance only. However, the author of the drawings can condition the copyright by including a proviso that the supplied information cannot be used for a new design on other projects. Furthermore, the author may wish to state that the proviso extends to the manufacture of spare parts or components unless required for the repair of works under the contract. The costs associated with the preparation of manuals and as-built information are usually included within the preliminaries.

The supply of as-built information does not need to include confidential manufacturing details or design knowledge used to create parts of a building or the calculations used, which if requested by a client and supplied would be for information only. This is to permit the building owner to appraise installed components, enabling the owner to amend, add or substitute works in the future which are not part of the contract with the owner relying on the supplied calculations that could influence any new design.

The supply of as-built information is usually required before the completion date and should accompany the operating manuals. However, if some information is missing at this time, it should not hinder the issue of a completion statement with the balance being a defect. When becoming a defect, the contractor must attend to the matter within the defects liability/rectification period.

7.3   Defects

A defect occurs when the standard of completed materials and/or workmanship or obligations to perform under contract fails to meet the specified requirements. Due to the nature of construction works, contract forms contain provisions that aim to address defects and the responsibilities of the parties to the contract within a defects liability/rectification period. This should not be confused with any project reference to a maintenance period, as it can mean maintenance responsibilities a building owner undertakes when following instructions in maintenance and operating manuals. It can also refer to a specific contract clause in relation to a project or part of a project when the contractor maintains parts of a project for a defined duration commencing from the expiry date of the defects liability/rectification period (e.g. mechanical equipment). Where parts of a project are handed over in sections, sectional defect periods are stated in the contract; this means a number of periods of varying lengths may be ongoing simultaneously. Under industrial terms, and possibly worded in the construction contract, defects applies to: the works themselves; the design and/or specifications reproduced in the works; and missing or rejected information such as a warranty that requires to be provided or updating.

7.3.1   Patent Defects

Patent defects are defective items discovered from observation of the works or confirmed following an investigation by test results which define a shortfall in the requirements. This can be as obvious as a leaking roof noticed from physical inspection, or discovered by an intrusive method of taking samples of concrete detaching itself from a structural component to check if the material complies with the specification.

The rectification of patent defects is generally a contractor's responsibility and, if a project is procured as a design-and-build contract, it will involve design responsibilities for the contractor. In general, a contractor is responsible for repairing defects in works they have carried out, that occur during the defects liability/rectification period, at their own expense. This will also apply to frost damage and damage caused by the presence of damp if found to be the root cause of a defect caused by freezing, thawing or lack of air replacement before practical completion. However, a client would be responsible for such undesirable defects if occurring after this date, because a building owner is responsible for ensuring the structure is adequately heated and ventilated as part of general maintenance.

The contract may include a clause giving the client the right to approach the contractor any time during the defects liability/rectification period to carry out repairs by issuing a notice of defect(s) to which the contractor must respond and repair the works within a stated timeframe. This may apply to emergency works only as, in reality, the process usually involves the client's agent preparing a defects list close to the end of the defects liability/rectification period which is issued as a notice to the contractor for action. A usual notice is one month to give the contractor time to make a presence and rectify the works. A contract may also state that a contractor must rectify defective works on or before a specified date, which is the end of the defects liability/rectification period plus a flexible allowance in days to allow the contractor to arrange access, gather a workforce and procure materials.

Where there is fair wear and tear of an installed part, the rectification should put the finish back to a condition it would be in if it was not for the defect, as the client cannot expect the rectification to put the finish to the required standard at the time of sectional/practical completion. If a defect is a result of the design and the project was a construct-only contract, the contractor is not usually obliged to make the adjustment unless the client issues a variation. It is therefore important for the contractor to understand its design obligations with defects; if a client considers a defect is due to a design flaw or an undesirable specification, the contractor can seek a variation if works are executed in accordance with the issued design and documentation under a construct-only contract. However, it could be a different scenario with a design-and-build contract where the contractor does not inherit the problem from an initial design prepared by a client's consultants, and must rectify the works at their own expense (subject to the terms and conditions of contract). Where client-engaged contractors are involved, the contractor will usually coordinate and supervise the works only, without the need to purchase anything.

During the defects liability/rectification period, the contract is very much alive and a contractor's lack of response to requests to carry out rectifications may be grounds for the client to terminate the contract. At such a late stage this would be exceptional conduct, as the client will be anxious to resolve defective works issues without any fuss and not be usually willing to engage others to rectify the works unless absolutely necessary. A client cannot employ independent contractors to repair defective works that are the contractor's responsibility and deduct money from a final account unless there is prior notice that the contractor consents to. However, if the contractor fails to comply with a request to repair a defect of their own creation, the client's agent can appoint others to rectify the works. If electing to appoint others because of the contractor's default, the contractor must be given notice in accordance with the contract or to a reasonable timeframe if nothing is stated. If the contractor consents to the works being carried out by others, the contractor will need to be advised of the pending costs before the works commence to avoid a surprise. The quantity surveyor should aim to mitigate this scenario by assessing what the price should be before the works commence or gain an approximation of the expense to be incurred, so the client can deduct an expected amount from the monies owed. Of course, in such a scenario it makes common sense for the contractor to repair their own defects. If the project is in a remote location however, with the contractor specifically locating themselves for the construction phase only, it may make commercial sense to negotiate the charges.

A clause of a contract may include a provision to extend the length of the defects liability/rectification period. Where applicable, this must be for a specific duration and could apply when replacement or the renewal of parts is necessary following the 'teething' of new mechanical plant or equipment. Where applying, it will usually trigger a separate liability period commencing on the date of completing the last installation of replacement parts.

At the end of the defects liability/rectification period, and subject to the works being rectified to the client's or client's agent's satisfaction, a certificate of making good is issued to the contractor. The contractor must retain this document as it is required for the release of final cash retention or return of the last bank guarantee. If the cash retention is not released or the bank guarantee returned automatically, the quantity surveyor should make a formal request when in receipt of the certificate.

7.3.2 Latent Defects

While patent defects are visibly obvious or discovered through investigation, latent defects are not so obvious. These defects are inherited by flaws in the design, workmanship, materials or a mixture that only become apparent over time which were either missed or not foreseen as requiring remedial action at the end of the defects liability/rectification period. As a procedure, the contractor is not usually contractually obliged to attend to defects occurring beyond the end of the defects liability/rectification period stated in the contract. In fact, after this period has lapsed and the certificate of making good has been issued, the contractor can refuse to carry out any rectification works even if the materials and workmanship are part of the contract. However, the refusal to repair anything produced by the contractor because the contract appears to provide relief of responsibilities serves no purpose to the contractor. On balance however, a defect can occur due to extraordinary wear and tear of a component or part of a building because of the client/end-user's activities that may be abnormal (e.g. a hardwood floor in a residential hallway used commercially as a public dance floor), with the contractor attending to the defect during the liability period and ceasing to attend to it upon expiry of the defects liability/rectification period. If the contractor refuses to attend to a defect it has produced that occurs after the period has expired, when it becomes a latent defect, the client/end-user can employ others to complete the repairs. Unfortunately, this would be at their own expense if the final account is agreed and paid, the certificate of making good issued and the conditions of contract discharged. To recover such expense, the client/end-user would not need to resort to legal action because the dispute resolution mechanisms linked to the contract would still apply if the defect is found to originate from the contract works and the client/end-user has not contributed to the defect (e.g. if the product has been used for the intended purpose and is suitably maintained). For effectiveness, this would need to be proven if not so obvious, for example, if it was the contractor that carried out the works that have become defective, or if the works were carried out by a client-engaged contractor. However, a contractor's liability does not cease at this point if the latent defect is a breach of contract because of the contractor's negligence. Here, the client will have remedy at common law because of the limitation period for the commencement of legal actions, which under the Limitation Act 1980 applicable to England and Wales is 6 years from the date of the breach under a Simple contract, and 12 years for a contract under seal or a Speciality contract. However, a client/end-user's common law remedy will be ineffective if the contractor is insolvent.

7.4 Final Accounts

In the lead up to a date of practical completion, and if site based on a project, the quantity surveyor will usually relocate to the contractor's head office to coordinate the certification process and the supply of operating manuals and as-built information. During this transition, the project team will organise the termination of temporary services, off-hiring of rented equipment, demobilisation of the site accommodation and reinstatement of the area disturbed by the presence of the accommodation. At this time, there is also a requirement for the quantity surveyor to consider final accounts and ascertain a final cost of the project.

7.4.1 Main Contract Final Account

The quantity surveyor will need to understand contractual provisions regarding the timely issue of a final account. This is because the issue of anything after the date of practical completion could be understood as complying with the contract; this in turn activates an assessment procedure for the issue of a final payment certificate, where in fact it may be the last progress claim prior to the issue of a final account. In other words, the contract may only permit one submission after the date of practical completion which is the final account.

The contract will usually specify a duration in days or weeks after the date of practical completion when the contractor must submit a final account. Alternatively, the contract might state a timeframe for the presentation of a final account after the issue of the making good certificate. An intention of issuing a final account is for it to lead to the issue of a penultimate and final payment certificate, the former being agreement to the final account including releasing half the cash retention. This certificate is issued at or soon after the date of practical completion, with the latter the release of the retention balance following the satisfactory completion of any rectifications with the certificate issued at or soon after expiry of the defects liability/rectification period. Where bank guarantees are issued instead of cash retention, the final payment certificate is the response to the issue of a final account with the guarantees released under separate cover subject to the works, and any rectifications being completed to the satisfaction of the client/client's agent. The final payment certificate is also a statement that recognises the project client's financial liability under the contract, or possibly (yet rarely) the contractor's liability to the client. Presentation of a final account should be as straightforward as possible and ideally constitute a summary statement comprising:

- the contract sum;
- the omission of any client contingencies included in the contract sum if not required;
- adjustment of the contract sum due to variations;
- omission of provisional and PC sums included in the contract sum reassessed as variations;
- loss and expense claims;
- agreed permanent deductions; and
- price fluctuations if the project is not a fixed price.

Where a project includes amounts that adjust the contract sum which are approved and paid, there is no need to re-present them as this would create unnecessary administration. Only new items or items under query should therefore have the necessary evidence or substantiation included with a final account submission.

A suitable contract clause will describe the presentation of a final account which should be cumulative to give a gross amount, less amounts already paid to produce a net amount due. Any cash retention or bank guarantees in lieu should be requested for release in isolation to the final account submission, and should be referred to in the consignment as a matter of course. Submission of the final account must be to the title named in the contract; on a construct-only project this may be the client's agent who may instruct the contractor to copy the details to the client's quantity surveyor/cost manager who will assess the submission on behalf of the client/client's agent. Table 7.1 shows a suitable final account summary that a contractor could submit as part of a consignment.

Table 7.1 Final account summary

FINAL ACCOUNT SUMMARY

PROJECT TITLE:

	OMISSIONS	ADDITIONS	
Agreed Contract Sum:			£3,000,000.00
Less Client Contingencies			−£ 50,000.00
			£2,950,000.00
Provisional sums adjustment:	£130,000,00	£15,000.00	
Prime cost sum adjustment:	£ 5,000.00	NIL	
Variations summary:	NIL	£65,000.00	
Additional works summary:	NIL	£30,000.00	
Contractors claims summary:	NIL	£15,000.00	
Total:	**£135,000.00**	**£125,000.00**	
Total OMISSIONS			−£135,000.00
Total ADDITIONS			£125,000.00
Total Final Account			**£2,940,000.00**
Less Retention (Bank guarantee supplied)			NIL
Less Architect Certificate Nrs 1 to 15			−£2,827,500.00
Amount due to the contractor			£ 112,500.00

Figures exclude VAT

The contractor's quantity surveyor must make note of any review period stated in the contract for the timely issue of a response from the client's agent after submitting a final account. During the review period, the client's quantity surveyor/cost manager (or other assessor) may coordinate with the contractor's quantity surveyor and request additional information which the contractor must endeavour to supply. A request for additional information would not be deemed a response, as it is merely the seeking of supplementary details needed to assess the final account, meaning the response must be the quantity surveyor/cost manager's statement of the final account and nothing less. If a contract states the review period leading to a response is for a period 'up to xx months' or 'xx days from receipt of the contractor's assessment' and a statement is issued beforehand, it would be a valid response. However, if a contract states that the period for a response is 'after a minimum period of xx months from receipt of the contractor's assessment' or 'after a period of xx days but not exceeding xx days', it means the contract favours a more thorough review period and a response before the minimum period will mean the statement would be invalid. Once any defined period for a response as stated in the contract lapses, the assessor must issue a statement of a final account or the assessor will have breached the client's contract with the contractor. The statement may be similar to the contractor's final account submission, including a cumulative sum stating the client's quantity surveyor's/cost manager's final account less payments made to produce an amount due to or from the contractor. To demonstrate, Table 7.2 provides a statement a contractor may receive from the client's agent or quantity surveyor/cost manager seeking the acceptance of a final account.

Table 7.2 Statement of final account

STATEMENT OF FINAL ACCOUNT

PROJECT TITLE:

Client: (Name and address)   Contractor: (Name and address)

Client's Agent: (Name and address)

Date of issue of the Statement: xx/xx/xxxx

Pursuant to the terms and conditions of the contract, and in respect of this statement, I/We certify that the sum of £2,940,000.00 (TWO MILLION NINE HUNDRED AND FORTY THOUSAND POUNDS EXACTLY) is the gross final account and the gross sum of £112,500.00 (ONE HUNDRED AND TWELVE THOUSAND FIVE HUNDRED POUNDS EXACTLY) is the amount due to the contractor. In certifying this statement I/We understand:

a) The amount due is the final amount to which the contractor is entitled to other than any remaining retention amount (where applicable) or Value Added Taxation (if applicable) under the rules of taxation law relevant to the project. Furthermore, in consideration for payment of the amount due, the contractor hereby releases the client from all monies, accounts, actions, proceedings, demands, costs, claims and expenses that the contractor had or has previously requested or stated.

b) The amount due is a full and final settlement that may be adjusted by liquidated damages and an audit by the client or client's agent.

Furthermore, I/We certify that the works have been carried out in accordance with the terms and conditions of the contract and the construction design and documentation.

Signed:

For and on behalf of: (Name of Contractor)

Witnessed by:

Date:

Ideally, the contractor's final account submission should agree with the statement, plus or minus amounts the contractor considers trivial, which an authorised signatory of the contractor must sign and return for it to become binding. If the contractor disagrees with the statement (and not necessarily the contractor's quantity surveyor), the document should not be signed and the contractor should attempt to negotiate the matter with the client/client's agent to reach an amicable settlement. Alternatively, the contractor may officially reject the statement and resort to dispute resolution if the client's quantity surveyor/cost manager or other assessor objects and affirms the statement. However, when electing for this decision, the contractor must consider what it may lose by not agreeing, as it may mean the cessation of further payments until there is an agreement or any dispute is resolved.

7.4.2   Issue and Effect of the Final Payment Certificate

A final payment certificate and the payment a contractor receives should be to the amount on the final account statement. However, a statement may include conditions giving the client the right to make amendments to the amount that could increase or reduce the value on the certificate, resulting in the payment of a different amount. Once receiving payment, and if differing from the amount on the certificate with the contractor owed money, an area of disagreement may arise regarding the deduction of liquidated damages. A disagreement may also occur when any financial incentives paid for early sectional completions are wiped out by liquidated damages because of a delay to the agreed date of practical completion. This can come as a sting in the tail to the contractor if agreeing to the statement to only later discover the amount due will be reduced, which may not be shown on the statement. The assessor issuing the statement may have no authority to deduct liquidated damages even if there is an obvious delay to the project, and may avoid advising of any deduction if the extent of the delay is arbitrary. This can mean the communicator of any deduction may be the client's agent that alters the value of the final payment certificate or may in fact be the client who decides to pay a different amount. The right to deduct liquidated damages is dealt with in the same manner as interim payments certificates, meaning the contractor must be informed of the deduction before payment is due by the payer issuing a 'pay less notice'. A contractor that agrees a final account and later discovers or understands that liquidated damages will be deducted should wait to receive payment until deciding to challenge the deduction, as there may be a possibility nothing will be paid.

If the contractor is advised of an intention to deduct liquidated damages, they must be to that stated in the contract. If they are different, or the contract is silent on liquidated damages, any deduction would be considered a penalty. In general, a penalty is a fine or punishment which is unenforceable under common law and applies where a client may benefit, meaning the contractor may resort to the law courts to recover any deducted amount. If a contract specifically states liquidated damages as nil a client cannot deduct anything and, if wishing to recover losses because of a delay, the delay must be a breach of contract by the contractor; this means the client would need to resort to the courts to recover their loss and seek an award in unliquidated damages.

Where liquidated damages are stated in a contract and deducted from the payment, and the length of delay is not contested by either party, the contractor may choose to query the calculation of the rate which a client may not wish to divulge because of business sensitivity. Stated liquidated damages must be specific, for example a rate per day or calendar week, and not vague. If choosing to query the damages included in the contract, the contractor must go through a court that will expose the calculations to understand their basis. In general, if a court considers the damages are a reasonable representation of an amount of business loss because of a delay, it may decide to leave them intact. If considered excessive due to uncertainty or because they include contingencies, the court may construe them as a penalty, thus making them unenforceable in full or in part. In the absence of negotiations between the parties on the matter, the contractor's challenge may be a long-drawn-out process to see if the liquidated damages are in fact a penalty. Separate to this is the question of the delay itself triggering the damages; if the contractor's request for an extension of time is rejected in part, it means there is no agreement to the end date with the project's duration rendered 'time at large'. For liquidated damages to be enforceable, the delay would need to be rejected in full, thus cancelling 'time at large'. This links to what is referred to in legal terms as the Prevention Principle, where a party cannot benefit from its own breach of contract; that is, one party cannot insist on the performance of the other if it caused the non-performance itself. In other words, if a project client acknowledges their part in a delay but not to the extent claimed by the contractor, the client still carries a degree of responsibility. If the client agrees to some extension of time and is less than the contractor seeks, it would still render 'time at large', meaning a neutral third party may be requested to intervene to determine a reasonable extension of time the parties could attempt to agree upon. In the meantime, if the parties are unable to agree there is an inability to apply stated liquidated damages.

7.4.3 Supply Chain Final Accounts

The contractor's quantity surveyor would be wise to address supply chain final accounts as soon as possible after the date of practical completion, as it is an important aspect of commercial and contract management of a construction project. This also assists in understanding the status of variations and/or claims under the main contract that may be dependent on submissions from the supply chain as part of the main contractor's submission. Furthermore, agreement and reconciliation of an amount to include in each final account is good practice, as it provides cost certainty regarding the contractor's liabilities.

The formal agreement to a final account with any supply chain member is seldom mandatory unless the contractor writes a suitable clause into an agreement. Without such a clause supply chain members will not need to agree a final account, and a method of determining the contractor's liabilities is for the quantity surveyor to request supply chain members to issue a final account in a collaborative manner irrespective of it being an obligation of the agreement. However, the accounting systems of some supply chain members may not be as robust as the contractor's and they may decline to carry out the request; a solution in that case is for the quantity surveyor to issue a statement in a Deed of Release. A Deed of Release is usually issued to subcontractors or consultants contracted under a lum-sum price for a fixed scope of works or services. The Deed represents an amount due under the agreement and is based upon the value of works/ services, plus/minus works changes and/or variations, less amounts paid that produces a liability. Where a stated amount on the Deed is the contractor's liability, it is subsequently invoiced, and once paid is the final financial liability excluding release of any withheld cash retention. Regarding subcontractors, and subject to the terms and conditions of each agreement, the Deed of Release will usually include a note to the effect that half of the retention will be released in a penultimate payment with the balance due at the end of the defects liability/rectification period (i.e. mirroring the main contract with a similar arrangement in place for managing the release of any bank guarantees). However, a contractor may agree to different timings with each subcontractor for the first release if their completion date is too distant from the date of practical completion (e.g. civil works). It is unlikely a contractor would agree to pay a second release before the end of the defects liability/rectification period, as it would mean security for any recall would be lost.

When assessing an amount to include in a Deed of Release, the contractor cannot deduct a penalty monetary amount as it cannot be legally enforced; if wishing to make a claim against the other party, it must be in the form of a financial claim (see Chapter 6, Section 6.7.7 for details of these types of claims). Furthermore, the contractor can only deduct liquidated damages if they are written specifically into an agreement; in any event these would usually be capped as subcontractors are not generally expected to accept the full allocation of liquidated damages under the main contract as it would not make commercial sense. Neither would it make sense for the contractor to apply damages where a time at large argument exists between the contractor and subcontractor where the latter is considered culpable for the contractor incurring liquidated damages, as such a claim would be superfluous. On the other hand, and for specified damages to be levied, a supply chain member's lack of performance must be proven as the sole reason the contractor misses the completion date resulting in liquidated damages being applied under the main contract, and can only apply to one entity. This could occur when a project is on target to complete on time but is hindered by the insolvency of a subcontractor when there is no programme float available. Here, the contractor could include the amount incurred as liquidated damages as part of a notional or final account for issue to the insolvency practitioner in an attempt to claim an amount from the estate of the insolvent business.

A Deed of Release can be formatted and worded in a similar fashion to that issued in a statement from the client's quantity surveyor/cost manager (or other assessor) under the main contract as shown in Table 7.2 above. For the Deed to be effective, it will need to be dated and signed by the supply chain member and contractor's quantity surveyor or project manager, with both parties retaining a copy of the dual signed document. If a supply chain member disagrees with the sum due on a Deed of Release, the parties must attempt to negotiate. If negotiations drift from an amicable agreement with the contractor formally objecting, it is up to the supply chain member to either accept the contractor's decision, or trigger the dispute resolution methods under the contract or seek intervention as a statutory right.

When dealing with plant hire companies to agree a final account, there may be surprises in store for the quantity surveyor if hired plant is considered still on-hire when it is clearly not on site. This will need clarifying as either an administration oversight, or one where the plant has been transferred to another project with the recording of the transfer managing to bypass the plant hire register. Theoretically, of course, all hire periods for plant should cease upon completion of a project unless they are retained for carrying out rectification works.

Rarely will material suppliers and plant hire companies sign off a final account when they have an invoicing system driven by their accounts procedures. However, some larger companies may have a unique account reference for a project and may liaise with the contractor's personnel during the construction phase and sign off a final account if requested to do so (e.g. scaffolding companies), to ensure hire periods are agreed.

7.4.4 Final Project Costs

The income the contractor receives for a completed project should exceed the cost incurred by a percentage resembling the offsite overheads/profit margin included in the tender. In order to assess the final margin, the final account under the main contract must be known together with the final project costs. The assessment of the final project costs is aided by the CMS that shows the value of orders placed as well as amounts paid in comparison with remaining liabilities on the balance of order values. At this stage, it is necessary to review accruals and decide if they will become actual costs or are residual amounts from approximate order values entered into the CMS. For example, if there is an order on the system to a supplier for the supply of bricks worth £50,000 that includes an allowance for waste of £5,000 with a final cost incurred of £45,000 and no further liabilities expected, the accrual can be released. In effect, this will reduce the final project costs for the brick supply by £5,000. This process should be repeated for other material supply orders procured to approximate values in addition to subcontractor's awards based on approximate quantities that become firm with the works prices valued and paid on agreed schedules of rates. However, care is required when releasing accruals as any release must only be for orders placed to approximate values and not fixed amounts. This is because a supply chain member may seek the balance of an order at any time, as time barring the issue of invoices is generally unacceptable and not ethical business practice.

The quantity surveyor must be satisfied a project's costs include accruals for amounts stated on minor work agreements and/or single purchases which are unpaid. As with purchase orders of approximate values, invoices can be issued at any time which, upon receipt, would become an automatic debt to the contractor. It would therefore be wise to spend time reviewing site instruction books and purchase order forms to become aware of the possibility of late invoicing. If the CMS has an accruals facility, the value of outstanding sums against purchase orders may be shown on the system and would be included in the cost by default. Where this facility is not a feature of the CMS or no CMS is available, the quantity surveyor would be wise to view financial statements from material suppliers and plant hire companies for verification, a procedure unfortunately not always adhered to. Financial statements are not favoured by quantity surveyors because they include a summary of invoices issued to a contractor's business over a defined period, and may not include project references. Furthermore, these statements usually include cumulative amounts owed by the contractor and deduct paid amounts from the bottom line to show company liability and not project liability. A project's financial liability is of course of interest when ascertaining a project's final expense, and it would be wise to contact any supplier in receipt of an award on a project undergoing a final account to identify unpaid invoices and understand what, if anything, is owed.

Beyond the date of practical completion, the process of ascertaining the final cost is regularly reviewed because the contractor still has responsibilities under the contract and may incur expenses through the defects liability/rectification period until receipt of the making good certificate. The intention of this periodic review is to identify expense incurred which can be considerable over the duration and is a part of project cost management to ensure the expense is suitably controlled. Normally, the preliminaries budget includes an amount to cover these costs that includes the cost of labour, plant, materials and supervision, and is recognition of the liabilities instead of an accurate assessment of the cost (as the expense is an unknown factor when preparing a tender and setting budgets).

A different scenario exists with national house builders that rely on income from the sale of properties. The largest type of house building company may find itself in a situation where they have a completed residential estate with unsold housing stock which needs to be maintained to promote the last sales. Here, the quantity surveyor will carry out a regular 'cost-to-complete' exercise with the aim of forecasting the final cost until the remaining stock is sold and the income received. The cost to complete is a forecast of expense required to see works through to completion in addition to the cost already incurred, which produces a projected final account. The cost-to-complete ideology may also be implemented by a main contractor during the defects liability/rectification period if there is a consensus the cost requires managing because of the rectification works involved, and if works variations are in progress.

7.5   Project Closure

7.5.1   Feedback

A contractor should consider a project it undertakes a success once practically complete, and in order to verify the result should seek confirmation and opinions from the
client and client's team where a client's team is engaged. Naturally, the contractor's project team and business will have worked hard to achieve the desirable outcome and
would welcome a pat on the back for a job well done. However, a contractor may be in
their infancy in business or recently gained the trust of a new client, which may have
resulted in shortcomings regarding deliverables while fostering a new working relationship. Moreover, the contractor may have constructed a type of building they have never
carried out before or procured the project in a new manner (e.g. design and build for the
first time), and may seek opinions regarding performance. Here, a contractor may welcome comments and any constructive criticisms of weaknesses in order to bolster
improvements for future projects. Feedback to a contractor is essential on completed
projects, even if the contractor considers themselves an expert in the construction of
certain types of buildings. For example, a contractor's prominent business may be
school building and has completed a unique and complex school project that tested
innovation. Here, a contractor may wish to learn if the innovation it provided met the
standards expected, which the client can provide with feedback.

With feedback, project clients may adopt their own key performance indicators as a
record of a contractor's performance which they can either divulge to the contractor or
preserve for their own use. In isolation, a contractor may wish to record feedback with
the use of a questionnaire that can be issued to clients for completing and returning to
the contractor. The type of questioning should focus on customer service that a client
can score with a rating, (e.g. 0 to 5 with 5 being the most favourable) and the questioning styled so that it takes 15–30 minutes to complete. It may also include sections that
warrant a written response instead of ratings, although this should be limited to preserve
the estimated time to complete. Questions should focus on the overall satisfaction
scored for the service provided commencing from the invitation to tender through to
the date of practical completion. The fact that there may be a dispute post-construction
that may be unresolved should not prohibit the contractor from issuing the questionnaire as it is feedback on the overall performance and particularly the end product that
is sought, which a client will usually understand. The questionnaire should conclude
with a tick box that, if ticked by the client, permits testimonials included in the questionnaire to be included on the contractor's website as well as permission to include any
photographs.

If the contractor wishes to issue a questionnaire to members of the client's team for
feedback, the client should be given an opportunity to accept or reject the request and
the contractor should only issue the forms upon the client's acceptance. Such questionnaires should focus on collaboration and may be paraphrased to suit the appropriate
discipline instead of being generic.

In addition to client's and client's team feedback, the quantity surveyor should provide
final project costs to the contractor's estimator for use with tendering new projects.
Information provided should include market labour rates, material costs and subcontractor prices as indicators of actual expense. The estimator will also be interested in the

effectiveness of the contractor's site management and constraints involved in the building process that impact the preliminaries, as this has an impact on time and the price for running the project and is useful information for future tendering.

7.5.2 Archiving and Retrieval

Once a project is practically complete, generated hard copy documents and files will need to be archived for the contractor's use and to comply with the law. The Companies Act 2006, Section 388(4) makes it mandatory for accounts records to be preserved for a period of 3 years by a private company and 6 years by a public company from the date they were made. In addition, the contractor must consider its closing position on a project and the limitation period of liabilities under legislation, i.e. 6 and 12 years for simple and sealed contracts respectively, and 15 years for negligence if applicable under the Latent Damage Act 1986. Notwithstanding this, the contractor will need to consider the location for the storage, the length of time the information is to be kept in store and if the contents will be transferred at any known time. To standardise the procedure, a contractor may have an established in-house arrangement describing the process for archiving information for easy retrieval, which may be defined in the company's policies and procedures manual. Options for storage may include containers or rooms within the contractor's premises or an independent storage facility, any of which will need a project completion record. A typical completion record will comprise the following.

- *Approvals*: A summary of authority approvals including planning and building regulations.
- *Compliance certificates*: schedule and copy of building services, testing and compliance certificates issued as part of the handover process.
- *Handover certificates*: client's agent's certificates of sectional and/or practical completion and the certificate(s) of making good.
- *Survey*: as installed strata plan survey.
- *Financial certificates*: as issued by the client/client's agent.
- *Final accounts*: applicable to the main contract and Deed of Release with each supply chain member.
- *Progress claims*: interim applications for payment under the main contract.
- *Supply chain payments*: supply chain member's files including tender information, pre-contract negotiations and forms of agreement, payments and correspondence.
- *Consultant correspondence*: to and from the contractor, including reports and payment details.
- *Client/architect instructions*: listed chronologically.
- *Client/architect correspondence*: to and from the contractor.
- *Minutes of meetings*: relevant to meetings with the client and any of the client's team.
- *General correspondence*: letters, public notices, etc. issued for any reason, to and from the contractor.
- *Site instructions*: issued to the contractor's supply chain.
- *Health and safety*: Construction and Occupational phase files and CDM plans including registers.
- *Other site administration*: a schedule of registers including plant and the site diary.

- *Project reports*: internal progress and cost reporting and reports to the client.
- *Retention accounts*: final cost management report produced by the cost management system.
- *Pre-construction file*: the handover file and details issued by the estimator at commencement.
- *Project performance summary*: a schedule including the contract construction programme and updates during the construction phase, delay register and trade attendance summary by month, if not included in the health and safety file. The summary can also include the client-completed feedback questionnaire on the contractor's performance.
- *Master building manual*: the contractor's copy of operating and maintenance manuals and as-built information, and one copy of each manual issued for separable portions including copies of collateral warranties.
- *Project contact list*: this includes the contact information for the client's team, the contractor's supply chain and project team members. The list should include a primary contact detail for the client and post-construction contacts (i.e. names of tenants or owners if made available to the contractor).
- *Contract and construction documents*: form of contract, specifications, drawings, etc. including photography during the construction phase and afterwards if to be used for commercial promotion.
- *Outstanding issues*: a list of any unresolved matters including disputes and planning conditions to be discharged as applicable at the time of creating the archive content. Any defects list at the time of archiving the files should also be included.

Collecting and archiving this information can be a mini-project in itself, with the hard copy option of storage consuming time and space on the contractor's premises. With the use of alternative storage systems and the growth of information technology, specialist companies can offer paper and paperless options as methods of storing information as well as a retrieval service. Services include:

- *Document warehouse*: a place to store hard copy information charged as an ongoing fee based on the volume stored. With this arrangement, the contractor has unrestricted access which usually involves an authorisation process to protect confidentially. Even although these warehouses may be described as waterproof and fireproof, a savvy contractor would keep an electronic back-up stored on a server at its head office in the event of irreversible damage.
- *Document reduction*: traditional folders and files can be replaced with vacuum packing which is a method of reducing the bulk factor of contents by as much as 50%. Here, documents are sealed with plastic wrapping in accessible pouches with the air removed. Pouches can be colour-coded for subject reference and can have security coding embossed in seals for more detailed identification.
- *Bulk document and large format scanning*: standard documents up to A3 in size can be scanned at the rate of thousands per day created as *.pdf, *.tiff or *.jpeg files using portable scanners. Once scanned, scheduled and suitably identified, information is supplied by the scanning company on compact disc/USB. Format scanning is also available for documents greater than A3 that involves the use of larger scanners. Once the documents are scanned, the contractor has the option to shred hard copies that a provider may offer as an additional service.

- *Electronic storage*: a contractor may elect to archive information on a computer server for general viewing instead of storing hard copy information, which frees up physical space and is an advantage of the system. Access to such electronic storage is usually password protected and restricted to members of the contractor's business, and located on a common shared drive (usually using any designated letter other than C:/, D:/ or E:/). Information retained by project team members on their own C:/, D:/ or E:/ drives on workstations or laptops which are relevant to a completed project must be saved and transferred to the shared drive to ensure all information is stored on a single server.
- *Electronic web viewing*: this has the same advantage as storage on a server, and comprises an electronic vault created by cloud computing. Here, information is uploaded and stored to a host cloud arranged by the contractor for access online using passwords for downloading. Some service providers may also offer onsite scanning services that involves creating a temporary network which can be set up at the contractor's offices so that scanned documents are uploaded to the cloud or host service provider.

Archiving is the last process for the successful delivery of a project and an important aspect of project administration. Once complete, time is released that permits the contractor's quantity surveyor to reflect on the experience of the completed scheme and use that experience to help deliver the next project.

Further Reading

Anumba, C., Carrillo, P. and Egbu, C. (2005) *Knowledge Management in Construction*, Wiley-Blackwell.

Atkinson, A. (2015) *JCT Contract Administration Pocket Book*, Routledge Pocket Books.

Best, R. and de Valence, G. (2002) *Design and Construction - Building in Value*, Butterworth-Heinemann.

Breach, M. (2011) *Essential Maths for Engineering and Construction*, Taylor and Francis.

Brook, M. (2017) *Estimating and Tendering for Construction Work*, Fifth Edition, Routledge.

Cartlidge, D. (2013) *Estimator's Pocket Book*, Routledge Pocket Books.

Cartlidge, D. (2017) *Quantity Surveyor's Pocket Book*, Third Edition, Routledge Pocket Books.

Chappell, D. (2012) *Understanding JCT Standard Building Contracts*, Ninth Edition, Taylor and Francis.

Charrett, D. and Loots, P. (2009) *Practical Guide to Engineering and Construction Contracts*, CCH Australia Ltd.

Chudley, R. and Greeno, R. (2016) *Building Construction Handbook*, Routledge.

CIOB (2009) *Code of Estimating Practice*, Seventh Edition, Blackwell Publishing Professional.

Cooke, B. and Williams, P. (2009) *Construction Planning, Programming and Control*, Third Edition, Wiley-Blackwell.

Emmitt, S. and Gorse, C. (2015) *Communication in Construction Teams* (Spon Research), Routledge.

Gibson, R.C. (2008) *Construction Delays: Extensions of Time and Prolongation Claims*, Taylor and Francis.

Hackett, M. and Statham, G. (2016) *The Aqua Group Guide to Procurement, Tendering and Contract Administration*, Wiley Blackwell.

Hughes, S., Mills, R. and O'Brien, P. (2008) *Payment in Construction: A Practical Guide*, RICS Books.

Hughes, W., Champion, R. and Murdoch, J. (2015) *Construction contracts, Law and Management*, Routledge.

Packer, A. (2016) *Building Measurement: New Rules of Measurement*, Second Edition, Routledge.

Pryke, S. (2009) *Construction Supply Chain Management*, Wiley-Blackwell.

RICS (2012) *NRM 2 - Detailed Measurement for Capital Building Works: NRM 2 (New Rules of Measurement)*, RICS Books.

Construction Quantity Surveying: A Practical Guide for the Contractor's QS, Second Edition. Donald Towey.
© 2018 John Wiley & Sons Ltd. Published 2018 by John Wiley & Sons Ltd.

Seeley, I.H. (1998) *Building Quantities Explained*, Palgrave Macmillan.

Shen, G.Q.P. and Yu, A.T.W. (2015) *Value Management in Construction and Real Estate: Methodology and Applications*, Routledge.

Towey, D. (2012) *Construction Quantity Surveying: A Practical Guide for the Contractor's QS*, First Edition, Wiley-Blackwell.

Towey, D. (2013) *Cost Management of Construction Projects*, Wiley-Blackwell.

Index

Printed and bound by CPI Group (UK) Ltd, Croydon, CR0 4YY

27/10/2024